LEARNING SYSTEMS
AND
INTELLIGENT ROBOTS

Edited by
K. S. Fu
School of Electrical Engineering
Purdue University
West Lafayette, Indiana

and
Julius T. Tou
Center for Information Research
University of Florida
Gainesville, Florida

PLENUM PRESS • NEW YORK AND LONDON

Library of Congress Cataloging in Publication Data

U.S.-Japan Seminar on Learning Control and Intelligent Control, 2d, Gainesville, Fla., 1973.
Learning systems and intelligent robots.

Includes bibliographical references.
1. Artificial intelligence—Congresses. 2. Information theory—Congresses.
I. Fu, King Sun, 1930- ed. II. Tou, Tsu-lieh, ed. III. Title.
Q334.U54 1973 001.53'5 74-11212
ISBN 0-306-30801-0

Proceedings of the U.S.-Japan Seminar on
Learning Control and Intelligent Control, held at
Gainesville, Florida, October 22-26, 1973

Seminar Organizers

K. S. Fu, *Purdue University, U.S.A.*
K. Nakamura, *Nagoya University, Japan*

Local Arrangements

J. T. Tou, *University of Florida, U.S.A.*

A Division of Plenum Publishing Corporation
227 West 17th Street, New York, N.Y. 10011

United Kingdom edition published by Plenum Press, London
A Division of Plenum Publishing Company, Ltd.
4a Lower John Street, London, W1R 3PD, England

Printed in the United States of America

Preface

This book contains the Proceedings of the Second U.S.-Japan Seminar on Learning Control and Intelligent Control. The seminar, held at Gainesville, Florida, from October 22 to 26, 1973, was sponsored by the U.S.-Japan Cooperative Science Program, jointly supported by the National Science Foundation and the Japan Society for the Promotion of Science. The full texts of the twenty-one presented papers are included. The papers cover a variety of topics related to learning control and intelligent control, ranging from pattern recognition to system identification, from learning control to intelligent robots.

During the past decade, there has been a considerable increase of interest in problems of machine learning, systems which exhibit learning behavior. In designing a system, if the a priori information required is unknown or incompletely known, one approach is to design a system which is capable of learning the unknown information during its operation. The learned information will then be used to improve the system's performance. This approach has been used in the design of pattern recognition systems, automatic control systems and system identification algorithms. If we naturally extend our goal to the design of systems which will behave more and more intelligently, learning systems research is only a preliminary step towards a general concept of integrated intelligent systems. One example of this class of systems is the intelligent robot, which integrates pattern recognition, learning and problem-solving into one intelligent system.

Credit for any success in this seminar must be shared with many people who contributed significantly of their time and talents. In organizing the seminar the co-chairman, Kahei Nakamura and K. S. Fu received considerable help from J. E. O'Connell of the National Science Foundation and the staff of the Japan Society for the Promotion of Science. It is the authors of the individual papers whose contributions made possible the seminar and the subsequent post-seminar proceedings. As the editors of the proceedings, we would like to express our deep appreciation for the help received from Terry Brown and Phyllis Straw during the preparation of this publication.

K. S. Fu
J. T. Tou
February 1974

Contents

THE CONCEPT OF A LINGUISTIC VARIABLE AND ITS APPLICATION TO APPROXIMATE REASONING

L. A. Zadeh

University of California

Berkeley, California 94720

One of the fundamental tenets of modern science is that a phenomenon cannot be claimed to be well understood until it can be characterized in quantitative terms.[1] Viewed in this perspective, much of what constitutes the core of scientific knowledge may be regarded as a reservoir of concepts and techniques which can be drawn upon to construct mathematical models of various types of systems and thereby yield quantitative information concerning their behavior.

This work was supported in part by the Navy Electronic Systems Command under Contract N00039-71-C-0255, the Army Research Office, Durham, N.C., under Grant DA-ARO-D-31-124-71-G174, and the National Science Foundation under Grant GK-10656X3. The writing of the paper was completed while the author was participating in a Joint Study Program with the Systems Research Department, IBM Research Laboratory, San Jose, California.

[1]As expressed by Lord Kelvin in 1883 [1], "In physical science a first essential step in the direction of learning any subject is to find principles of numerical reckoning and practicable methods for measuring some quality connected with it. I often say that when you can measure what you are speaking about and express it in numbers, you know something about it; but when you cannot measure it, when you cannot express it in numbers, your knowledge is of a meager and unsatisfactory kind: it may be the beginning of knowledge but you have scarcely, in your thoughts, advanced to the state of science, whatever the matter may be".

Given our veneration for what is precise, rigorous and quantitative, and our disdain for what is fuzzy, unrigorous and qualitative, it is not surprising that the advent of digital computers has resulted in a rapid expansion in the use of quantitative methods throughout most fields of human knowledge. Unquestionably, computers have proved to be highly effective in dealing with mechanistic systems, that is, with inanimate systems whose behavior is governed by the laws of mechanics, physics, chemistry and electromagnetism. Unfortunately, the same cannot be said about humanistic systems,[2] which-so far at least- have proved to be rather impervious to mathematical analysis and computer simulation. Indeed, it is widely agreed that the use of computers has not shed much light on the basic issues arising in philosophy, psychology, literature, law, politics, sociology and other human-oriented fields. Nor have computers added significantly to our understanding of human thought processes-excepting, perhaps, some examples to the contrary that can be drawn from artificial intelligence and related fields [2], [3], [4], [5].

It may be argued, as we have done in [6] and [7], that the ineffectiveness of computers in dealing with humanistic systems is a manifestation of what might be called the principle of incompatibility-a principle which asserts that high precision is incompatible with high complexity.[3] Thus, it may well be the case that the conventional techniques of system analysis and computer simulation-based as they are on precise manipulations of numerical data-are intrinsically incapable of coming to grips with the fuzziness of human thought processes and decision-making. The acceptance of this premise suggests that, in order to be able to make significant assertions about the behavior of humanistic systems, it may be necessary to abandon the high standards of rigor and precision that we have become conditioned to expect of our mathematical analyses of well-structured mechanistic systems, and become more tolerant of approaches which are approximate in nature. Indeed, it is entirely possible that only through the use of such approaches could computer simulation become truly effective as a tool for the analysis of systems which are too complex or too

---

[2]By a humanistic system we mean a system whose behavior is strongly influenced by human judgement, perception or emotions. Examples of humanistic systems are: economic systems, political systems, legal systems, religious systems, etc. A single individual and his thought processes may also be viewed as a humanistic system.

[3] Stated somewhat more concretely, the complexity of a system and the precision with which it can be analyzed bear a roughly inverse relation to one another.

ill-defined for the application of conventional quantitative techniques.

In retreating from precision in the face of overpowering complexity, it is natural to explore the use of what might be called linguistic variables, that is, variables whose values are not numbers but words or sentences in a natural or artificial language. The motivation for the use of words or sentences rather than numbers is that linguistic characterizations are, in general, less specific than numerical ones. For example, in speaking of age, when we say "John is young," we are less precise than when we say, "John is 25." In this sense, the label young may be regarded as a linguistic value of the variable Age, with the understanding that it plays the same role as the numerical value 25 but is less precise and hence, less informative. The same is true of the linguistic values very young, not young, extremely young, not very young, etc. as contrasted with the numerical values 20, 21, 22, 23, ...

If the values of a numerical variable are visualized as points in a plane, then the values of a linguistic variable may be likened to ball-parks with fuzzy boundaries. In fact, it is by virtue of the employment of ball-parks rather than points that linguistic variables acquire the ability to serve as a means of approximate characterization of phenomena which are too complex or too ill-defined to be susceptible of description in precise terms. What is also important, however, is that by the use of a so-called extension principle, much of the existing mathematical apparatus of systems analysis can be adapted to the manipulation of linguistic variables. In this way, we may be able to develop an approximate calculus of linguistic variables which could be of use in a wide variety of practical applications.

The totality of values of a linguistic variable constitute its term-set, which in principle could have an infinite number of elements. For example, the term-set of the linguistic variable Age might read

Age = young + not young + very young + not very young + very very young + ... + old + not old + very old + not very old + ... + not very young and not very old + ... + middle-aged + not middle-aged + ... + not old and not middle-aged + ... + extremely old + ... (1)

in which + is used to denote the union rather than the arithmetic sum. Similarly, the term-set of the linguistic variable Appearance might be

Appearance = _beautiful_ + _pretty_ + _cute_ + _handsome_ + _attractive_ + _not beautiful_ + _very pretty_ + _very very handsome_ + _more or less pretty_ + _quite pretty_ + _quite handsome_ + _fairly handsome_ + _not very attractive and not very unattractive_ + ...

In the case of the linguistic variable _Age_, the numerical variable _age_ whose values are the numbers 0, 1, 2, 3, ..., 100 constitutes what may be called the _base variable_ for _Age_. In terms of this variable, a linguistic value such as _young_ may be interpreted as a label for a _fuzzy restriction_ on the values of the base variable. This fuzzy restriction is what we take to be the _meaning_ of _young_.

A fuzzy restriction on the values of the base variable is characterized by a _compatibility function_ which associates with each value of the base variable a number in the interval [0,1] which represents its _compatibility_ with the fuzzy restriction. For example, the compatabilities of the numerical ages 22, 28 and 35 with the fuzzy restriction labeled _young_ might be 1, 0.7 and 0.2, respectively.

The conventional interpretation of the statement "John is young," is that John is a member of the class of young men. However, considering that the class of young men is a fuzzy set, that is, there is no sharp transition from being young to not being young, the assertion that John is a member of the class of young men is inconsistent with the precise mathematical definition of "is a member of". The concept of a linguistic variable allows us to get around this difficulty in the following manner.

The name "John" is viewed as a name of a composite linguistic variable whose components are linguistic variables named _Age_, _Height_, _Weight_, _Appearance_, etc. Then, the statement "John is young," is interpreted as an _assignment equation_

Age = _young_

which assigns the value _young_ to the linguistic variable _Age_. In turn, the value _young_ is interpreted as a label for a fuzzy restriction on the base variable _age_, with the meaning of this fuzzy restriction defined by its compatibility function.

There are several basic aspects of the concept of a linguistic variable that are in need of elaboration.

First, it is important to understand that the notion of compatibility is distinct from that of probability. Thus, the statement that the compatibility of, say, 28 with _young_ is 0.7, has no relation to the probability of the age-value 28. The correct interpretation of the compatibility-value 0.7 is that it is merely a subjective indication of the extent to which the age-value 28 fits one's conception of the label *young*. As we shall see in later sections, the rules of manipulation applying to compatibilities are different from those applying to probabilities, although there are certain parallels between the two.

Second, we shall usually assume that a linguistic variable is *structured* in the sense that it is associated with two rules: (i) a *syntactic rule*, which specifies the manner in which the linguistic values which are in the term-set of the variable may be generated. In regard to this rule, our usual assumption will be that the terms in the term-set of the variable are generated by a context-free grammar.

The second rule, (ii), is a *semantic rule* which specifies a procedure for computing the meaning of any given linguistic value. In this connection, we observe that a typical value of a linguistic variable, e.g., *not very young and not very old*, involves what might be called the primary terms, e.g., *young* and *old*, whose meaning is both subjective and context-dependent. We assume that the meaning of such terms is specified *a priori*.

In addition to the primary terms, a linguistic value may involve connectives such as *and*, *or*, *either*, *neither*, etc.; the negation *not*; and the hedges such as *very*, *more or less*, *completely*, *quite*, *fairly*, *extremely*, *somewhat*, etc. As we shall see in later sections, the connectives, the hedges and the negation may be treated as operators which modify the meaning of their operands in a specified, context-independent, fashion. Thus, if the meaning of *young* is defined by the compatibility function whose form is shown in Fig. 1, then the meaning of *very young* could be obtained by squaring the compatibility function of *young*, while that of *not young* would be given by subtracting the compatibility function of *young* from unity (Fig. 2). These two rules are special instances of a more general semantic rule which is described in [9].

Third, when we speak of a linguistic variable such as *Age*, the underlying base variable, *age* is numerical in nature. Thus, in this case we can define the meaning of a linguistic value such as *young* by a compatibility function which associates with each age in the interval [0,100] a number in the interval 0, 1 which represents the compatibility of that age with the label *young*.

On the other hand, in the case of the linguistic variable *Appearance*, we do not have a well-defined base variable; that is, we do not know how to express the degree of beauty, say, as a function of some physical measurements. We could still associate with each member of a group of ladies, for example, a grade

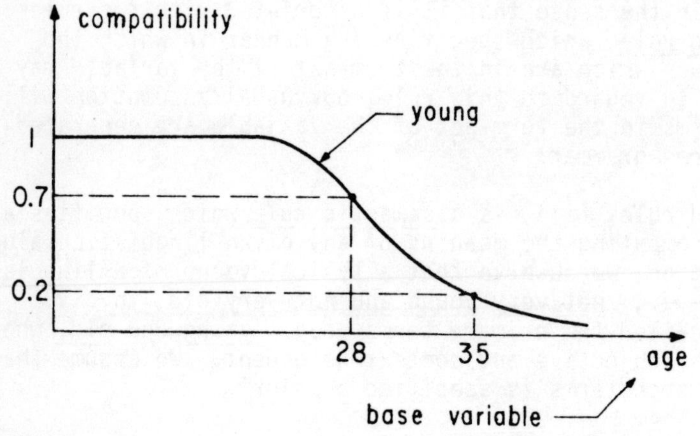

Fig. 1. Compatibility function for young.

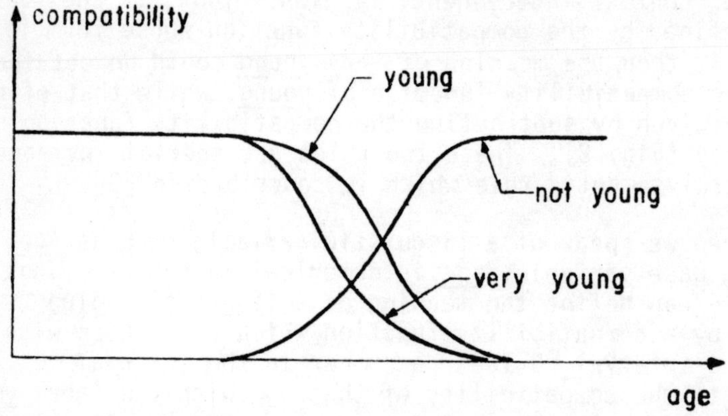

Fig. 2. Compatibilities of young, not young, and very young.

of membership in the class of beautiful women, say 0.9 with Fay, 0.7 with Adele, 0.8 with Kathy and 0.9 with Vera, but these values of the compatibility function would be based on impressions which we may not be able to articulate or formalize in explicit terms. In other words, we are defining the compatibility function not on a set of mathematically well-defined objects, but on a set of labeled impressions. Such definitions are meaningful to a human but not - at least directly - to a computer.[4]

As we shall see later, in the first case, where the base variable is numerical in nature, linguistic variables can be treated in a reasonably precise fashion by the use of the extension principle for fuzzy sets. In the second case, their treatment becomes much more qualitative. In both cases, however, some computation is involved - to a lesser or greater degree. Thus, it should be understood that the linguistic approach is not entirely qualitative in nature. Rather, the computations are performed behind the scene and, at the end, linguistic approximation is employed to convert numbers into words.

A particularly important area of application for the concept of a linguistic variable is that of approximate reasoning, by which we mean a type of reasoning which is neither precise nor very imprecise. As an illustration, the following inference would be an instance of approximate reasoning:

x is small

x and y are approximately equal

therefore

y is more or less small.

The concept of a linguistic variable enters into approximate reasoning as a result of treating Truth as a linguistic variable whose truth-values form a term-set such as shown below

Truth = true + not true + very true + completely true +
more or less true + fairly true + essentially true
+...+ false + very false + neither true nor false +...

The corresponding base variable, then, is assumed to be a number in the interval 0,[1], and the meaning of a primary term such as

---

[4]The basic problem which is involved here is that of abstraction from a set of samples of elements of a fuzzy set. A discussion of this problem may be found in [8].

true is identified with a fuzzy restriction on the values of the base variable. As usual, such a restriction is characterized by a compatibility function which associates a number in the interval [0, 1] with each numerical truth-value. For example- the compatibility of the numerical truth-value 0.7 with the linguistic truth-value very true might be 0.6. Thus, in the case of truth-values, the compatibility function is a mapping from the unit interval to itself.

Treating truth as a linguistic variable leads to a fuzzy logic which may well be a better approximation to the logic involved in human decision processes than the classical two-valued logic. Thus, in fuzzy logic it is meaningful to assert what would be inadmissibly vague in classical logic, e.g.,

> The truth-value of "Berkeley is close to San Francisco," is quite true.

> The truth-value of "Palo Alto is close to San Francisco," is fairly true.

Therefore, the truth-value of "Palo Alto is more or less close to Berkeley," is more or less true.

Another important area of application for the concept of a linguistic variable lies in the realm of probability theory. If probability is treated as a linguistic variable, its term-set would typically be:

> Probability = likely + very likely + unlikely + extremely likely + fairly likely + ... + probable + improbable + more or less probable + ...

By legitimizing the use of linguistic probability-values, we make it possible to respond to a question such as, "What is the probability that it will be a warm day a week from today," with an answer such as fairly high, instead of, say, 0.8. The linguistic answer would, in general, be much more realistic, considering, first, that warm day is a fuzzy event, and, second, that our understanding of weather dynamics is not sufficient to allow us to make unequivocal assertions about the underlying probabilities.

In the following, the concept of a linguistic variable and its applications to approximate reasoning will be briefly discussed. To place the concept of a linguistic variable in a proper perspective, we shall begin our discussion with a formalization of the notion of a conventional (nonfuzzy) variable. For our purposes, it will be helpful to visualize such a variable as a tagged valise with rigid (hard) sides. Putting an object into the valise corresponds to assigning a value to the variable, and the restriction on what can be put in corresponds to a subset of the universe of discourse which comprises those points which can be assigned as values to the variable. In terms of this analogy, a fuzzy variable may be likened to a tagged valise with soft rather than rigid sides. In this case, the restriction on what can be put in is fuzzy in nature, and is defined by a compatibility function which associates with each object annumber in the interval [0, 1] representing the degree of ease with which that object can be fitted in the valise. For example, given a valise named X, the compatibility of a coat with X would be 1, while that of a record-player might be 0.7.

An important concept in the case of fuzzy variables is that of noninteraction, which is analogous to the concept of independence in the case of random variables. This concept arises when we deal with two or more fuzzy variables, each of which may be likened to a compartment in a soft valise. Such fuzzy variables are interactive if the assignment of a value to one affects the fuzzy restrictions placed on the others. This effect may be likened to the interference between objects which are put into different compartments of a soft valise.

A linguistic variable is defined as a variable whose values are fuzzy variables. In terms of our valise analogy, a linguistic variable corresponds to a hard valise into which we can put soft valises, with each soft valise carrying a name tag which describes a fuzzy restriction on what can be put into that valise.

The concept of a linguistic variable can be applied to probabilities, and it is shown that linguistic probabilities can be used for computational purposes. However, because of the constraint that the numerical probabilities must add up to unity, the computations in question involve the solution of nonlinear programs and hence are not as simple to perform as computations involving numerical probabilities.

The so-called compositional rule of inference is interpreted as the process of solving a simultaneous system of so-called relational assignment equations in which linguistic values are assigned to fuzzy restrictions. Thus, if a statement such as "x is small" is interpreted as an assignment of the linguistic value small to the fuzzy restriction on x, and the statement "x and y are approximately equal," is interpreted as the assignment of a fuzzy relation labeled approximately equal to the fuzzy restriction on the ordered pair (x,y), then the conclusion "y is more or less small," may be viewed as a linguistic approximation to the solution of the simultaneous equations

R(x) = small

R(x,y) = approximately equal

in which R(x) and R(x,y) denote the restrictions on x and (x,y), respectively.

The compositional rule of inference leads to a generalized modus ponens, which may be viewed as an extension of the familiar rule of inference: If A is true and A implies B, then B is true. Interesting examples include a fuzzy theorem in elementary geometry and the use of fuzzy flowcharts for the representation of definitional fuzzy algorithms.

REFERENCES

1. Sir William Thomson, Popular Lectures and Addresses, McMillan and Co., London, 1891.
2. E. Feigenbaum, Computers and Thought, McGraw-Hill Book Co., New York, 1963.
3. M. Minsky and S. Papert, Perceptrons: An Introduction to Computational Geometry, The M.I.T. Press, Cambridge, Mass., 1969.
4. M. Arbib, The Metaphorical Brain, Wiley-Interscience, New York, 1972.
5. A. Newell and H. Simon, Human Problem Solving, Prentice-Hall, Englewood Hills, N.J., 1972.
6. L. A. Zadeh, "Fuzzy Languages and Their Relation to Human and Machine Intelligence," Proc. of Inter. Conference on Man and Computer, Bordeaux, France, pp. 130-165, S. Karger, Basel, 1972.
7. L. A. Zadeh, "Outline of a New Approach to the Analysis of Complex Systems and Decision Processes," IEEE Trans. on Systems, Man and Cybernetics, Vol. SMC-3, pp. 28-44, Jan. 1973.
8. R. E. Bellman, R. E. Kalaba and L. A. Zadeh, "Abstraction and Pattern Classification," Jour. Math. Analysis and Appl., Vol. 13, pp. 1-7, 1966.
9. L. A. Zadeh, "The Concept of a Linguistic Variable and Its Application to Approximate Reasoning," Memorandum No. ERL-M411, Oct. 1973, Electronics Research Laboratory, University of California, Berkeley.

FUNDAMENTAL CONCEPTS AND SOCIAL CONSEQUENCES OF ARTIFICIAL INTELLIGENCE

Shuhei Aida

University of Electro-Communications

Tokyo 182, Japan

1. CONCEPT OF ARTIFICIAL INTELLIGENCE

The definition and the domain of "artificial intelligence" is still obscure. This means the science of artificial intelligence is interdisciplinary and still at the frontier stage. It is of interest to scholars in the sense that future development is expected in connection with the information science and other related sciences.

However, as for using the term, artificial intelligence, there must be a certain general aspect in the domain covered by the term, and that aspect should have close connection with the meaning of the term. In this paper, the domain of "artificial intelligence" is regarded as a space for considering the general aspect just mentioned.

Based upon the understanding that a function of artificial intelligence consists essentially of manipulation of information, my discussion starts with examination of the concept of information. If we are to take up what we call information, information means a physical stimulus in terms of certain time and space or a physical constituent state. As material of information, we can think of matter and energy. If we extract a space here that consists of matter, energy, and information as a space composing this world, this space can be regarded as a partial space of the space for considering artificial intelligence because information physically originates in matter and energy. Then is there any reason to use the term information, if it is a physical stimulus in terms of time and space, or physical constituent state composed of matter and energy? There must exist certain entity that regards it as information. The entity mentioned at this point does not have to be a living creature; it could be a machine. We call this entity, which regards the physical stimulus or physical constituent state as information, a subject. The essence of a subject is the act of "regarding as information". However, what induces this act is problem sense, which the entity has his own or is given by other subjects. As for computers, this problem sense is given by man through programs. Man may impose problem sense on himself or it may be given by others.

There are two kinds of entity in the environment of a subject; one being the object of a problem sense of the subject, and the other which links the subject to the entity in terms of information. The former should be defined as an object and the latter as a medium. We have now gotten a partial space composed of subject, medium and object as a passage of information. This partial space actually is dependent upon problem sense of the subject. As elements, subject, medium and object do have unique characteristics in the space of information and information is following these characteristics. Therefore, the problem sense of the subject clarifies the structure penetrating the two partial spaces of matter, energy and information on the one hand, and subject, medium and object on the other. Under these circumstances, the subject searches for the solution to his own problems.

The information itself possesses the three aspects as follows; countability, morphological aspect, and descriptiveness. The space consisting of these three aspects may be defined as "information space" (Figure 1). Although each of these aspects is to be explained individually, these aspects cannot be considered as being independent of each other.

1) Countability

Countability of information can be obtained through correspondence between the physical stimulus or the physical constituent state and the set of numbers, such as natural numbers, real numbers, and vectors. Information aimed at countability should be called countable information. The subject having relation with the countable information should have the function of memorizing and manipulating numbers.

2) Morphological Aspect

Morphological aspect of information may be defined as a form or a geometrical picture of extention of physical stimulus or physical constituent state in terms of time and space. For instance, morphological information can be observed as a pattern

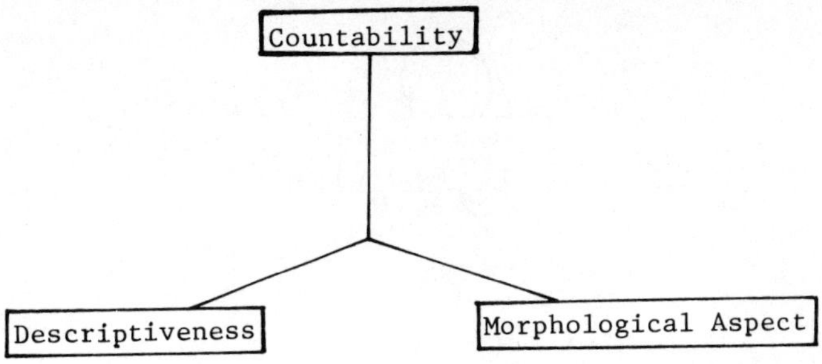

Fig. 1. Information space.

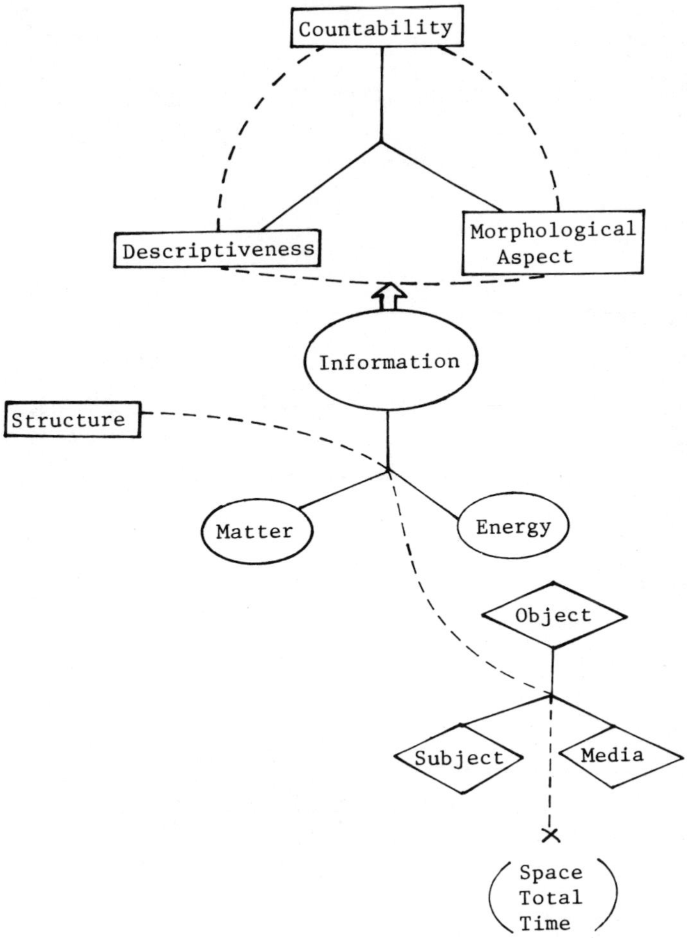

Fig. 2. Image of the thinking space.

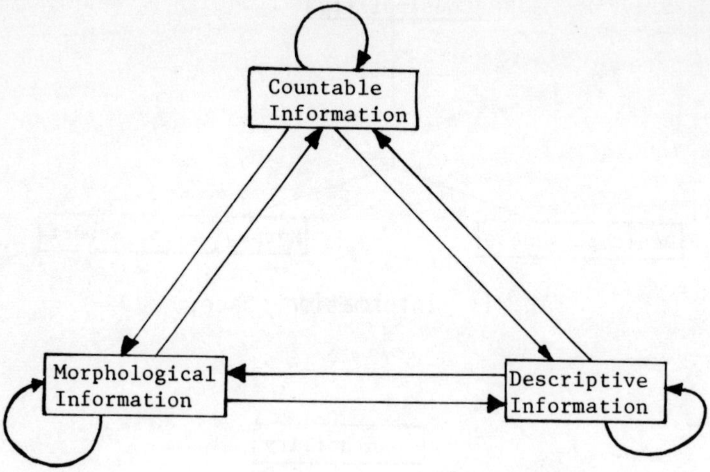

Fig. 3. Process of the transformation.

Table 1

Countability	X	-5	-4	-3	-2	-1	0	1	2	3	4
	Y	25	16	9	4	1	0	1	4	9	16
Morphological aspect	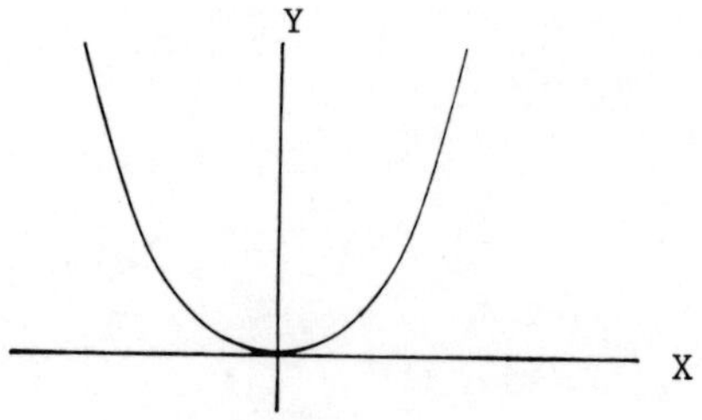										
Descriptiveness	$Y = X^2$										

written in ink on the paper with regard to recognition of letters. As for the movement of a rocket, the morphological information may be its shape or the orbit drawn. The information aiming at the morphological aspect is defined as morphological information. The subject dealing with morphological information should be able to preserve the extention in terms of space and time characteristic to morphological information.

3) Descriptiveness

Descriptiveness can be obtained through description of physical stimulus or physical constituent state in terms of the arrangement of symbols, such as formal language, logical equation, mathematical equation, and natural language. Information aiming at descriptiveness should be called descriptive information. The subject dealing with descriptive information should have the function of manipulating symbols, logically.

The three aspects of information do not exist independent of each other as already mentioned. For example, it is possible to regard countability as a special case of descriptiveness. Besides, we use morphological information when we write figures or symbols on paper. It is also possible to point out countability, morphological aspect, and descriptiveness as shown in Figure 1 if we examine the definition of a certain function in terms of information. The countable definition shows the correspondence of an independent variable to a dependent variable, in quantitative terms. A morphological definition is given in terms of a graph as co-ordinates. A descriptive definition is given as an equation such as $y=x^2$. When a subject regards a certain physical stimulus and a physical constituent state as information and tries to manipulate them, it is important to clarify which aspect he has an eye on. Figure 2 shows the entire image of a space of thinking obtained by the discussion thus far.

One of the manipulations of information as a function of artificial intelligence is transformation of information between these three aspects or within one of the aspects. Figure 3 shows the process of the transformation. All the possible ways of transformation can be listed as follows;

countable information → countable information
countable information → morphological information
countable information → descriptive information

morphological information → countable information
morphological information → morphological information
morphological information → descriptive information

descriptive information → countable information
descriptive information → morphological information
descriptive information → descriptive information

Various manipulations of information, such as management, observation, control and evaluation of information may be classified as one of the transformations of information or as being in line with one of them.

The artificial intelligence, functioning as a subject, forms a model of the object within itself utilizing its original internal structure and the information of objects obtained from the environment through a medium. Since the model itself is a kind of information, it has three aspects; countable model, morphological model, descriptive model or the model possessing two or three of these aspects. Within the artificial intelligence, the object is identified with the model, and the attitude of artificial intelligence toward the object is determined according to this model. Therefore, the model theory, modelling theory and the theory for manipulation of models should be attached importance at this point.

The heuristics, attracting the attention of scholars as a matter concerning artificial intelligence at present, is a method for solving the complicated problem or the problem with an enormous number of possible phases as in chess, for example. It utilizes a concept of state. Each state is expressed in terms of descriptive information using a symbol. Symbols thus defined are manipulated through a change of state. This method seems to be useful for the process of transformation between the aspects of information. The concept of fuzzy set may be effective under the circumstance when the artificial intelligence as a subject is given subjectivity. Moreover, in some cases, problem sense may be expressed as fuzzy set if it is not made clear. Examination of inter-relations between the fuzzy of the three aspects of information is necessary with regard to management of fuzzy information.

2. UNIT OF RATIONAL BEHAVIOR (URB)

Unit of Rational Behavior (URB) as a basic element of expressing the social system was proposed by Prof. S. Kumon (University of Tokyo). It generalized the idea of "homo-economics" which played an essential role in the economic theories. URB has an internal structure of its own and a linking structure with nature and other URBs. URBs may be defined as individuals, or organizations such as enterprises or the government offices, for convenience in expressing the social system in question. URB obtains information from the environment through linking structure, recognizes and evaluates the world and at the same time, tries to maintain or to achieve the desirable condition, through the internal structure. In order to attain this objective, URB sends out all sorts of information to the environment. Matter and energy are mutually exchanged between the URB and the environment.

The internal structure and the linking structure of URB is to be explained in the following.

2-1 Internal Structure of URB

Internal structure of URB may be considered as composed of two parts; an executive part and the controlling part. Figure 4 shows an example of the internal structure of the executive part. The action of the controlling part as shown in Figure 4 may be explained as follows;

1) The "detecting block" selects the information from the executive part and the external world at its discretion to send out to "recognition block".

2) At the "recognition block", newly obtained information is handled and world image is re-formed based on the principles of recognition (including the memory of world images previously formed). Then according to this image formed, assuming the set of possible means available, a model of cause and effect is (re-) formed as to how the condition of the world may be changed (or maintained) as a result of certain steps. The information regarding the pairs of cause and effect thus obtained is sent to the "evaluation block".

3) The "evaluation block" evaluates and arranges the various conditions of the world which may be achieved, in the order of desirability, using the set of objectives of its own. Then, it is determined as to which step should be taken and what kind of condition of the world should be realized, according to the standard of selection (such as standard of maximization, standard of satisfaction).

4) The "enforcement block" receives the information from the "evaluation block" and sends it out in the form of a signal having significance for the executive part (and the external world).

The active selecting action is characteristic to the action of the controlling part described above. URB may form different images of the world regarding the identical information, or may form different models of cause and effect concerning the same world image. Different evaluations are possible for the same set of pairs of cause and effect, or different means are possible

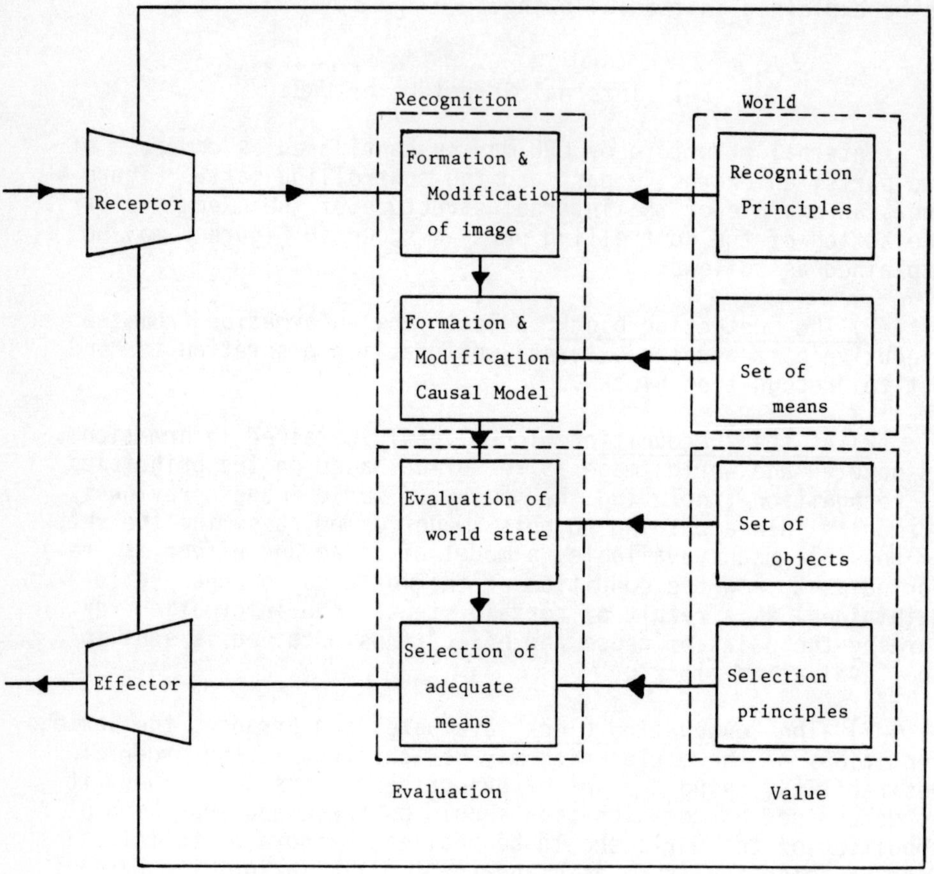

Fig. 4. Internal structure of the executive part.

with regard to the same evaluation. Of course these are supposed to be possible in terms of "technique". In the actual process, URB makes selections with the help of informations from active elements contained in the "world view block" and the "value system block". As for the moment when the choice is made, a single output information corresponds to a single input information. Four active elements shown in Figure 4 are not free of external influence. They may be transformed through learning based on the past experience or through approaches by other URBs (persuasion, education, etc.). Connotation of the rationality of the URB mentioned here, is to examine the condition of the world and the possibility of change, and to select the proper action giving consideration to the objectives and available means.

URB possesses a world image of its own. Constituent factors of conditions of the world, that it picks up as objects of interest, at the evaluation of the image of the world, may be called as commodity (for it). Commodity can be classified by various standards. For example, commodity for a certain URB may be classified as follows:

1) inner commodity: The value of which can be directly changed by the action of URB.

2) outer commodity: The value of which can be changed only through the decision of other URB (through collective decision of a number of URBs).

3) environmental commodity: The value of which cannot be changed by any of URBs. The change of its value is caused only by the function of "nature" or "environment".

Commodity may also be classfied by the following: some commodities can cause the change of the value of another commodity or of its own at its active state, while others cannot. The former may be defined as expedient commodity. The act of placing the expedient commodity to an active state mentioned above may be defined as to use. The commodities which cannot become expedient commodity may be called end commodity. End commodity may correspond to what K.E. Boulding has termed as "psychic capital".

Now, let me explain the act of "using". Firstly, we can think of "use in a narrow sense". "Use" in a narrow sense consists of production and consumption; production aiming at creation and addition of other expedient commodity, while consumption aiming at creation and addition of end commodity. With regard to the use of expedient commodity in a narrow sense, consideration should be given to concurrence, that is, when a certain URB is using a certain expedient commodity, other URBs are eliminated from using a certain expedient commodity, other URBs are eliminated from using it in the same sense. If a certain URB excludes the others and justifies his monopolistic use, this may be defined so as to set up the right of using regarding the commodity.

Based upon the assumption stated above with regard to commodity, let me explain the executive part of URB. It is convenient to consider it as a space containing both 1) the end commodity for URB and 2) expedient commodity for which URB possesses the proprietary right (or unlimited right of using) or temporary right of using. The executive part of URB receives instructions from the controlling part as to which expedient commodity should be used and how. The value of commodity which exists in nature or executive part, changes as expedient commodity is being used in practice. In some cases, this change accompanies physical movement of commodity itself, between the executive part and the nature, and between the executive parts, while in other cases, not. Regarding the former, the movement of commodity between the executive parts is called transfer of commodity and; as for the latter, the movement of commodity between the executive part and the nature is called metabolism of commodity.

The relation between the controlling part and the executive part of URB, described thus far, may be shown as in Figure 5. The transformation of commodity here means formation or extinction of commodity as a result of use of commodity in a narrow sense.

2-2 Linking Structure of URB

URB tries to create or to maintain the desirable condition of commodity (or condition of the world) utilizing the executive part or the expedient commodity existing externally. Each URB rarely acts independently of other URBs. Generally, there are various interactions played between URBs (formation of links). As basic patterns of links as a result of interaction, we can think of two types of links; the link through commodity and the link through information.

Examples of the links through commodity are as follows;

1) such act as desertion, suicide, subordination, expulsion, murder, and conquest which result in the extinction of the existence of URB as URB.

2) exchange: that is, (1) transfer of the right of commodity. (2) transfer of the right of using. (3) change of value of an inner commodity for URB which exists as outer commodity for the partner.

3) presentation or robbery: that is, arbitrary alteration

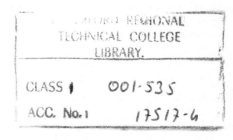

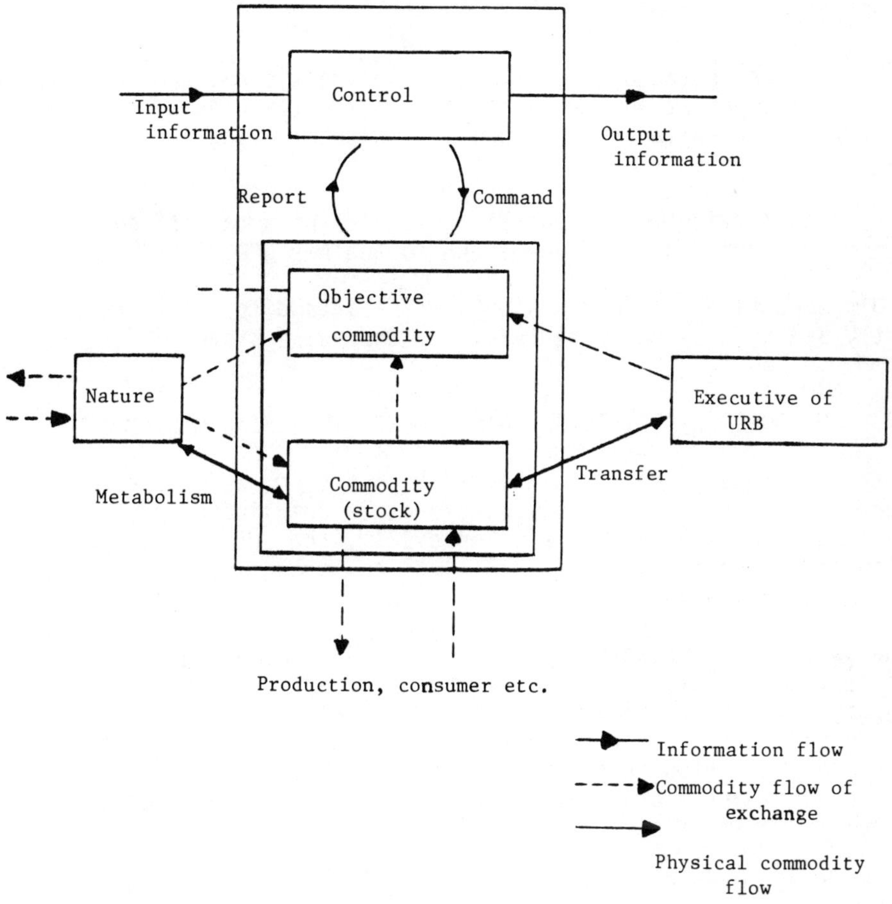

Fig. 5. Unit of Rational Behavior. (URB)

of the value of inner commodity which exists as outer commodity for the partner, or transfer of the (right of) commodity or the right of using, at URBs discretion.

Link through information is the most important method of reaching at consensus, with regard to the decision making (action) of each other, for a number of URBs in opposition. Examples are listed below.

1) such as mediation, arbitration, voting system, and law: which try to reach at an agreement with regard to rules of conduct or the process to attain consensus, without going into the content of the world view or the value system of each URB.

2) "instruction" or "report": which leaves a part of evaluating or recognizing function of URB to another URB.

3) "education" or "persuasion"; which intend to influence by directly acting upon the world view, or the value system.

3. THE ROLE OF ARTIFICIAL INTELLIGENCE IN SOCIAL SYSTEM

When we regard the social system as a system, we become aware of the difficulty with control because it is complicated as it contains all sorts of elements. Especially, the social system working under the influence of political and economic systems bears various problems within its activity. These problems as such from collapse of family to global problem of population, accompany the danger of growing worse without clarifying the ambiguity, since they have close connection with the inner level of the consciousness of man. The problems of the social system must be solved because they are directly affecting man. At this point, we give consideration to two ways of thinking, regarding the relation between artificial intelligence or intelligence in general and social system or system in general.

The first method considers artificial intelligence as a strategy for solution of problems and utilizes it to solve the problems within the social system. It may be a position of the application of artificial intelligence. Such issues require consideration as to uder which circumstances the method of artificial intelligence is necessary, or as to what kind of method is required, or in what cases a certain method is applicable. The other interprets the concept of intelligence in an enlarged scale. A system in general is regarded as having unique intelligence. This method aims at explaining structure of the system in terms of its intelligence. The intelligence mentioned here may be defined as system intelligence. Related to this is a method of describing the nature of the system through extracting intelligent factors from the system.

Let me explain the two methods mentioned. The first way is to utilize artificial intelligence as means of solving problems. In this case, the subject who solves the problem is URB within the social system, and it uses the method of heuristics. The existing condition of our bureaucracy leads us to anticipate that the internal structure of URB, exchanging information and commodity with the external world through linking structure, may become complicated if the social system becomes extremely complicated. The application of the method of heuristics to the complicated political issues is promising under these circumstances. Actually, the two processes;

1) formation and modification of the image of the world by principles of recognition

2) formation and modification of the model of cause and effect by the set of means

have to be taken into account. However, the expression of each content of the four blocks contained in 1) and 2), must be clearly transformed into the form of information in advance for the application of heuristics. That is, in terms of countable information, morphological information, and descriptive information.

The second method, based upon system intelligence, is intended to explain the structure of the system. Since we have assumed that the social system is composed of the links of URB, the function the social system carries out depends upon the linking structure and the internal structure of URB.

Therefore, consideration should be given to the change of linking structure and external structure of URB if we take up issues such as stability, variation and self organization of the social system. Since URB, the constituent of the linkage, is causing the change of linking structure, it is necessary to study the behavior of URB.

Let me consider URB in the domain of artificial intelligence. Like artificial intelligence, URB exchanges matter, energy and information with the environment. As for artificial intelligence, matter and energy carry information. However, as for URB, matter and energy are defined as commodity, being heterogenious to information. URB has to deal with all sorts of information appearing in the social system since it composes the social system. The point is clear if we consider the example of multi-national enterprises. Since URB is careless at handling information, highly efficient artificial intelligence is needed in order to achieve precision. The artificial intelligence will monitor itself and the environment so that the society will always maintain the best condition.

Based on the viewpoints thus developed, the concept of artificial intelligence shall be utilized as future strategy to solve the problems and to define self contradiction of the society of the human race, concerning energy, diplomacy, armament, etc.

4. UNIT OF ENVIRONMENTAL POLLUTION (UEP)

A hierarchical system composed of a central structure controlling the whole area and a network structure for communicating information is effective as a system for monitoring, predicting, and controlling the environment of an extensive area efficiently. However, there may be considerable difficulty in grasping the environment of a wide area continuously and equally. Even if it is limited to the air pollution, the condition of pollution differs locally, and it is difficult to monitor the entire area by just one system.

Therefore in order to monitor the air pollution constantly, to take immediate measures in times of emergencies, and to predict, it is effective to divide up the area into a number of unit spaces, as monitoring, prediction and control carried out in each unit and management concentrated at the center. Hence, it is necessary to set up pollution units, each containing a uniform effect characteristic to its locality, regarding urban structure, energy, and information. This should be a minimum unit for grasping the condition of pollution and for taking immediate measures in accordance with local characteristics. At the same time, pollution measures should be taken for the whole area being influenced by the unit.

This divided area unit is named as "Unit of Environmental Pollution"(UEP), concerning around the air pollution. The objectives of setting up UEP are listed below.

1) Monitoring, prediction, and control becomes easier as functions of the system.

2) Ambiguity characteristic to the air pollution if coped with by this method.

3) It manages the information concerning the air pollution such as urban structure, energy data, physiological or psychological data for man and other creatures.

4) Determination of environmental capacity is easy.

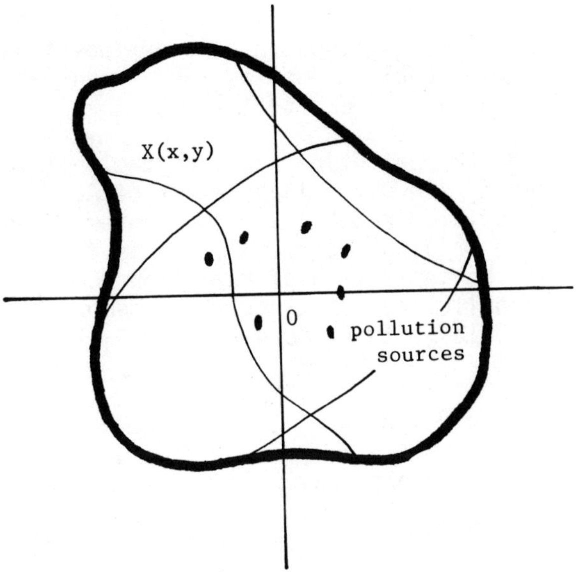

Fig. 6. Unit of Environmental Pollution.

4-1 Determination of UEP

At the primary stage UEP is defined as an area of high concentration centering around the specific pollution source. However, it is extremely difficult to determine the border of the area because of the ambiguity and variation of the distribution of pollution in terms of space and time, depending upon the meteorological conditions or change in the structure, energy, or distribution of information.

Therefore, the determination of the boundary utilizing the fuzzy set should be considered here.

Given the characteristics of the set in terms of concentration of a specific polluting substance, every position in the area above the threshold level, α_0 ppm as for concentration of pollution, is defined as an element of the set.

Now, X stands for the pollution space of the whole area and A_i represents the set of ith UEP. The element of A_i is provided as a co-ordinate of position x(X,Y,) and the center of pollution source is given as the original point. A_i stands for a set of positions showing the concentration of pollution above α_0 with regard to a specific polluting substance. The membership function showing the degree X belongs to A_i is given as $\mu A_i(X)$. At this point,

$$\mu A_i : x \to M$$

Although M is a membership space, it is assumed that M= [0,1]. That is,

$$\mu A_i : x \to [0,1]$$

When the value of $\mu A_i(x)$ is close to 1, the degree of x belonging to A_i is high. Conversely, it is low when close to 0. μA_i is a function determined by meteorological condition W(x), change of internal elements I(x), and external effects O(x).

However, at the primary level, it is defined according to the observed data or statistical data obtained at the border area in advance, or, by general tendency. Boundary can be given as a value of x corresponding to the point of partition, which divides the membership function by a certain threshold. As diagrammatized in Figure 6, UEP is defined as a summation of sets of each polluting substance.

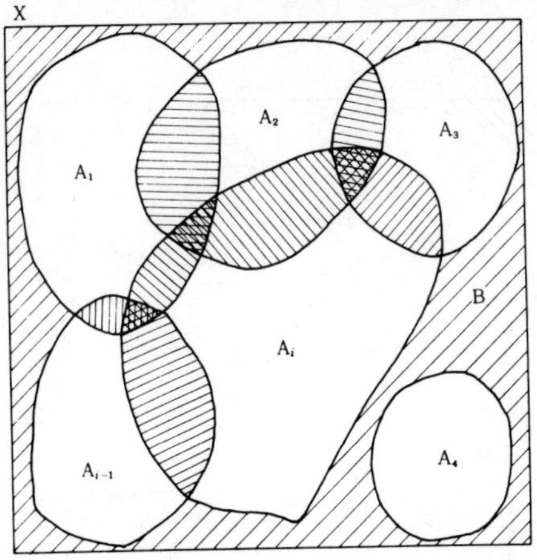

(a) Arrangements of UEPs.

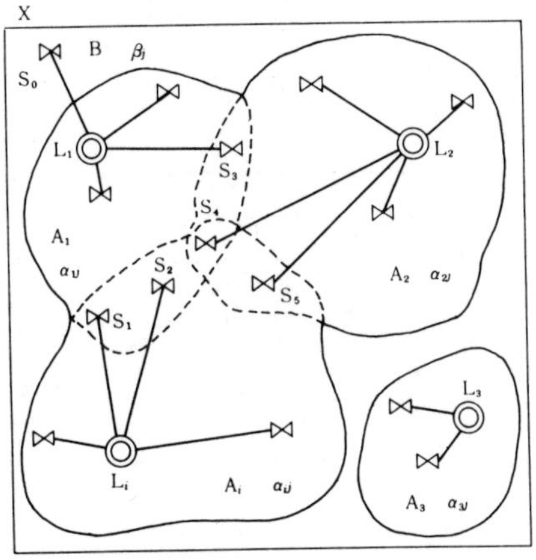

(b) Arrangements of sensors and local stations in UEPs.

$$\bigcup_{i=1}^{n} A_i \quad \& \quad \left(\bigcup_{i=1}^{n}\right) A_i \qquad a$$

Fig. 7. Arrangements of UEPs.

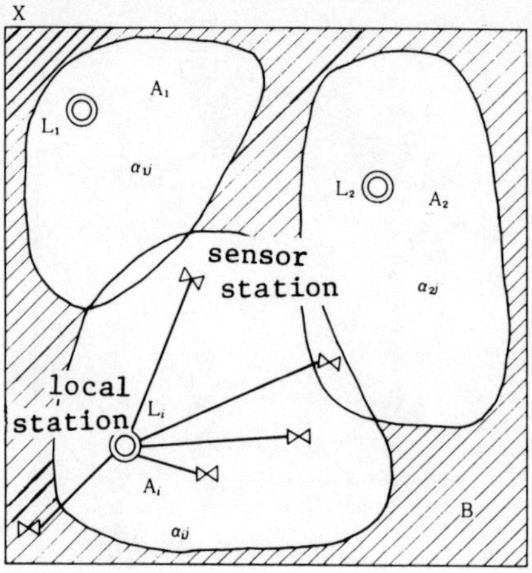

(c) Arrangements of sensors and local stations.

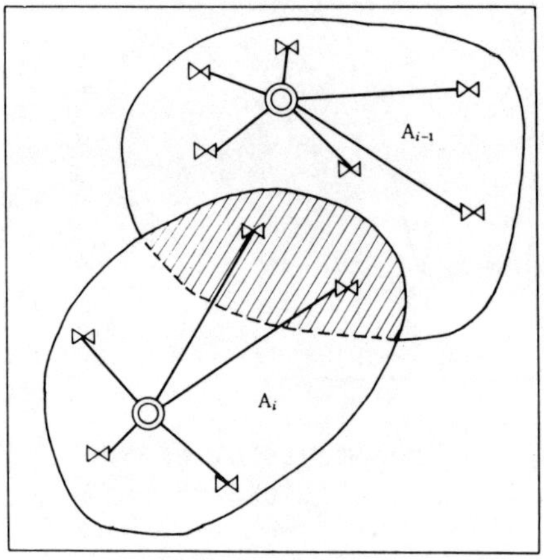

(d) $A_{i-1} \in A_i$

Fig. 7. Arrangements of UEPs (cont'd.).

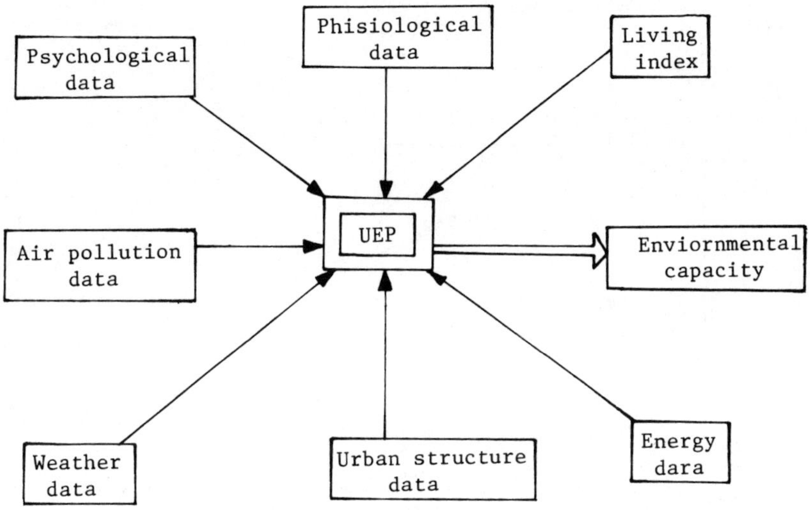

Fig. 8. Air pollution index.

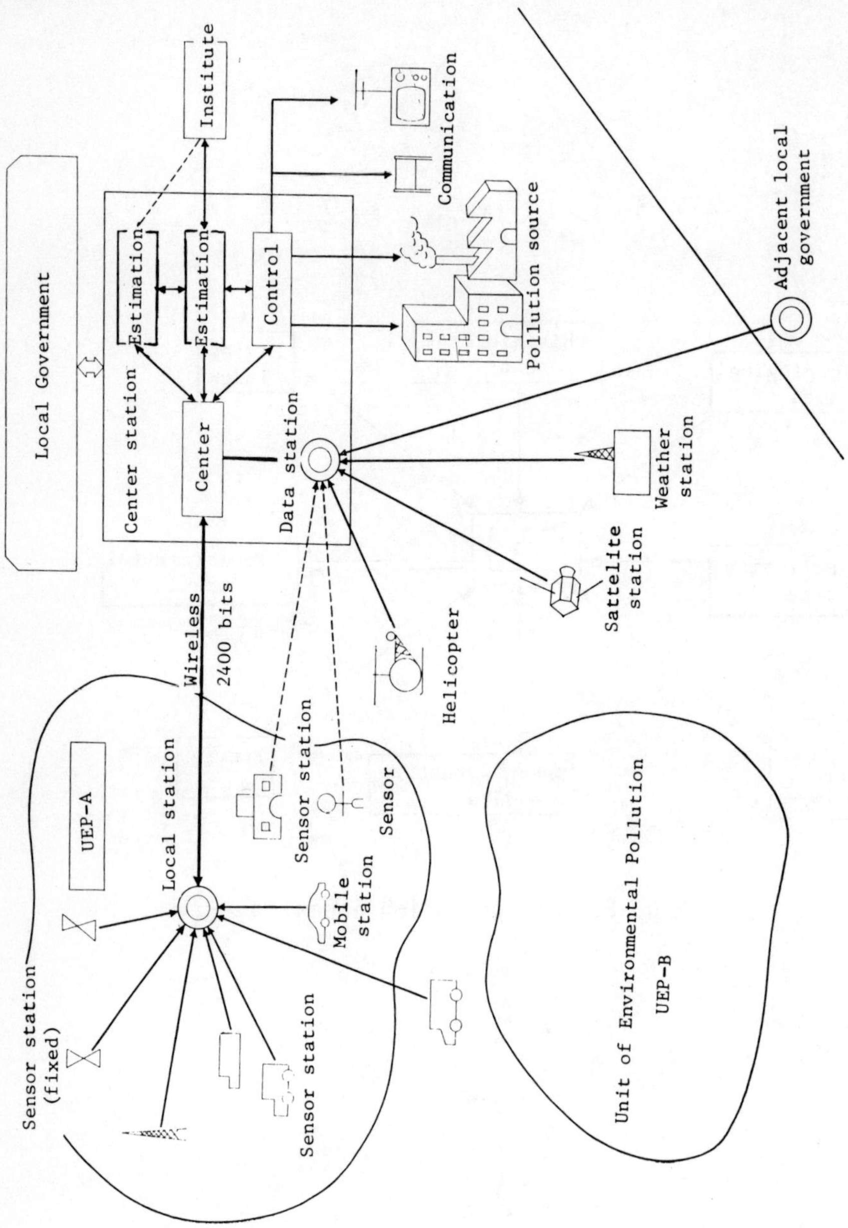

Fig. 9. Air pollution control system using the concepts of UEP.

4-2 Characteristics of UEP and Air Pollution

This intends to determine the environmental capacity based not only on the physical data but also on the ecological viewpoint giving consideration to the medical and social aspects of the air pollution as to how in fact, creatures are physiologically influenced by the air pollution, or, as to how human emotion or consciousness is psychologically affected. Figure 8 shows an example of the structure of air pollution controlling system. UEP has a function as a data distributing sub-system. That is, a filtering function of excluding the unnecessary part of the numerous data accumulated and filing only the useful data. Figure 9 shows an example of the arrangement of local station and sensor stations in UEP.

In case there is some overlapping of UEP as diagrammatized, sensor stations should be located without overlapping.

In the following stage, most suitable UEP should be determined by simulation using various methods.

BIOROBOTS FOR SIMULATION STUDIES OF LEARNING AND INTELLIGENT CONTROLS

Tosio Kitagawa, Honorary Professor Fukuoka

Kyushu University

Kyushu, Japan

TABLE OF CONTENT

1. INTRODUCTION AND SUMMARY

The purpose of this paper is to give a basic consideration for introducing a mathematical formulation of biorobots for simulation studies of ecosphere with an intension to give light to some essential features of learning and intelligent controls, as illustrated in Figure 5.2 in Section 5.

Various theories and techniques, such as sophistications and elaborations of optimal and adaptive controls, regarding learning controls have been mostly developed along the general traditional framework of control theory and control engineering. Intelligent controls have been discussed by many authors in pursuit along the prolongation of this line of development and by others in the framework of the current computer science. Nevertheless, it seems to us that there are urgent needs for deeper considerations upon the notions of learning and intelligence in order to realize more adequate formulation of learning control and intelligent control. In Section 2 we shall enumerate the five principles which should be adopted in our approach to the present topics. The key idea in our approach is to consider a certain type of intelligence which we may coin as biological intelligence, besides the current notions of human intelligence and mechanical one, as enumerated in Principle 1. Interdisciplinary cooperative approach ranging over biological science, information science, and mathematical science is indispensable for the study of biological intelligence, an enumerated in Principle 2. Principle 3 emphasizes the biological science aspect of our approaches by pointing out that the notions of ecosphere and evolution are indispensable to discuss biological intelligence. On the other hand, Principle 4 indicates that the whole aspect of logic of information science should be taken into our consideration in dealing with biological intelligence. The last Principle 5 is concerned with the role of the mathematical science approach in cooperation with experiments and technologies, both of which are also indispensable components of our researches on biological intelligence.

Section 3 is devoted to a brief illustration of what we call the logic of information science. There are three coordinations in our formulation of the logic of information science. It is remarked that, although we are now in this Seminar concerned with learning control and intelligent control, the two other aspects of feasibility as shown in the third coordination, namely, eizon and creation, besides the authentic aspect of control, are indispensable to appreciate the deep inplications of learning and intelligence.

In Section 4 we explain our notion of biological intelligence by showing the interralationship between the logic of information science enumerated in Section 3 and genetic epistemology of J. Piaget [11], [12]. After giving five basic observations on the Piaget theory, we shall show the implications of the three coordinations given in Section 3 with reference to the genetic epistemology.

Section 5 is devoted to mathematical formulations of biorobots and their implications to the studies of learning controls and intelligent controls. The notion of biorobots was introduced by the present author in his recent paper [10] in connection with his cell space approaches to biomathematics. The first four subsecions give an outline of the notion of biorobot and ecosphere in which these biorobots are participating members of the game to be played there. In Subsection 5.5 we shall refer to the work of L.T. Fogel, A.J. Owens and M.J. Walsh [4] , and we shall point out our emphasis on biorobots and ecosphere, in comparison with their works.

A few comments are given here on our previous works [6] [7], [8], which are more or less concerned with some features of sophiscated control procedures. It is remarked that although most of them are lacking in giving any explicit reference to learning and intelligence, they can be considered to be located in the neighborhood of learning and intelligent controls. In fact, some features of these types of control can be traced in the formulations of these previous papers. It is the view of the present author that, in view of biological intelligence explained in the first four Sections, we have to modify each of the formulations given in our previous papers [6], [7], [8], so as to take into our consideration the notion of biological intelligence, biorobots and ecosphere discussed in this paper.

2. FIVE PRINCIPLES IN OUR APPROACHES TO LEARNING CONTROLS AND INTELLIGENT CONTROLS

Principle 1. We should set up a systematic research of (i) oberservational surveys, (ii) hypothesis testing investigations and (iii) logico-mathematical deductive studies of biological intelligence which includes animal learning processes as well as plant adaptations.

Principle 2. We should appeal to an interdisciplinary approach which ranges over the area partly covering (a) biological science, (b) information science and (c) mathematical science.

Principle 3. We should have a systematic description of the real pictures of ecosphere in which various biological organisms attain their existence through competitions and cooperations and in which historical processes of biological organisms including genesis, growth, breakdown, and disintegration of each species have been involved.

Principle 4. We should refer to the whole feasibility aspects of information science including control, eizon and creation. In particular, biological intelligences should be discussed with reference to three functions of information, namely, cognition, direction, and evaluation.

Principle 5. We should set up a logico-mathematical formulation of biological intelligences as a theoretical framework, which will work under the cooperation of the two other indispensable components of approaches, namely, experimental works and technological realizations regarding biological intelligences.

3. LOGIC OF INFORMATION SCIENCE APPROACHES

In order to make clear the logical formulation of information science approaches, a system of three coordinations was introduced in one of our previous papers [9] . The first coordination is concerned with three aspects of information structure: (a) objectivity, (b) subjectivity, and (c) practices, while the second coordination with three aspects of information functions: (α) cognition, (β) direction and (γ) evaluation. Table 1 shows that each of these six aspects (axises) consists of three fundamental notions. For instance (a) objectivity consists of (a_1) pattern, (a_2) operation and (a_3) transformation; (b) subjectivity consists of (b_1) operation, (b_2) adaptation, and (b_3) strategy; (c) practices consists of (c_1) optimalization, (c_2) stability, and (c_3) learning. Similarly for (α) congnition, (β) direction, and (γ) evaluation as shown in Table 1.

Now the third coordination is concerned with the three feasibility aspects called (III_1) control, (III_2) eizon, and (III_3) creation. Each of these (III_i) (i=1,2,3) has six notions consisting of (a_i), (b_i), (c_i), (α_i), (β_i), and (γ_i), where i=1 for control, i=2 for evolution, and i=3 for creation.

Besides the two fundamental notions of (β_1) direction and (β_3) creation there comes another one coined as "eizon" which seems to be indispensable in formulating all the essential aspects of direction. The terminology "eizon" is a new Japanese word which was first introduced by the author in his monograph in October, 1969. The word "eizon" is a combination of two Chinese characters used in Japanese pronounciation, "ei" and "son (=zon," where "ei" comes from the current Japanese word "keiei" corresponding to "management" in English, while "zon" comes from the sophisticated notion "jitsuzon" in Japanese which has been introduced to translate existence in the sense of existentialism in the French school of philosophy). The reason we introduce the direction aspect of "eizon" comes from the real situations where our direction aspect cannot and should not be formulated by either control or creation aspects. In the first place, in these cases we cannot and should not have either any definitely prescribed single aim or a prescribed set of actions from which we can choose a suitable subset and/or a subsequence of actions for our direction. This shows a sharp distinction with the current control aspect. Secondly, in these cases we cannot and should not be so positive in our direction aspect as to create any set or actions which will suit our situations. This again shows a clear distinction with the so called creation aspect. In fact, we are essentially concerned with the feasibility of the existence; and we cannot and should not be adhering to any unchangeable aims for our direction, all of which are essentially subordinate to the feasibility of the existence itself. The supreme coordination principle for management should also be concerned with the feasibility of existence.

Table I - Three coordinations (I), (II) and (III) is the logic of information.

	(III_1)Control	(III_2)Eizon
(I_1)Objectivity	(a_1)Pattern	(a_2)Chaos
(I_2)Subjectivity	(b_1)Operation	(b_2)Adaptation
(I_3)Practices	(c_1)Optimalization	(c_2)Stability
(II_1)Cognition	(α_1)Deduction	(α_2)Induction
(II_2)Direction	(β_1)Control	(β_2)Eizon
(II_3)Evaluation	(γ_1)Efficiency	(γ_2)Reliability

(III_3)Creation
(a_3)Transformation
(b_3)Strategy
(c_3)Learning
(α_3)Abduction
(β_3)Creation
(γ_3)Plasticity

Without entering into any further detailed explanations of these eighteen component notions given in Table 1, which was fully illustrated in our previous work [9], we are content to point out the following three remarks:

(1^0) In considering the topics regarding learning control and intelligent control, it is an authentic attitude to rely upon all three of the feasibility aspects uncluding eizon and creation aspects besides control aspect. This assertion is due to the observation that adaptation should be referred to the framework of eizon aspect and that intelligence beyond adaptation should be discussed with reference to creation aspect.

(2^0) In order to have a systematic research of observational surveys, hypothesis testing investigations, and logico-mathematical studies on biological intelligence, the second coordination referring to the three information function aspects are most useful, if not indispensable, in the sense that it will give us a genetic epistemological channel connecting mechanical, biological and human intelligences.

(3^0) The differences among control, eizon, and creation can also be observed by a comparison of the two frameworks given in our previous papers [5] and [9] , respectively.

4. BIOLOGICAL INTELLIGENCE AND LOGIC OF INFORMATION SCIENCE

The terminology of biological intelligence is neither currently used nor sufficiently well-defined until it has been given a solid consensus in many scientific circles. The purpose of this Section is to explain our notion of biological intelligence by illustrating the interrelationship between our understanding, or the logic of information science shown in Section 3, and the general framework of genetic epistemology due to the famous work of J. Piaget [11] and [12].

In fact, we shall show that each of the key notions introduced by J. Piaget [11], [12] in his genetic epistemological approaches can be understood as a prototype of the respective key notion in our formulation of information logic. For this purpose, let us start with a few comments on the general observations which are basic in our approach.

Observation 1. In order to have a deep understanding of intelligence which is both biological and logical, we have to consider both of affective life and cognitive one. The Japanese version of "information", JÔHÔ, consists of two Chinese characters JÔ and HÔ (in Japanese pronounciation), where the former indicates the affective aspect of information, while the latter the cognitive one.

Observation 2. Our notion of feasibility is concerned with feasibility of existence which is secured by an equilibrium between the action of the organism on the environment and vice versa.

According to J. Piaget [11], assimilation is used to describe the action of organism on surrounding objects, in so far as this action depends upon previous behaviours involving the same or similar objects; and accomodation is used by him to denote the action of environment on the organism, which leads to a simple modification of the action of the organism. (See [11] p.7). J. Piaget [10] defines an adaptation as an equilibrium between assimilation and accomodation which amounts to the same as an equilibrium of interaction between subject and object. (See [11] p.9-11).

Observation 3. According to J. Piaget [11], intelligence itself does not consist of an isolated and sharply differentiated class of cognitive processes. This denial involves a radical functional continuity between the higher forms of thought and the whole mass of lower types of cognitive and motor adaptation. Intelligence is defined in terms of the progressive reversibility of the mobile structures which it forms, and "intelligence constitutes the state of equilibrium towards which tend all the successive adaptations of a sensori-motor and cognitive nature, as well as all the assimilatory and accomodatory interactions between the organism and the environment" (See J. Piaget [11] p.11).

Observation 4. The mechanism of transition proposed in the Piaget theory is an equilibration process. Piaget believes that an equilibration-equilibrium model is particularly suited to the analysis of ontogenetic change of structures. (See J.H. Flavell [2]).

Observation 5. The assimilation-accomodation model, due to J. Piaget ([11], [12]), as a theory of intelligence has an important implication to the notion of play and imitation. (See E.W. Beth and J. Piaget [1] and J.H. Flavell [2]).

After these preparatory observations on the general framework of genetic epistemology due to J. Piaget, its correspondence to our formulation of logic of information science can be enumerated in the following way:

(1^0) The first coordination. Each of the three key notions in setting up the intelligence of the lower level such as (a') environment, (b') organisms, and (c') action is a prototype of each of the three constituent axes of the first coordination, (a) objectivity, (b) subjectivity and (c) practices, respectively. Figure 4.1 illustrates some aspects of information logic from genetic epistemology.

(2^0) The first and the second coordination. Each of the two main features of organisms, cognition and affections, is the origin of structuration and valuation respectively. In our formulation of the information logic, the information structure is formulated in the framework of the first coordination, can be recognized as an objective realization of the three cognitive functions of information, namely (α) cognition given in the first axis of the second coordination. On the other hand, valuation based upon affective life of organism is objectively described by means of the two functions of informations, (β)direction and (γ) evaluation in the second coordination.

(3^0) The third coordination. The notion of biological adaptation explained by Piaget [11], [12] in connection with development of intelligence which implies three features (i') assimulation, (ii') accomodation and (iii') equilibration, can be considered as a prototype of our fundamental notion called feasibility, more exactly, feasibility of existence. There are three basic features called (i) control, (ii) eizon (feasibility of existence by passive attitudes), and (iii) creation. In fact, there is a correspondence between (i^k) and ($i^{k'}$) by which we can consider ($i^{k'}$) as a prototype of (i^k) (k=1,2,3). See Figure 4.1.

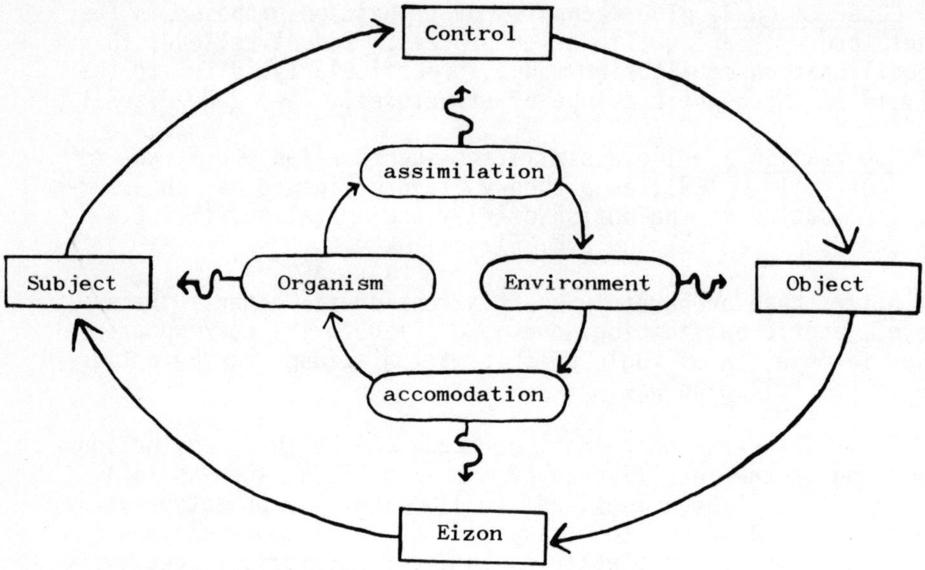

Fig. 4.1. The logic of information viewed from genetic epistemology.

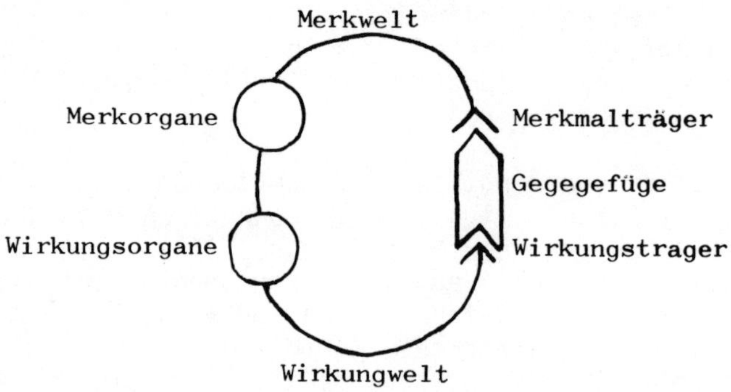

Fig. 5.1. Umwelt.

In this correspondence, stability and reliability in the eizon space in the second coordination can be considered to be crucially required in equilibrium. On the other hand, equilibration can be recognized to be assocaited with adaptation process which requires a new type of equilibrium after the stage of temporary disintegration. The process of disintegration of the old structure system and formation of a new structure system can be recognized as belonging to the realm of creation from the standpoint of subjective attitudes. In this connection, it is noted that plays and imitations, both of which have a deep implication in the assimilation-accomodation model due to J. Piaget, are also crucial in understanding the real functions of the creation process. This remark is also useful in understanding that two notions of learning and strategy are indispensable to distinguish the characteristic features of creation in comparison with the other two feasibility aspects, control and eizon. Abduction in the cognition axis and plasticity in the evaluation axis can also be recognized as being indispensable in our formulation of creation space.

(4^0) The first and the third coordination. After setting up the two-dimensional framework of the first and the third coordinations, the objectivity axis can be considered to have (a_1) pattern in control, (a_2) chaos in eizon and (a_3) transformation in creation. In the subjectivity axis, the primitive notion of action can be formulated as operation in the logic of information, which is also fundamental in a certain stage of development in the Piaget theory of genetic epistemology. Similar observations hold to be valid regarding the practice axis and the cognition axis.

5. MATHEMATICAL FORMULATION OF BIOROBOTS AND THEIR IMPLICATIONS TO THE STUDIES OF LEARNING CONTROLS AND INTELLIGENT CONTROLS

The terminology of biorobots was introduced in a recent paper [10] of the author in connection with cell space approaches to biomathematics. The purpose of the present Section is (1) to illustrate the implications of cell space approaches as mathematical models in which some of essential features of genetic epistemology illustrated in Section 4 can be formulated and analysed, and then (2) to point out the possibility of realizing at least part of the five principles of our approaches set up in Section 3.

5.1. Umweld and Basic Cell Space

It is crucial to have a local space representation which can represent the essential features of the local world in which a certain organism does have its living. The notion of "Umwelt" introduced by J. von Uexküll and G. Kriszat [13] does seem to give us an excellent description for this purpose. In fact, various key notions as shown in Figure 5.1 and their mutual interrelationship yield us the possibility of applying system approach to its study and that of introducing a notion of local space associated with each individual organism. One of the possible mathematical formulations which may well represent some essential features of such a local space associated with the organism will be given by a basic cell space in a cell space approach, in which a certain set of local transformation associated with each basic cell space is defined.

5.2 Biological Intelligence and Biorobots

The notion of current robots have been mostly, if not exclusively, introduced in order to represent a technological imitation of a certain set of human abilities including pattern recognitions, motion behaviors and learning processes. We have introduced in our recent paper [10] the notion of biorobots by which we aim to give a mathematical similation of a certain set of abilities of biological existences including the functions of informations as well as the mobility. Biorobots as mathematical models can yield us logico-deductive existences, and a coexistence of various kinds of biorobots provide a mathematical formulation of ecosphere. The real pictures of competitions and coexistences in the real ecosphere can be simulated by the mutual interferences among the existing biorobots. Animal robots, as well as plant robots, must be included as indispensable members of the ecosphere model. Here we are concerned not only with a food chain, which is mostly viewed from the standpoint of matter and energy, but also with an information interrelationship as one indispensable component of the whole aspect of the real ecosphere. Each individual biorobot is assumed to have a certain biological intelligence which we explained in Section 4. Consequently, our mathematical formulation of biosphere is concerned with the society of the various organisms. In fact, each individual organism is participating in the ecosphere where various ecological phenomena due to competitions, coexistences, and struggle for existence in general sense are being performed.

The abilities of the participants must be objectly defined regarding (α) cognitions, (β) direction, (γ) evaluation as well as (δ) mobility. Regarding (α) cognition, for instance, the ability to recognize patterns can be described with reference to (i) masking systems and (ii) type of mask functions used by the participant for pattern recognitions, which make it possible to have cell space formulation of computational geometry due to E.F. Codd [2]. The operational views of biological intelligence as illustrated in Section 4 can also be applied to (β) direction, (γ) evaluation and (δ) mobility.

5.3. Game Theoretic Formulation of Ecosphere

What we propose here as a mathematical model of an ecosystem is a game in which biological robots are being introduced. There are five leading principles upon which our formulation of ecogame should be based. (See [10]).

Principle 1. Our emphasis is to investigate common aspects of principles which are valid for different levels of biological existence in spite of the superfluous complexities of biological phenomena.

Principle 2. Each species that participates in the ecosphere has as its mathematical model a robot that has a certain set of biological intelligence and mobility. Each player in the game is not necessarily a single person, but it may be a family of the robots, each of which may play independently in a parallel way or cooperatively with their mutual relationship.

Principle 3. We introduce not a single game but a sequence of mutually connected games. Game rules should not be fixed but should be subject to change in such a way that they can be adapted to experiences.

Principle 4. In our approach three feasibility aspects, called control, eizon and creation should be taken into our consideration as guiding principles to observe the process of plays of the game. This implies, in particular, that we should not be confined to an evaluation scheme of a single criterium function.

Principle 5. The dynamical behaviors of the ecosphere are shown by tracing each individual play as a sample. Strategic behaviors of the players should be objectively formulated in our mathematical model with reference to biological robots associated with the ecogame. Computer simulations can contribute to investigating the dynamics of an ecosphere. An accumulation of play experiments of ecogame and their analysis from the standpoints of three feasibility aspects will give us the comparative study of various different levels of biological intelligence, each of which is possessed by a certain species of biorobot. The dynamic behaviors of the ecosphere and their final fortunes may be primarily judged from the supreme criterium as to whether the ecosphere can continue to exist.

5.4 Evolution of Biorobots Under the Feasibility Conditions of Ecosphere

In our game theoretic approach to ecosphere, we start with a certain set of various different biorobots whose biological intelligence are assigned respectively. Then the simulation

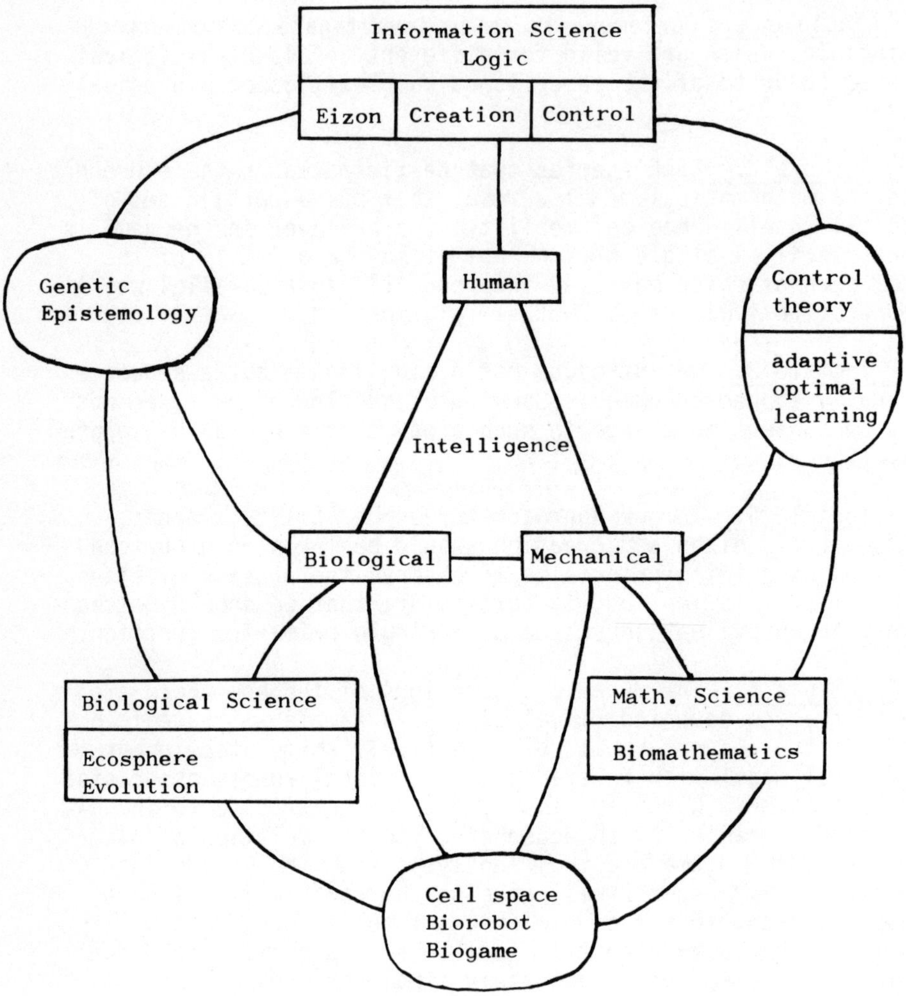

Fig. 5.2. Intelligence studies in interdisciplinary approaches.

experiments of the plays may sometimes show an equilibrium state in the sense that the play can continue furthermore without any ruin of the species of biorobots. But in other cases their fortune may turn out to be a ruin of the game; that is, some species of biorobots may become extinct as the play goes on. In such a case it is interesting to make a modification of biological intelligence of the species and to have another set of simulation experiments for the new set of biorobots. Such a modification process of the biological intelligences will give an insight about the evolution process of biological intelligence in the natural history. (See also [10]).

5.5 Artificial Intelligence Through Simulated Evolution of Biological Intelligence

In their use of evolutionary programming in creating artificial intelligence, L.J. Fogel, A.J. Owens, and M.J. Walsh [4] viewed intelligent behavior as a composite of an ability to predict the sensed environment, coupled with an ability to select a suitable response in the light of the prediction and the given goal. (See [4] p. 119). The problem of predicting each next symbol is reduced to the problem of finding the finite-state machine which plays the role of evolving organisms. The machine demonstrating greater worth is selected to serve as the new parent. Without going into any more detail of their approach, it is remarked that our main emphasis in this paper is a mathematical formulation of ecosphere where ontogenetic development of individual organism and evolution process of species have been realized through the history of their competition and their mutual coexistences.

In concluding this paper, Figure 5.2 will serve to illustrate our aims and our proposals for building up the connection routes among information science, biological science, and mathematical science, which will give us a clear picture describing the mutual relationship among control theory, genetic epistemology and our cell space approaches including biorobots and biogames.

REFERENCES

(1) Beth, E.W., and Piaget, J. : Mathematical epistemology and psychology (Translated by Mays, W.), D. Reidel Publishing Company, Dordrecht-Holland, 1966.

(2) Codd, E.J.: Cellular Automata, Academic Press, New York London, 1968.

(3) Flavell, J.H.: The Developmental Psychology of Jean Piaget, Van Nostrand Reinhold Comp., New York, Cincinnati, Toronto, London, Melbourne, 1963.

(4) Fogel, L.T., Owens, A.J., and Walsh, M.J.: Artificial Intelligence through Simulated Evolution, John Wiley & Sons, New York, London, Sydney, 1966.

(5) Kitagawa, T.: "The logical aspect of successive processes of statistical inferences and controls", Bull. Inter. Statist. Ins. 38(1961), 151-164.

(6) Kitagawa, T.: "Successive process of statistical controls", (3), Mem. Fac. Sci., Kyushu Univ., Ser. A, 14(1960), 1-33.

(7) Kitagawa, T.: "Successive process of optimizing pro edures", Proc. 4th Berkeley Symp. on Math. Statist. and Probability. I (1961), 21-71.

(8) Kitagawa, T.: "A mathematical formulation of the evolutionary operation program", Mem. Fac. Sci., Kyushu Univ., Ser. A, 15(1961), 21-51.

(9) Kitagawa, T.: "Three coordinate systems for information science approaches", Information Sciences, 5(1973), 157-169.

(10) Kitagawa, T.: "Cell space approaches in biomathematics", Math. Biosciences, 17(1973) (in press)

(11) Piaget, J.: The Psychology of Intelligence, (Translated from the French by Piercy, M. and Berklyne, D.E.), Poutledge & Kegan Paul LTD, London, 1950.

(12) Piaget, J.: Genetic Epistemology (Translated by Duckwork,E.) Columbia Univ. Press, New York, 1970.

(13) von Üexkull, J. and Kriszat, G.: Streifzüge durch die Umwelten von Tieren und Menschen, S. Fischer Verlag, Frankfurt am Main, 1970.

A MATHEMATICAL NEURON MODEL WHICH HAS A STAIRCASELIKE RESPONSE CHARACTERISTIC

Jin-ichi Nagumo

University of Tokyo

Tokyo, Japan

In 1961, Harmon proposed an electronic circuit as a neuron model and carried out a series of experimental investigations of its properties [1]. The neuron model is shown in Fig. 1, where two transistors, T_1 and T_2, together form a monostable circuit whose function is to produce a single pulse if the threshold is exceeded, and T_3 and T_4 form an amplifier and low impedance output source. The function of T_5 is to invert an otherwise excitatory input, and the inverted input is summed with the excitatory input to produce inhibition. The refractory function is obtained from the time constant associated with T_1 and T_2, whose pulse emitting activity may be inhibited by a graded change in their virtual threshold.

During the course of experimental studies with his neuron model, Harmon found an "unusual and unsuspected" phenomenon which is described below.

If a unit of the neuron model is used to drive another directly, and the pulse amplitude is sufficiently high, the firing of the second follows the first, pulse for pulse. However, if the output amplitude of the driving unit be monotonically decreased, firing in the second unit begins to skip. More precisely, as the driven unit receives progressively less excitation, there will be a critical point at which a given driving pulse is insufficient to fire the driven unit. However, it turns out that when the next driving pulse comes along, it is integrated with the preceding one to come firing, and the firing frequency of the driven unit is half that of the driver. One would expect that as the driving pulse amplitudes are reduced still more, every third pulse would be effective, then every fourth, and so on.

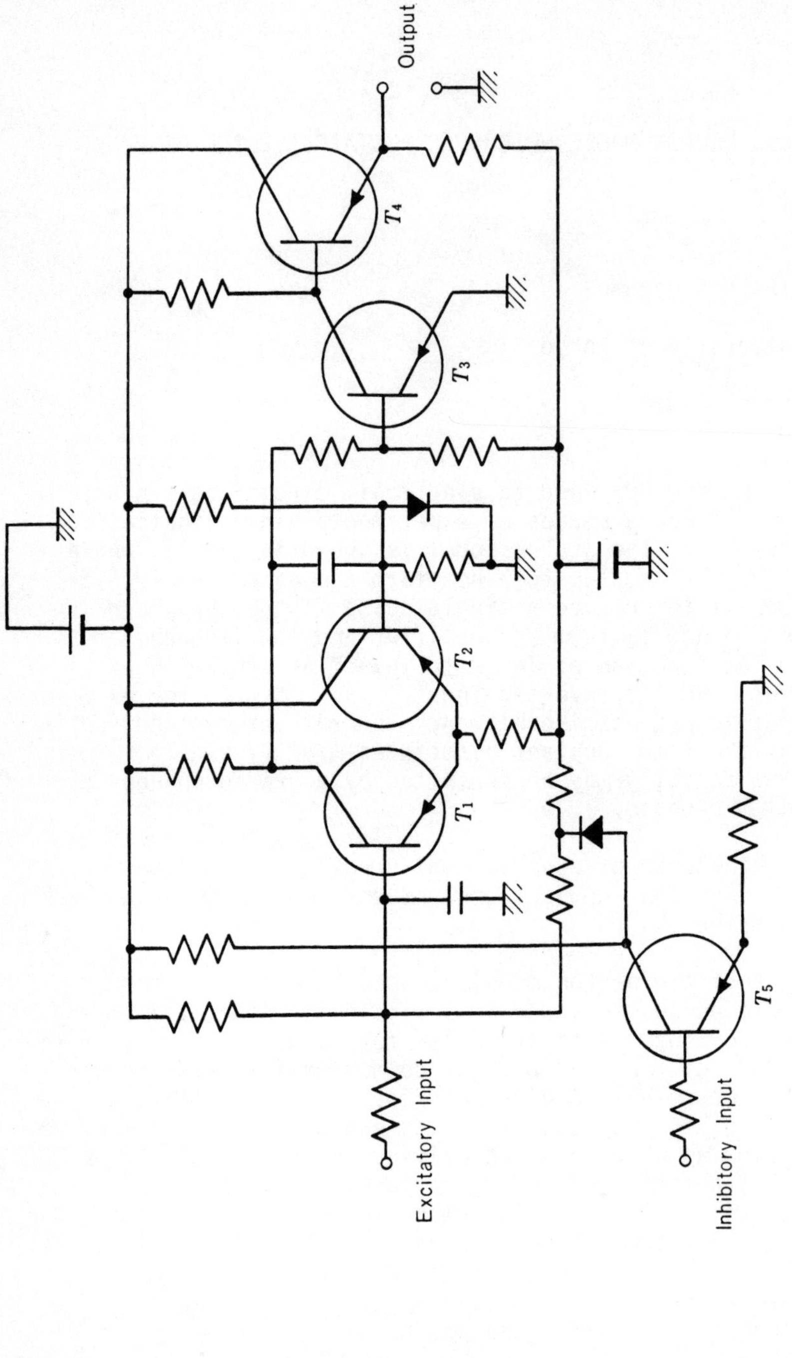

Fig. 1. Harmon's electronic neuron model.

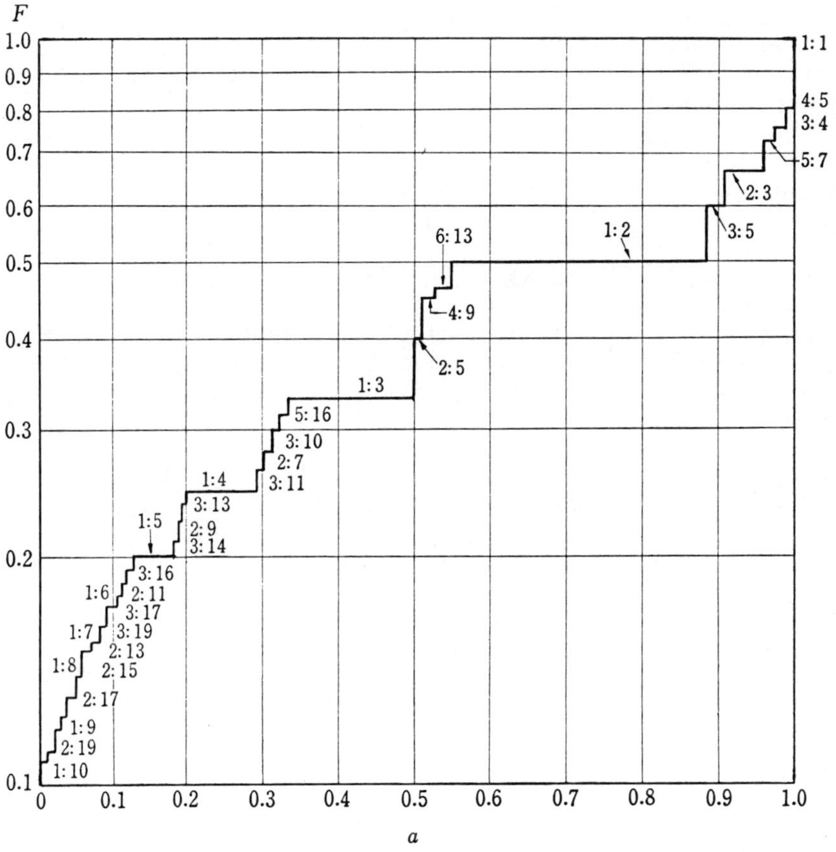

Fig. 2. Relationship between the amplitude of driving pulses (a) and the pulse frequency ratio of the driven unit to that of the driving unit (F). Reproduced from Harmon (1) with a minor modification.

This is what one can reasonably expect to happen, but in fact, it does not. What does occur is much more complicated. As the excitation of the second unit decreases, the integral steps expected show up, but considerably more complex behavior also appears. Thus, other than the predicted steps such as 1:1, 1:2, 1:3,..., 1:10, a much larger number of nonintegral steps, such as 3:5, 5:16, 3:19, etc., also appear. The relationship between the amplitude of driving pulses (let it be a) and the ratio of the pulse frequency of the driven unit to that of the driving unit (let it be F) is shown in Fig. 2, which was reproduced from Harmon [1] with a minor modification.

The purpose of this paper is to present a mathematical neuron model which can elucidate the above experimental results.

I. MATHEMATICAL NEURON MODEL AND AVERAGE FIRING RATE

From the functional point of view, a neuron can be regarded as a threshold element with a refractory period, where a threshold element takes, at each instant, either of two possible states which correspond to the resting and excited states. In the present model, the refractoriness is so assumed that the inhibitory influence of past firing upon the excitability of the neuron at the present instant decreases exponentially with time, and time is assumed to be discrete. Under these assumptions, the behavior of the neuron is expressed by a nonlinear difference equation [2]:

$$x_{n+1} = 1[A_n - \alpha \sum_{r=0}^{n} b^{-r} x_{n-r} - \theta], \qquad (1)$$

where $1[x] = 1\ (x \geqq 0), = 0\ (x < 0)$,

x_n : the state of the neuron at the instant n. The resting state is represented by 0, and the excited state by 1,

A_n : magnitude of the input stimulus applied at the instant n,

θ : threshold value,

$\alpha > 0,\ b > 1.$

Although our neuron model may be regarded as a discrete-time version of Caianiello and DeLuca's continuous-time neuron model [3],

$$x(t+\tau) = 1[A(t) - \alpha \int_0^t b^{-r} x(t-r)\, dr - \theta], \qquad (2)$$

the former has a much richer variety of forms of solution than the latter.

Now, introduction of a new variable y_n:

$$y_n = \alpha^{-1}(A_n - \theta) - \sum_{r=0}^{n} b^{-r} x_{n-r} \tag{3}$$

reduces (1) to

$$y_{n+1} = b^{-1} y_n + a_n - 1[y_n] , \tag{4}$$

where

$$a_n = \frac{1}{\alpha}[(A_{n+1} - \frac{A_n}{b}) - \theta(1 - \frac{1}{b})] , \tag{5}$$

and

$$x_{n+1} = 1[\alpha y_n] = 1[y_n]. \tag{6}$$

In case the magnitude of the input stimulus is constant, that is,

$$A_n = A \tag{7}$$

for all n, it follows that

$$a_n = \frac{A - \theta}{\alpha}(1 - \frac{1}{b}) = a \text{ (constant)}, \tag{8}$$

and (4) becomes

$$\begin{cases} y_n \geqq 0 : y_{n+1} = b^{-1} y_n + a - 1, \\ y_n < 0 : y_{n+1} = b^{-1} y_n + a. \end{cases} \tag{9}$$

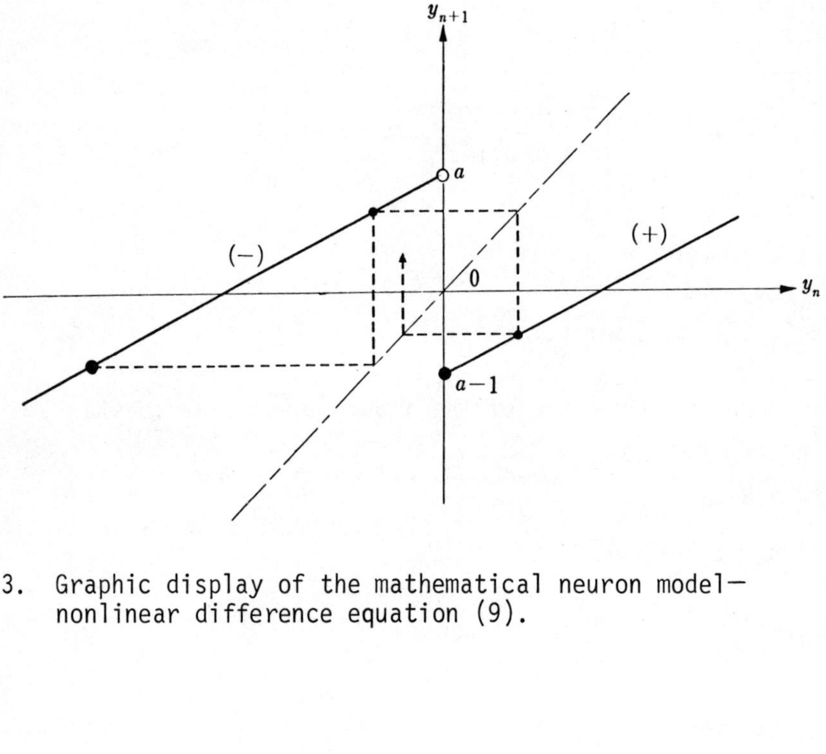

Fig. 3. Graphic display of the mathematical neuron model—nonlinear difference equation (9).

The former equation corresponds to the (+) branch, and the latter to the (-) branch in Fig. 3. Hereafter, the case of the constant input is considered, and a is regarded as representing the magnitude of the input stimulus.

Given the initial value y_0, a sequence of y:

$$y_0, y_1, y_2, y_3, \ldots\ldots \tag{10}$$

is determined from (9), and correspondingly a sequence of x:

$$x_1, x_2, x_3, \ldots\ldots\ldots \tag{11}$$

follows from (6).

In the sequel, our consideration is restricted to cases where (10) is a periodic sequence or a sequence which asymptotically approaches a periodic sequence. As will be seen later, this restriction does not affect the generality of our conclusion. In such cases, (11), after a finite number of steps of n, becomes a periodic sequence iterating $x_1^*, x_2^*, x_3^*, \ldots, x_\ell^*$ infinitely, and denoted by

$$\{x_1^* \; x_2^* \; x_3^* \; \ldots \; x_\ell^*\} \; . \tag{12}$$

For such periodic sequences, definition of the average firing rate F(a) is given by

$$F(a) = \frac{\text{number of 1 in } x_1^*, x_2^*, \ldots, x_\ell^*}{\ell} \; . \tag{13}$$

Simple considerations show that if $a \geqq 1$, $y_n \to y^* \geqq 0$ as $n \to \infty$. Hence $x_n = 1$ for $n > n_0$, and $F(a) = 1$. On the other hand, if $a < 0$, $y_n \to y^* < 0$. Hence $x_n = 0$ for $n > n_0$, and $F(a) = 0$. Thus our problem is to investigate the relationship between a and F for $0 < a < 1$.

It is worth mentioning that the replacements of a by 1-a, y_n by $-y_n$ in (9) cause an interchange of the (+) branch and the (-) branch, so that F is replaced by 1-F. In other words, the function F(a) is symmetrical with respect to the point(a=1/2,F=1/2).

II. SET OF PERIODIC SEQUENCES S HAVING CERTAIN SPECIAL FORMS

In what follows, our consideration will be further limited to such periodic sequences as have certain special forms, and denote the whole of such periodic sequences by S. It will become clear later that this limitation does not affect the generality of our conclusion. The set S is the totality of an infinite number of subsets S_1, S_2, S_3,..., each $S_i (i = 1,2,3,...)$ being a set of periodic sequences having the special form described below.

Denote a periodic sequence in which 0 appears consecutively n (a positive integer) times after 1 has appeared consecutively m (a positive integer) times by $\{1^m 0^n\}$. Then the set S_1 is the whole of periodic sequences in the form $\{1^m 0^n\}$ with m=1 or n =1.

The set S_2 is constructed from every pair of elements which are neighboring in the set S_1 by the same method as above. To cite an example, all periodic sequences of the form $\{(01)^m (001)^n\}$ (where m=1 or n=1), which are derived from two elements {01} and {001} neighboring in S_1, belong to the set S_2.

In the same way, the set S_3 is set up from every pair of elements which are neighboring in the set S_2. For example, all periodic sequences of the form $\{(01001)^m (01001001)^n\}$ (where m=1 or n=1), which are derived from two elements {01001} and {01001001} neighboring in S_2, belong to the set S_3.

Then the set S is defined as a union of all such subsets S_i $(i = 1,2,3,...)$ by

$$S = S_1 \cup S_2 \cup S_3 \cup \ldots \quad .$$

Obviously

$$S_i \cap S_j = \phi \text{ for } i \neq j.$$

Our next task then is to show that each element of the set S is a periodic solution of (9) having a particular value of a. Henceforth, the correspondence between the periodic sequence of S and the value of a in (9) will be investigated.

1. Elements of Set S_1

As an example of the element of S_1, $\{0^m 1\}$ $(m \geqq 1)$ is taken into consideration. In Fig. 4

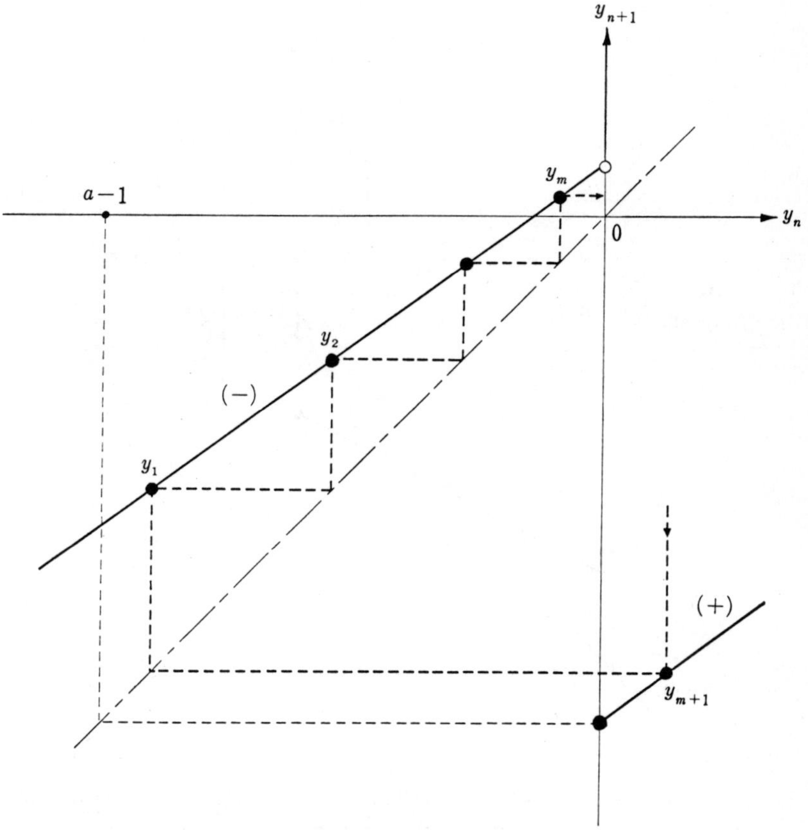

Fig. 4. Graphic display of periodic sequence $\{0^m1\}$ ($m \geqq 1$), which is an element of the set S_1.

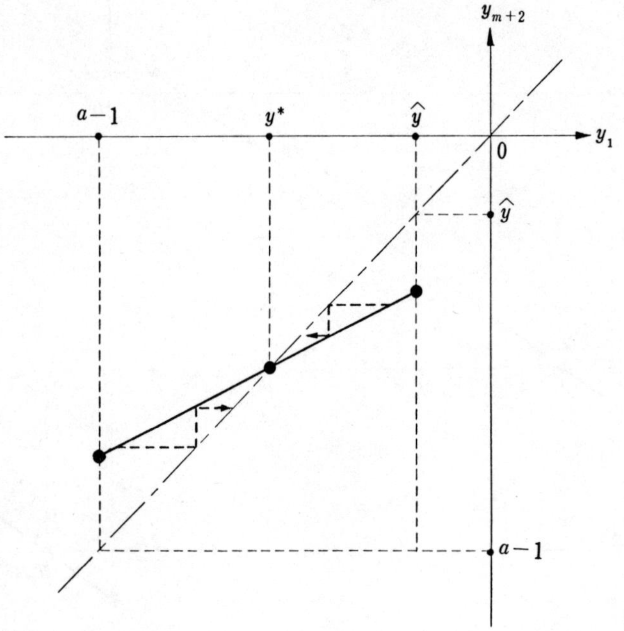

Fig. 5. If condition (18) is satisfied, the corresponding sequence of y approaches to a periodic sequence independent of the initial value.

$$y_1 < 0,$$

$$y_2 = b^{-1}y_1 + a < 0,$$

$$y_3 = b^{-2}y_1 + a(1+b^{-1}) < 0,$$

$$\vdots$$

$$y_{m+1} = b^{-m}y_1 + a\ (1+b^{-1}+b^{-2}+\ldots+b^{-m+1}) > 0.$$

Define

$$y_{m+2} = b^{-m-1}y_1 + a(1+b^{-1}+b^{-2}+\ldots+b^{-m}), \tag{14}$$

and put $y_{m+2} = y_1$. Then

$$y_1 = b\ \ \{a(b^m + b^{m-1}+\ldots+1) - b^m\}\ (b^{m+1}-1)^{-1}, \tag{15}$$

and

$$y_m = b\ \ \{a(b^m+b^{m-1}+\ldots+1) - b\}\ (b^{m+1}-1)^{-1}. \tag{16}$$

These y_1 and y_m must satisfy the inequalities

$$y_m < 0,\quad y_1 \geqq a-1, \tag{17}$$

which are rewritten as

$$\frac{b}{b^m+b^{m-1}+\ldots+1} > a \geqq \frac{1}{b^m+b^{m-1}+\ldots+1}. \tag{18}$$

Conversely, if condition (18) is satisfied, it follows from (14) that $y_{m+2} \geqq a-1$ for $y_1 = a-1$. Moreover, when $y_m=0$, or when $y_1=\hat{y}$, $\hat{y}_{m+2}<\hat{y}$, where $\hat{y} = -a(b^{m-1}+b^{m-2}+\ \ldots\ +b)$. Hence it is obvious from Fig. 5 that $y_{1+i(m+1)} \to y^*$ as $i \to \infty$, where $\hat{\hat{y}} > y^* \geqq a-1$, independent of the initial value y_0. Consequently the corresponding sequence of x becomes the periodic sequence $\{0^m1\}$ after a finite number of steps.

To summarize, it has been shown that (18) is a necessary and sufficient condition for the sequence of y to become a periodic sequence which corresponds to the periodic sequence $\{0^m1\}$ of x. The value of a which satisfies (18) is called the value (or interval) of a which corresponds to the periodic sequence $\{0^m1\}$. For this sequence it is apparent that $F=(m+1)^{-1}$.

It should be noticed that condition (18) can be directly derived from (4), as described below.

From (4),

$$
\begin{aligned}
y_1 &= b^{-1}y_0 + a - x_1,\\
y_2 &= b^{-1}y_1 + a - x_2,\\
y_3 &= b^{-1}y_2 + a - x_3,\\
&\vdots\\
y_\ell &= b^{-1}y_{\ell-1} + a - x_\ell\ ,
\end{aligned}
\tag{19}
$$

hence

$$y_\ell = b^{-\ell}y_0 + b^{-\ell+1}(a\mu - h_1), \tag{20}$$

where

$$\mu = b^{\ell-1} + b^{\ell-2} + \ldots + b+1,$$

$$h_\ell = b^{\ell-1}x_\ell + b^{\ell-2}x_{\ell-1} + \ldots + bx_2 + x_1 .$$

If the sequence of y is a periodic sequence with period ℓ, then

$$y_\ell = y_0, \tag{21}$$

from which

$$y_0 = \frac{b}{b^\ell - 1}(a\mu - h_1) \tag{22}$$

is obtained. Furthermore, from (22) and the first equation of (19)

$$y_1 = \frac{b}{b^\ell - 1}(a\mu - h_2), \tag{23}$$

where

$$h_2 = b^{\ell-1}x_1 + b^{\ell-2}x_\ell + \ldots + bx_3 + x_2.$$

Similarly,

$$y_2 = \frac{b}{b^{\ell} - 1}(a\mu - h_3),$$

$$\vdots$$

$$y_{\ell-1} = \frac{b}{b^{\ell} - 1}(a\mu - h_{\ell}),$$

where

$$h_3 = b^{\ell-1}x_2 + b^{\ell-2}x_1 + b^{\ell-3}x_{\ell} + \ldots + bx_4 + x_3,$$

$$\vdots$$

$$h_{\ell} = b^{\ell-1}x_{\ell-1} + b^{\ell-2}x_{\ell-2} + \ldots + bx_1 + x_{\ell}.$$

These equations are rewritten as

$$x_p = 1\left[a - \frac{h_p}{\mu}\right] \qquad (p = 1, 2, \ldots, \ell), \tag{24}$$

where

$$h_p = b^{\ell-1}x_{p-1} + b^{\ell-2}x_{p-2} + \ldots + bx_{p+1} + x_p.$$

Now let the periodic sequence of x under consideration be $\{x_1^* x_2^* \ldots x_{\ell}^*\}$.

Obviously, for such a p as $x_p^* = 1$,

$$a \geqq \frac{h_p}{\mu}$$

holds. On the contrary, for such a p as $x_p^* = 0$,

$$a < \frac{h_p}{\mu}$$

holds. Let the maximal h_p for which $x_p^* = 1$ be $h_{\bar{v}}$, and the minimal h_p for which $x_p^* = 0$ be h_L. Then it immediately follows that

$$\frac{h_L}{\mu} > a \geqq \frac{h_{\bar{v}}}{\mu}. \tag{25}$$

Application of (25) to periodic sequence $\{0^m 1\}$ yields (18).

It should be noted that the sequence of x becomes the periodic

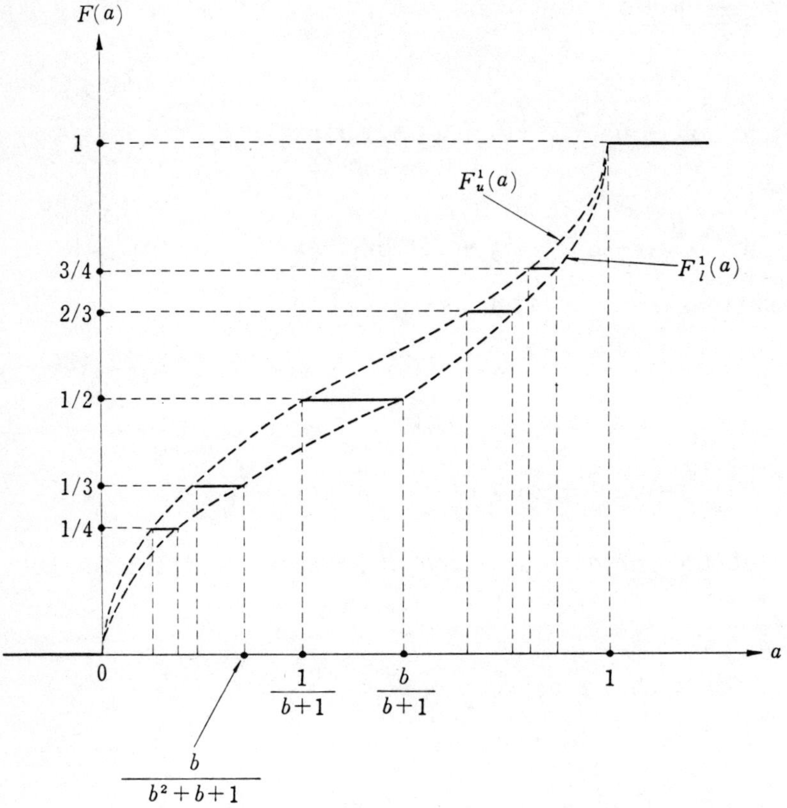

Fig. 6. The relation between a and F for periodic sequences of the set S_1. $F_u^1(a)$ and $F_\ell^1(a)$ are functions which give an upper bound and a lower bound, respectively.

$$\frac{b^3 + b^2 + b}{b^3 + b^2 + b + 1} \geqq a \geqq \frac{b^3 + b^2 + 1}{b^3 + b^2 + b + 1} \qquad F = \frac{3}{4}$$

$$\left(\frac{1\ 1\ 1\ 0}{1\ 1\ 1\ 1} \geqq a(b) \geqq \frac{1\ 1\ 0\ 1}{1\ 1\ 1\ 1}\right)$$

$$\frac{b^2 + b}{b^2 + b + 1} \geqq a \geqq \frac{b^2 + 1}{b^2 + b + 1} \qquad F = \frac{2}{3}$$

$$\left(\frac{1\ 1\ 0}{1\ 1\ 1} \geqq a(b) \geqq \frac{1\ 0\ 1}{1\ 1\ 1}\right)$$

$$\frac{b}{b + 1} \geqq a \geqq \frac{1}{b + 1} \qquad F = \frac{1}{2}$$

$$\left(\frac{1\ 0}{1\ 1} \geqq a(b) \geqq \frac{0\ 1}{1\ 1}\right)$$

$$\frac{b}{b^2 + b + 1} \geqq a \geqq \frac{1}{b^2 + b + 1} \qquad F = \frac{1}{3}$$

$$\left(\frac{0\ 1\ 0}{1\ 1\ 1} \geqq a(b) \geqq \frac{0\ 0\ 1}{1\ 1\ 1}\right)$$

$$\frac{b}{b^3 + b^2 + b + 1} \geqq a \geqq \frac{1}{b^3 + b^2 + b + 1} \qquad F = \frac{1}{4}$$

$$\left(\frac{0\ 0\ 1\ 0}{1\ 1\ 1\ 1} \geqq a(b) \geqq \frac{0\ 0\ 0\ 1}{1\ 1\ 1\ 1}\right)$$

Table I. The values of a and F which correspond to periodic sequences $\{1^m0\}$ and $\{10^m\}$ of the set S_1 for cases $m = 1,2$ and 3.

sequence $\{0^m 1\}$ even if $a = b(b^m + b^{m-1} + . . . +1)^{-1}$. Therefore, instead of (18),

$$\frac{b}{b^m + b^{m-1} +\ldots+1} \geqq a \geqq \frac{1}{b^m + b^{m-1} +\ldots+1} \tag{26}$$

is adopted as the interval of a which corresponds to the periodic sequence $\{0^m 1\}$.

Representation of (26) in the scale of b yields

$$\frac{0\,0\cdot\cdot\cdot 0\,1\,0}{1\,1\cdot\cdot\cdot 1\,1\,1} \geqq a\ (b) \geqq \frac{0\,0\cdot\cdot\cdot 0\,0\,1}{1\,1\cdot\cdot\cdot 1\,1\,1}, \tag{27}$$

where a (b) means the value of a represented in the scale of b. Numerators on the left and right-hand sides of (27) indicate the periodic sequence under consideration, and the numerator on the left is derived from that on the right by consecutive rotations of the numerals of the latter. Incidentally, the value of F is given by the ratio of the sum of the numerals in the numerator to that in the denominator.

Replacements of a by 1-a, F by 1-F in (26) yield results for periodic sequence $\{1^m 0\}$. The values of a and F, which correspond to the periodic sequences $\{1^m 0\}$ and $\{0^m 1\}$, are shown in Table I and in Fig. 6 for cases m=1, 2 and 3.

In Fig. 6, $F_u^1(a)$ and $F_\ell^1(a)$ are functions which give an upper bound and a lower bound respectively, and

$$\begin{aligned} F_u^1\,(a) &= \frac{1}{\log\,(1 + \frac{b-1}{a})} \quad (\frac{1}{b+1} \geqq a > 0), \\ &= 1 - \frac{1}{\log\,(1 + \frac{b(b-1)}{1-a})} \quad (1 > a \geqq \frac{1}{b+1}), \end{aligned} \tag{28}$$

$$\begin{aligned} F_\ell^1\,(a) &= \frac{1}{\log\,(1 + \frac{b(b-1)}{a})} \quad (\frac{b}{b+1} \geqq a > 0), \\ &= 1 - \frac{1}{\log\,(1 + \frac{b-1}{1-a})} \quad (1 > a \geqq \frac{b}{b+1}). \end{aligned} \tag{29}$$

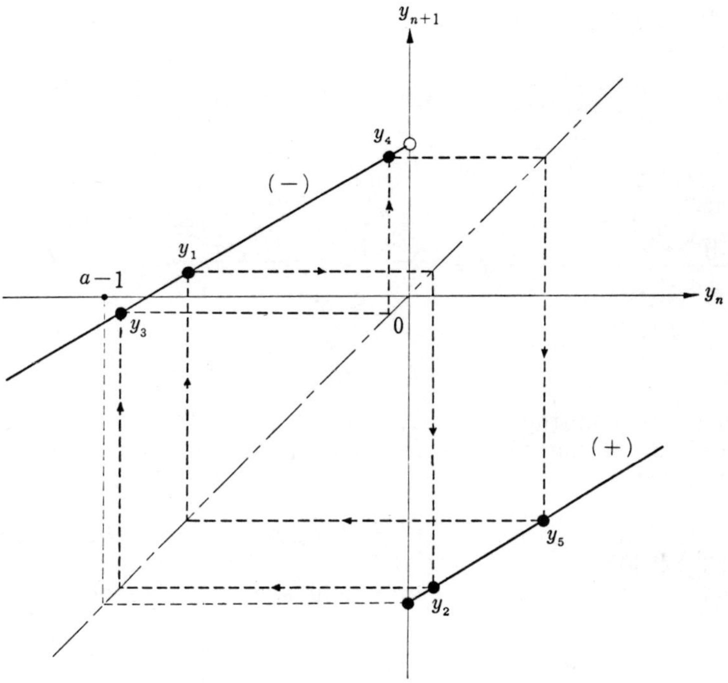

Fig. 7. Graphic display of periodic sequence {01001}, which is an element of the set S_2.

2. Elements of Set S_2

The values of a and F will be computed for {01001} which was previously cited as an example of the element of the set S_2 (Fig. 7). By a calculation like that before, the values y_1, y_2, ..., y_5 are obtained as functions of a and b. The condition corresponding to (17) turns out

$$y_4 < 0, \qquad y_3 \geqq a - 1, \tag{30}$$

from which

$$\frac{b^3 + b}{b^4 + b^3 + \cdots + 1} \geqq a \geqq \frac{b^3 + 1}{b^4 + b^3 + \cdots + 1} \tag{31}$$

is derived corresponding to (26). Conversely, if condition (31) is satisfied, corresponding sequence of x becomes periodic sequence {01001} after a finite number of steps. Clearly F=2/5.

Table II and Fig. 8 show the values of a and F which correspond to $\{(01)^m(001)\}$ and $\{(01)(001)^m\}$ for cases m=1,2 and 3. In Fig. 8, functions giving an upper bound and a lower bound are

$$F_u^2(a) = \frac{1}{3}\left[1 + \frac{1}{\log\left(1 + \dfrac{b-1}{a(b^2 + b + 1) - b}\right)}\right]$$

$$\left(\frac{b^3 + 1}{b^4 + b^3 + \cdots + 1} \geqq a > \frac{b}{b^2 + b + 1}\right),$$

$$= \frac{1}{2}\left[1 - \frac{1}{\log\left(1 + \dfrac{b^2(b - 1)}{1 - a(b + 1)}\right)}\right] \tag{32}$$

$$\left(\frac{1}{b + 1} > a \geqq \frac{b^3 + 1}{b^4 + b^3 + \ldots + 1}\right),$$

$$\frac{b^7 + b^5 + b^3 + b}{b^8 + b^7 + \cdots + 1} \geqq a \geqq \frac{b^7 + b^5 + b^3 + 1}{b^8 + b^7 + \cdots + 1} \qquad F = \frac{4}{9}$$

$$\left(\frac{0\,1\,0\,1\,0\,1\,0\,1\,0}{1\,1\,1\,1\,1\,1\,1\,1\,1} \geqq a(b) \geqq \frac{0\,1\,0\,1\,0\,1\,0\,0\,1}{1\,1\,1\,1\,1\,1\,1\,1\,1}\right)$$

$$\frac{b^5 + b^3 + b}{b^6 + b^5 + \cdots + 1} \geqq a \geqq \frac{b^5 + b^3 + 1}{b^6 + b^5 + \cdots + 1} \qquad F = \frac{3}{7}$$

$$\left(\frac{0\,1\,0\,1\,0\,1\,0}{1\,1\,1\,1\,1\,1\,1} \geqq a(b) \geqq \frac{0\,1\,0\,1\,0\,0\,1}{1\,1\,1\,1\,1\,1\,1}\right)$$

$$\frac{b^3 + b}{b^4 + b^3 + \cdots + 1} \geqq a \geqq \frac{b^3 + 1}{b^4 + b^3 + \cdots + 1} \qquad F = \frac{2}{5}$$

$$\left(\frac{0\,1\,0\,1\,0}{1\,1\,1\,1\,1} \geqq a(b) \geqq \frac{0\,1\,0\,0\,1}{1\,1\,1\,1\,1}\right)$$

$$\frac{b^6 + b^3 + b}{b^7 + b^6 + \cdots + 1} \geqq a \geqq \frac{b^6 + b^3 + 1}{b^7 + b^6 + \cdots + 1} \qquad F = \frac{3}{8}$$

$$\left(\frac{0\,1\,0\,0\,1\,0\,1\,0}{1\,1\,1\,1\,1\,1\,1\,1} \geqq a(b) \geqq \frac{0\,1\,0\,0\,1\,0\,0\,1}{1\,1\,1\,1\,1\,1\,1\,1}\right)$$

$$\frac{b^9 + b^6 + b^3 + b}{b^{10} + b^9 + \cdots + 1} \geqq a \geqq \frac{b^9 + b^6 + b^3 + 1}{b^{10} + b^9 + \cdots + 1} \qquad F = \frac{4}{11}$$

$$\left(\frac{0\,1\,0\,0\,1\,0\,0\,1\,0\,1\,0}{1\,1\,1\,1\,1\,1\,1\,1\,1\,1\,1} \geqq a(b) \geqq \frac{0\,1\,0\,0\,1\,0\,0\,1\,0\,0\,1}{1\,1\,1\,1\,1\,1\,1\,1\,1\,1\,1}\right)$$

Table II. The values of a and F which correspond to periodic sequences $\{(01)^m(001)\}$ and $\{(01)(001)^m\}$ of the set S_2 for cases m = 1,2 and 3.

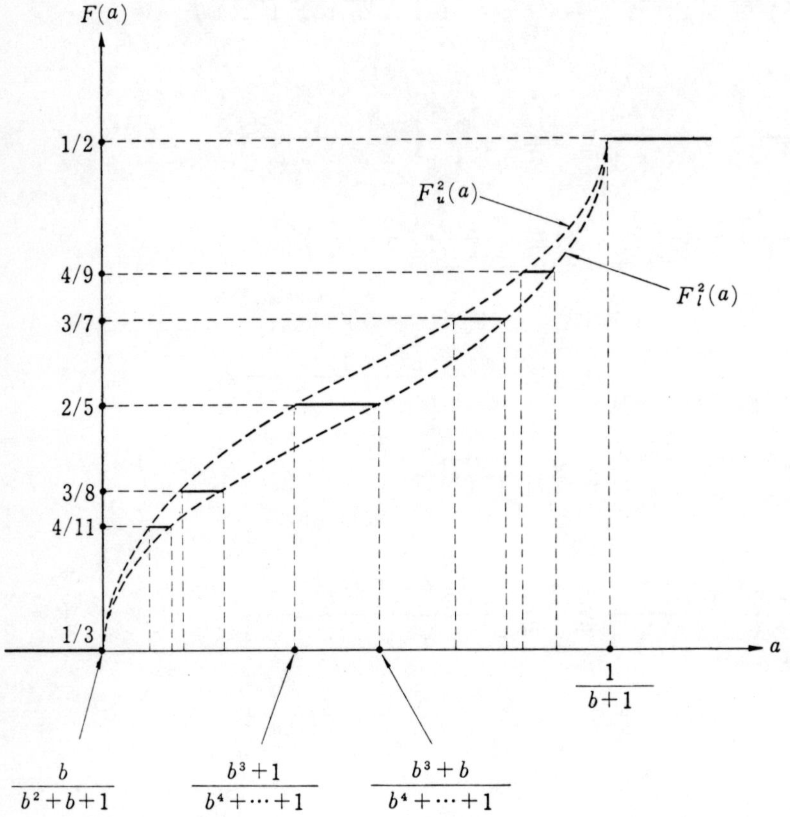

Fig. 8. The relation between a and F for periodic sequences $\{(01)^m(001)\}$ and $\{(01)(001)^m\}$ of the set S_2. $F_u^2(a)$ and F_ℓ^2 are functions which give an upper bound and a lower bound, respectively.

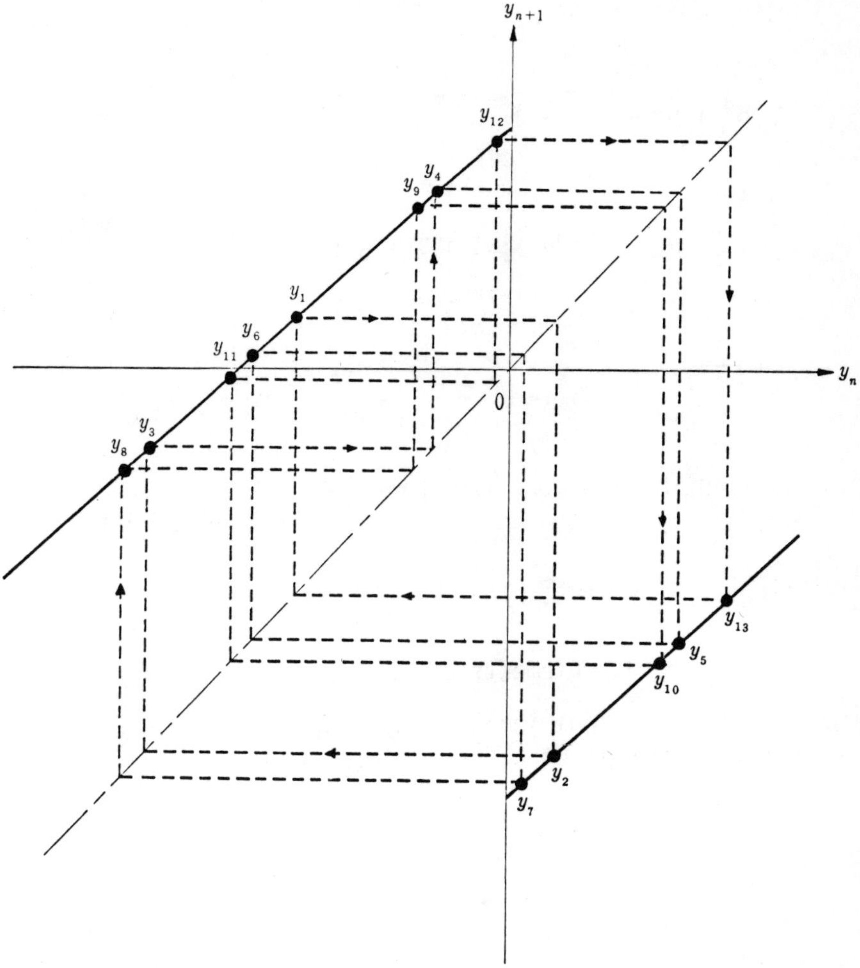

Fig. 9. Graphic display of periodic sequence {0100101001001}, which is an element of the set S_3.

$$F^2_\ell(a) = \frac{1}{3}\; 1 + \left[\frac{1}{\log\left(1 + \dfrac{b^3(b-1)}{a(b^2+b+1)-b}\right)}\right]$$

$$\left(\frac{b^3+b}{b^4+b^3+\cdots+1} \geqq a > \frac{b}{b^2+b+1}\right),$$

$$= \frac{1}{2}\left[1 - \frac{1}{\log\left(1 + \dfrac{b-1}{1-a(b+1)}\right)}\right] \qquad (33)$$

$$\left(\frac{1}{b+1} > a \geqq \frac{b^3+b}{b^4+b^3+\cdots+1}\right),$$

and the following inequalities hold for $\frac{1}{b+1} > a > \frac{b}{b^2+b+1}$.

$$\frac{1}{2} > F^1_u(a) > F^2_u(a) > F(a) > F^2_\ell(a) > F^1_\ell(a) > \frac{1}{3}.$$

3. Elements of Set S_3

For the example {0100101001001} of the element of S_3 previously cited, the condition corresponding to (17) is found to be

$$y_{12} < 0, \quad y_8 \geqq a - 1 \qquad (34)$$

or

$$\frac{b^{11}+b^8+b^6+b^3+b}{b^{12}+b^{11}+\cdots+1} \geqq a \geqq \frac{b^{11}+b^8+b^6+b^3+1}{b^{12}+b^{11}+\cdots+1} \qquad (35)$$

(See Fig. 9). Obviously F=5/13.

$$\frac{b^{16}+b^{13}+b^{11}+b^{8}+b^{6}+b^{3}+b}{b^{17}+b^{16}+\cdots\cdots+1} \geqq a \geqq \frac{b^{16}+b^{13}+b^{11}+b^{8}+b^{6}+b^{3}+1}{b^{17}+b^{16}+\cdots\cdots+1} \qquad F=\frac{7}{18}$$

$$\left(\frac{0\;1\;0\;0\;1\;0\;1\;0\;0\;1\;0\;1\;0\;0\;1\;0\;1\;0}{1\;1\;1\;1\;1\;1\;1\;1\;1\;1\;1\;1\;1\;1\;1\;1\;1\;1} \geqq a(b) \geqq \frac{0\;1\;0\;0\;1\;0\;1\;0\;0\;1\;0\;1\;0\;0\;1\;0\;0\;1}{1\;1\;1\;1\;1\;1\;1\;1\;1\;1\;1\;1\;1\;1\;1\;1\;1\;1}\right)$$

$$\frac{b^{11}+b^{8}+b^{6}+b^{3}+b}{b^{12}+b^{11}+\cdots\cdots+1} \geqq a \geqq \frac{b^{11}+b^{8}+b^{6}+b^{3}+1}{b^{12}+b^{11}+\cdots\cdots+1} \qquad F=\frac{5}{13}$$

$$\left(\frac{0\;1\;0\;0\;1\;0\;1\;0\;0\;1\;0\;1\;0}{1\;1\;1\;1\;1\;1\;1\;1\;1\;1\;1\;1\;1} \geqq a(b) \geqq \frac{0\;1\;0\;0\;1\;0\;1\;0\;0\;1\;0\;0\;1}{1\;1\;1\;1\;1\;1\;1\;1\;1\;1\;1\;1\;1}\right)$$

$$\frac{b^{19}+b^{16}+b^{14}+b^{11}+b^{8}+b^{6}+b^{3}+b}{b^{20}+b^{19}+\cdots\cdots+1} \geqq a \geqq \frac{b^{19}+b^{16}+b^{14}+b^{11}+b^{8}+b^{6}+b^{3}+1}{b^{20}+b^{19}+\cdots\cdots+1} \qquad F=\frac{8}{21}$$

$$\left(\frac{0\;1\;0\;0\;1\;0\;1\;0\;0\;1\;0\;0\;1\;0\;1\;0\;0\;1\;0\;1\;0}{1\;1\;1\;1\;1\;1\;1\;1\;1\;1\;1\;1\;1\;1\;1\;1\;1\;1\;1\;1\;1} \geqq a(b) \geqq \frac{0\;1\;0\;0\;1\;0\;1\;0\;0\;1\;0\;0\;1\;0\;1\;0\;0\;1\;0\;0\;1}{1\;1\;1\;1\;1\;1\;1\;1\;1\;1\;1\;1\;1\;1\;1\;1\;1\;1\;1\;1\;1}\right)$$

Table III. The value of a and F which correspond to periodic sequences $\{(01001)^m(01001001)\}$ and $\{(01001)(01001001)^m\}$ of the set S_3 for cases m - 1 and 2.

The values of a and F are shown in Table III which correspond to periodic sequences $\{(01001)^m(01001001)\}$ and $\{(01001)(01001001)^m\}$ for cases m=1 and 2.

III. TOTAL LENGTH OF THE INTERVALS OF a WHICH CORRESPOND TO THE ELEMENTS OF S

Our next task then is to calculate the total length of intervals of a which correspond to the elements of S.

For the element $\{10\} = \{01\}$ of S_1, the corresponding interval is

$$\frac{b}{b+1} \geqq a \geqq \frac{1}{b+1},$$

so that the length of the interval is equal to $(b-1)^2(b^2-1)^{-1}$. Similarly, the length of the intervals for $\{110\}$ and $\{001\}$ is $(b-1)^2(b^3-1)^{-1}$, one for $\{1110\}$ and $\{0001\}$ is $(b-1)^2(b^4-1)^{-1}$. Hence the total length of intervals which correspond to the elements of S_1 is given by $(b-1)^2L_1$, where

$$L_1 = \frac{1}{b^2-1} + 2\left(\frac{1}{b^3-1} + \frac{1}{b^4-1} + \frac{1}{b^5-1} + \ldots\right).$$

Next, the total length of intervals of the elements of S_2 will be considered. The intervals for the elements of S_2, which lie between $\{01\}$ and $\{001\}$, are as follows. From Table II, it is seen that the interval for $\{(01)(001)\}$ is

$$\frac{b^3+b}{b^4+b^3+\ldots+1} \geqq a \geqq \frac{b^3+1}{b^4+b^3+\ldots+1},$$

so that the length of the interval is given by $(b-1)^2(b^5-1)^{-1}$. The lengths of the intervals for $\{(01)^2(001)\}$ and $\{(01)^3(001)\}$ are $(b-1)^2(b^7-1)^{-1}$ and $(b-1)^2(b^9-1)^{-1}$, respectively. On the other hand, the lengths of intervals for $\{(01)(001)^2\}$ and $\{(01)(001)^3\}$ are $(b-1)^2(b^8-1)^{-1}$ and $(b-1)^2(b^{11}-1)^{-1}$, respectively. Hence the total length of intervals for elements of S_2 which lie between $\{01\}$ and $\{001\}$ is given by $(b-1)^2L_2^{(5)}$, where

$$L_2^{(5)} = \frac{1}{b^5-1} + \left(\frac{1}{b^7-1} + \frac{1}{b^9-1} + \frac{1}{b^{11}-1} + \cdots\right) + \left(\frac{1}{b^8-1} + \frac{1}{b^{11}-1} + \frac{1}{b^{14}-1} + \ldots\right).$$

The length of the intervals corresponding to the elements of S_2 which lie between {110} and {10} is also equal to $(b-1)^2 L_2^{(5)}$.

Likewise, the length of intervals corresponding to the elements of S_2, which lie between {001} and {0001} and between {1110} and {110}, is given by $(b-1)^2 L_2^{(7)}$, where

$$L_2^{(7)} = \frac{1}{b^7 - 1} + \left(\frac{1}{b^{10} - 1} + \frac{1}{b^{13} - 1} + \frac{1}{b^{16} - 1} + \cdots\right) + \left(\frac{1}{b^{11} - 1} + \frac{1}{b^{15} - 1} + \frac{1}{b^{19} - 1} + \cdots\right).$$

In the same way, $L_2^{(9)}$, $L_2^{(11)}$, $L_2^{(13)}$,... are obtained, and the total length of intervals corresponding to the elements of S_2 is given by $(b-1)^2 L_2$, where

$$L_2 = 2\,(L_2^{(5)} + L_2^{(7)} + L_2^{(9)} + \ldots).$$

Next, the total length of intervals which correspond to elements of S_3 is calculated and denoted by $(b-1)^2 L_3$. To cite an example, the total length of intervals corresponding to the element of S_3 which lie between {(01)(001)} and {(01)(001)^2} is, referring to Table III, given by $(b-1)^2 L_2^{(13)}$, where

$$L_3^{(13)} = \frac{1}{b^{13}-1} + \left(\frac{1}{b^{18}-1} + \frac{1}{b^{23}-1} + \frac{1}{b^{28}-1} + \ldots\right) + \left(\frac{1}{b^{21}-1} + \frac{1}{b^{29}-1} + \frac{1}{b^{37}-1} + \ldots\right).$$

After all, the length of intervals corresponding to the elements of S is given by $(b-1)^2 L$ with

$$L = \sum_{i=1}^{\infty} L_i,$$

where $(b-1)^2 L_i$ is the total length of intervals corresponding to the elements of S_i $(i=1,2,3,\ldots)$.

The above-mentioned results are summarized in Table IV, where the integers $n (n \geqq 2)$ indicate the terms $(b^n-1)^{-1}$, C_i $(i=1,2,3)$

C_1	C_2	C_3
2	...	
	13	
	11	
	9	
	7	...
		12
	5	...
		13
	8	...
	11	
	...	
3	...	
	13	
	10	
	7	
	11	
4	...	
	...	
	13	
	9	
	...	
5	...	
	11	
	...	
6	...	
	13	
	...	
7		
.		
.		
.		

Table IV. Positive integer n implies the term $(b^n-1)^{-1}$, $(b-1)^2$ times of which is the length of the interval carresponding to an element of S. C_i is the set of integers n which correspond to the elements S_i. Only a half is shown in this table because it is symmetrical.

n	C_1	C_2	C_3	Σ	$\varphi(n)$
2	1	0	0	1	1
3	1	0	0	1	2
4	1	0	0	1	2
5	1	1	0	2	4
6	1	0	0	1	2
7	1	2	0	3	6
8	1	1	0	2	4
9	1	2	0	3	6
10	1	1	0	2	4
11	1	4	0	5	10
12	1	0	1	2	4
13	1	4	1	6	12
⋮	⋮	⋮	⋮	⋮	⋮

Table V. This table contains numbers of the integer n which belongs to each C_i $(i = 1,2,3)$, the sume of them Σ, and double the sum (n) (except when $n = 2$).

represents the set of the integers n which correspond to the elements of S_i, and only a half of the whole is shown in this table because it is symmetrical. Table V contains numbers of the integer n which belongs to each C_i (i=1,2,3), the sume of them Σ, and double the sum $\phi(n)$ (except for the case n = 2). Since $\phi(n)$ is the number of terms $(b^n-1)^{-1}$ which are involved in L,

$$L = \sum_{n=2}^{\infty} \frac{\phi(n)}{b^n-1} . \tag{36}$$

It is proved that $\phi(n)$ is nothing but Euler's function (see Appendix), namely, the number of positive integers not greater than and prime to n.

Now

$$\sum_{n=1}^{\infty} \frac{\phi(n)}{b^n-1} = \sum_{n=1}^{\infty} \phi(n) \sum_{r=1}^{\infty} b^{-rn}$$

$$= \sum_{m=1}^{\infty} b^{-m} \sum_{s|m} \phi(s), \tag{37}$$

where the last sum is taken over every divisor S of m. But since [4]

$$\sum_{s|m} \phi(S) = m, \tag{38}$$

it immediately follows that

$$\sum_{n=1}^{\infty} \frac{\phi(n)}{b^n-1} = \sum_{m=1}^{\infty} mb^{-m} = b(b-1)^{-2} , \tag{39}$$

and hence,

$$L = b(b-1)^{-2} - (b-1)^{-1} = (b-1)^{-2}. \tag{40}$$

Thus, it is shown that the total length of the intervals which correspond to the elements of S is equal to unity.

IV. FUNCTION F(a)

As has been seen in the previous section, the function F(a) is defined over an enumerably infinite number of intervals which are everywhere dense in the interval $0 \leqq a \leqq 1$, and the total length of the intervals is equal to unity. It is clear that this function can be uniquely extended to a function which is defined throughout the interval $0 \leqq a \leqq 1$ in a natural way. The extended function (denoted by F(a) anew) is continuous, nondecreasing, flat ($F'(a) = 0$) almost everywhere in $0 \leqq a \leqq 1$; nevertheless, $F(0) = 0$, $F(1) = 1$. Thus, the conclusion that the function F(a) is an extended Cantor's function [5] is achieved.

As described in the preceding section, our consideration of the solution of (1) was restricted to periodic solutions having certain special forms. It has now become clear that this restriction does not affect the generality of our conclusion. In other words, it is 'almost' sufficient to consider only periodic solutions with such special forms instead of considering all kinds of solutions of (1). The conclusion can elucidate the essential feature of Harmon's experimetnal finding shown in Figure 2.

The physiological meaning of the staircaselike characteristic is as follows. There are two points of view to be considered.

In the first viewpoint, a is regarded as the magnitude of an external stimulus, and F(a) as the f ring frequency of the neuron. Regarded from this point of view, the author is not aware of any experimental result where a real neuron did exhibit such a staircaselike characteristic. Since the behavior of a real neuron is usually 'dull' compared with that of the electronic model neuron, it might be somewhat difficult to obtain such a sharp characteristic as a staircaselike one on the behavior of real neurons. Nevertheless, it does not seem impossible if an experiment is carefully carried out with that intention.

The second viewpoint is as follows: As is known, the synaptic conductance between two neurons is presumed to change due to use of the synapse and this gives the physiological basis of the theory of learning and/or the learning machines.

For an excitatory synapse, it will be reasonable to assume that the synaptic conductance between two neurons increases if the post synaptic neuron fires due to an impulse from the presynaptic neuron. Since in our case the presynaptic neuron fires at every

Harmon	F	m	n	Harmon	F	m	n
1 : 1	F_1	∞	—	2 : 9	F_3	3	1
4 : 5	F_1	4	0	3 : 14	F_4	3	2
3 : 4	F_1	3	0	1 : 5	F_4	4	0
5 : 7	F_1	2	1	3 : 16	F_3	4	2
2 : 3	F_1	2	0	2 : 11	F_3	4	1
3 : 5	F_1	1	1	3 : 17	F_4	4	2
1 : 2	F_1	1	0	1 : 6	F_4	5	0
6 : 13	F_3	1	5	3 : 19	F_3	5	2
4 : 9	F_3	1	3	2 : 13	F_3	5	1
2 : 5	F_3	1	1	1 : 7	F_3	6	0
1 : 3	F_4	2	0	2 : 15	F_3	6	1
5 : 16	F_3	2	4	1 : 8	F_4	7	0
3 : 10	F_3	2	2	2 : 17	F_3	7	1
2 : 7	F_3	2	1	1 : 9	F_4	8	0
3 : 11	F_4	2	2	2 : 19	F_3	8	1
1 : 4	F_4	3	0	1 : 10	F_4	9	0
3 : 13	F_3	3	2				

Table VI. Comparison of the values of F between Harmon's experimental results and our theoretical results.

instant, the firing frequency of the postsynaptic neuron corresponds to the amount of increment of the synaptic conductance. Hence, if a is regarded as the synaptic conductance, F(a) may be regarded as representing the increment of the synaptic conductance when the latter is equal to a.

V. CORRESPONDENCE TO HARMON'S EXPERIMENTAL RESULTS

The values of F of the periodic sequences which belong to set S_2 and lie between $\{1^{m+1}0\}$ and $\{1^m0\}$ are given by

$$F_1 = \frac{n(m+1) + m}{n(m+2) + (m+1)} \qquad (n \geqq 1),$$

or

$$F_2 = \frac{(m+1) + nm}{(m+2) + (m+1)} \qquad (n \geqq 1),$$

and those between $\{0^m1\}$ and $\{0^{m+1}1\}$ are given by

$$F_3 = \frac{n + 1}{n(m+1) + (m+2)} \qquad (n \geqq 1),$$

or

$$F_4 = \frac{1 + n}{(m+1) + n(m+2)} \qquad (n \geqq 1).$$

A detailed correspondence of these values of F to Harmon's experimental results is shown in Table VI. Since a in (9) is not exactly the same as that in Figure 2, correspondence of the values of a cannot be made.

VI. AN ANALOGUE CIRCUIT OF THE MATHEMATICAL NEURON MODEL

Let us consider the very simple circuit in Figure 10 which is composed of a delay line and a negative resistance element. Let the characteristic impedance of the dealy line be Z, the characteristic of the negative resistance element be as shown in Figure 11, and assume $Z > R$.

Let the series inductance and parallel capacitance per unit length of the delay line, assumed to be lossless, be L and C, respectively. Then the relation between the voltage ν across the line and the current i through the line is given by

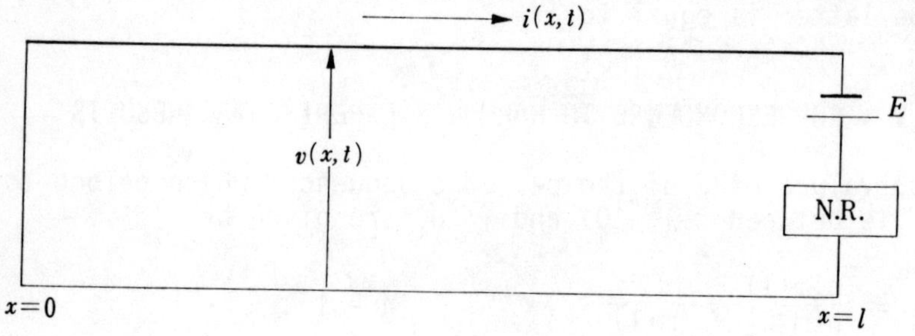

Fig. 10. An analogue circuit of the mathematical neuron model.

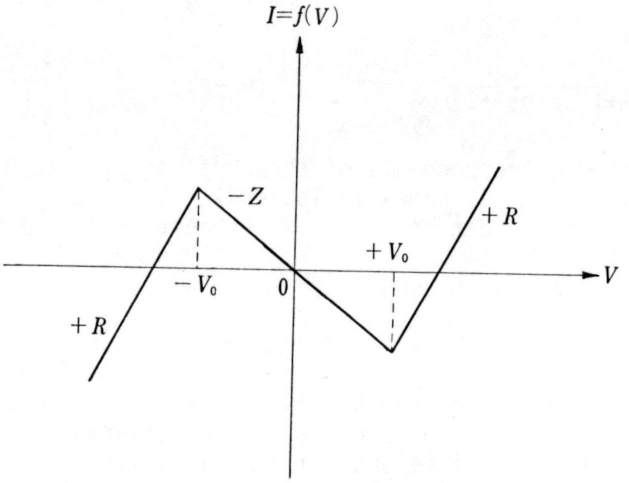

Fig. 11. The characteristic of the negative resistance element in the analogue circuit, where Z is the characteristic impedance of the delay line and Z > R.

$$\frac{\partial v}{\partial x} = -L\,\frac{\partial i}{\partial t}\,,\quad \frac{\partial i}{\partial x} = -\,C\,\frac{\partial v}{\partial t}\,. \tag{41}$$

As the line is shorted at $x = 0$ and connected with the negative resistance element at $x = \ell$, boundary conditions become

$$v(0,t) = 0, \tag{42}$$

$$i(\ell,t) = f(v(\ell,t) + E), \tag{43}$$

where E is the dc bias voltage, and $I = f(V)$ is the expression representing the characteristic curve of the negative resistance element (Fig. 11).

It is assumed that the voltage along the line at $t = 0$ is $\alpha(x)$ and the current is $\beta(x)$. Therefore, the initial condition may be written as

$$v(x,0) = \alpha(x) \qquad (0 < x < \ell), \tag{44}$$

$$i(x,0) = \beta(x) \qquad (0 < x < \ell), \tag{45}$$

where $\alpha(0) = 0$.

D'Alembert's solution of (41) is of the form

$$v(x,t) = \phi_1(t - \frac{x}{w}) + \phi_2(t + \frac{x}{w}), \tag{46}$$

$$i(x,t) = \frac{1}{Z}\{\phi_1(t - \frac{x}{w}) - \phi_2(t + \frac{x}{w})\}\,, \tag{47}$$

where $w = (LC)^{-\frac{1}{2}}$ is the propagation velocity of waves through the line, $Z = (L/C)^{\frac{1}{2}}$ is the characteristic impedance of the line, and ϕ_1 and ϕ_2 are arbitrary functions to be determined.

The combination (46) and (42) gives

$$\phi_1(t) + \phi_2(t) = 0. \tag{48}$$

With the substitution of (48) into (46) and (47), the results are

$$v(x,t) = \phi_1(t - \frac{x}{w}) - \phi_1(t + \frac{x}{w}), \tag{49}$$

$$i(x,t) = \frac{1}{Z}\{\phi_1(t-\frac{x}{w}) + \phi_1(t + \frac{x}{w})\}\,. \tag{50}$$

From (43), (49) and (50)

$$\phi_1(t - \frac{T}{2}) + \phi_1(t + \frac{T}{2}) = Zf(\phi_1(t - \frac{T}{2}) - \phi_1(t + \frac{T}{2}) + E) , \quad (51)$$

where $T/2 = \ell/w$: the propagation time of waves through the line.

Equation (51) may be written as

$$\phi_1(t + \frac{T}{2}) = g[\phi_1(t - \frac{T}{2})]. \quad (52)$$

This equation is a difference equation with the difference T. If $\phi_1(t)$ is given for $-T/2 < t < T/2$, $\phi_1(t)$ may be successively determined for $T/2 < t < 3T/2$, $3T/2 < t < 5T/2$,....

Now, from (44) and (49)

$$\phi_1(-\frac{x}{w}) - \phi_1(\frac{x}{w}) = \alpha(x) \qquad (0 < x < \ell) , \quad (53)$$

and from (45) and (50)

$$\phi_1(-\frac{x}{w}) + \phi_1(\frac{x}{w}) = Z\,\beta(x) \qquad (0 < x < \ell). \quad (54)$$

Accordingly

$$\phi_1(-\frac{x}{w}) = \frac{1}{2}\{\alpha(x) + Z\,\beta(x)\} \qquad (0 < x < \ell), \quad (55)$$

$$\phi_1(\frac{x}{w}) = -\frac{1}{2}\{\alpha(x) - Z\beta(x)\} \qquad (0 < x < \ell). \quad (56)$$

Equation (55) gives $\phi_1(t)$ for $-T/2 < t \leqq 0$, and (56) for $0 \leqq t < T/2$. The value of $\phi_1(t)$ for $-T/2 < t < T/2$, therefore, is determined by combining (55) with (56). Thus our problem is reduced to the difference equation (51) or (52).

By the introduction of a new variable $\psi(t)$:

$$\psi(t) = \psi_1(t - \frac{T}{2}),$$

(52) becomes

$$\psi(t + T) = g[\psi(t)]. \quad (57)$$

Furthermore, making use of new variables

$$\begin{aligned} \xi &= \frac{1}{\sqrt{2}}\{\psi(t) - \psi(t + T)\} , \\ \eta &= \frac{1}{\sqrt{2}}\{\psi(t) + \psi(t + T)\} , \end{aligned} \quad (58)$$

(51) is rewirtten as

$$\eta = \frac{Z}{\sqrt{2}}\, f(\sqrt{2}\,\xi + E). \quad (59)$$

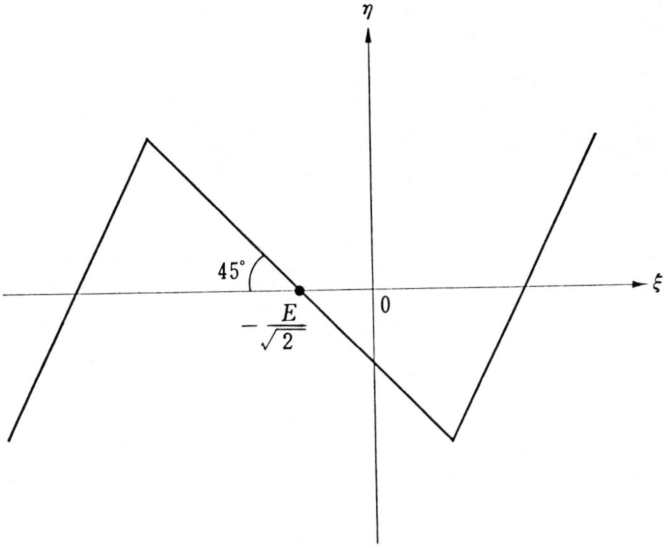

Fig. 12. Graphic display of the relation between ξ and η in (58).

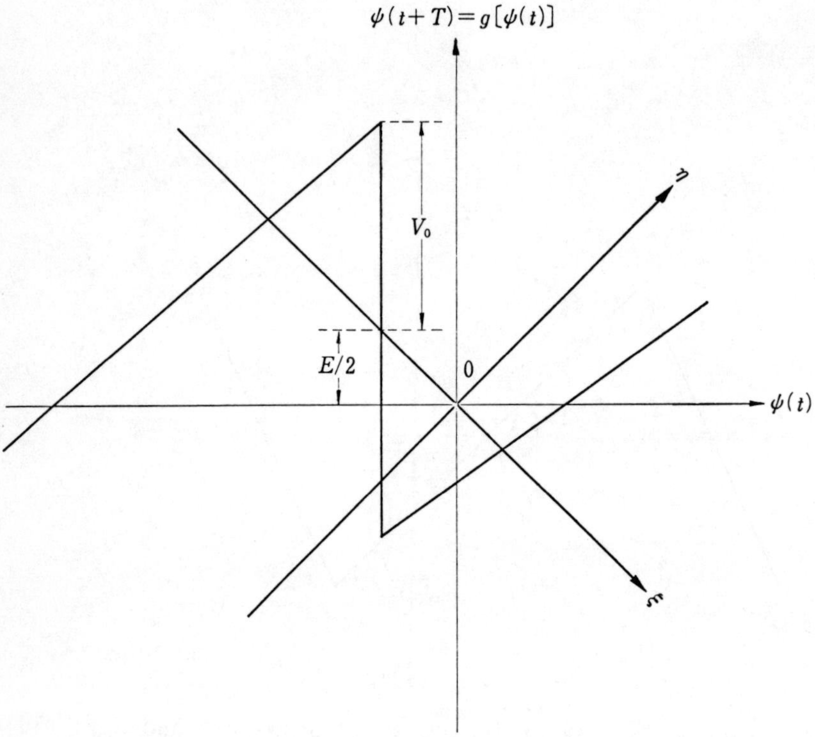

Fig. 13. Comparison of this figure with Fig. 3 leads to the conclusion that the circuit in Fig. 10 is an analogue of the mathematical neuron model.

Obviously

$$\xi = \frac{1}{\sqrt{2}} v(\ell,t), \qquad \eta = \frac{Z}{\sqrt{2}} i(\ell,t). \tag{60}$$

From Fig. 11, (59) becomes as shown in Fig. 12. Hence, the relation between $\psi(t)$ and $\psi(t+T)$ is obtained by rotating the curve in Fig. 12 45^0 to the right. This is shown in Fig. 13. By comparing this figure with Fig. 3, it is found that the relation between $\psi(t)$ and $\psi(t+T)$ takes the same form as that between y_n and y_{n+1}. Furthermore, a simple computation gives

$$a = \frac{1}{2}(1 + \frac{E}{\overline{V}_0}), \tag{61}$$

$$b = \frac{Z + R}{Z - R} > 1. \tag{62}$$

Therefore, the condition $-\overline{V}_0 < E < \overline{V}_0$ implies $0 < a < 1$.

After all, it has been shown that

$$x_n = 1[v(\ell,nT)] \qquad (n = 1,2,3,\ldots), \tag{63}$$

where T is double the delay time of the delay line.

Some experimental results of this circuit, when a tunnel diode is used as the negative resistance element, was given in Reference 6.

VII. REMARKS

So far, our consideration is limited to the case where the amplitude of the input pulse sequence is constant. The subsequent problem will be the discussion on those cases where the pulse amplitude of the input sequence is not constant but changes periodically with period ℓ.

Up to the present, two special cases were solved. One is the case where the input sequence is of the type $\{0^{\ell-1}A\}$, where A is a positive constant and ℓ is an arbitrary integer greater than unity. The result of this case is similar to that of the constant input case mentioned above, except that the characteristic of this case is composed of the lower left part of the characteristic of the constant input case, as shown in Fig. 14. See Ref. 7.

The other case which was solved is that the input sequence is of the type {0AA}, where A is a positive constant and $\ell=3$. The result of this case is fairly complicated. See Ref. 8.

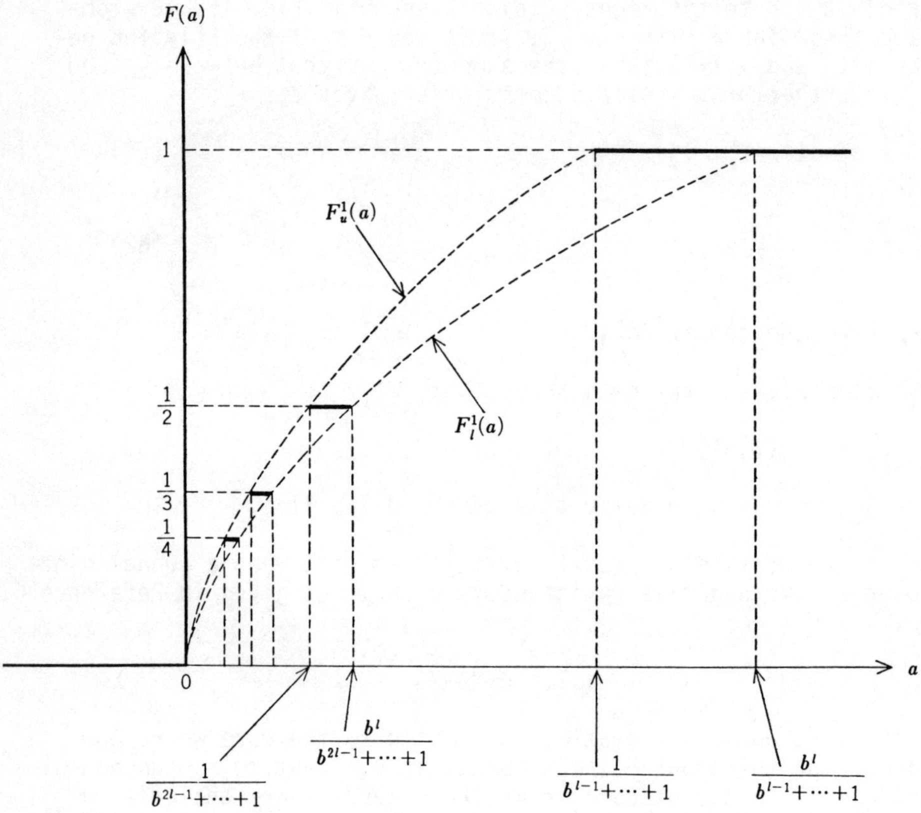

Fig. 14. The relation between a and F for the periodic input of the type $\{0^{\ell-1}A\}$, where A is a positive constant and $\ell \geqq 2$.

ACKNOWLEDGEMENTS

The proof in Appendix was given by Prof. S. Furuya, Dept. of Mathematics, University of Tokyo, for which the author wishes to express his gratitude. Part of this paper was published in Ref. 9. An alternative approach to the same problem was given by Shunsuke Sato, Osaka University. See Ref. 10.

APPENDIX - PROOF THAT $\phi(n)$ IS EULER'S FUNCTION

In this appendix, the constitution of Table IV will be investigated. First of all,

$C_1 = \{2,3,4,\ldots\}$.

The element of C_2 which lies between p_1 (positive integer not less than 2) and p_1+1 takes the form $n = p_2p_1+q_2(p_1+1)$, where p_2 and q_2 are positive integers and $p_2=1$ or $q_2=1$.

In general, the positive integer of the form $n = pp_1+q(p_1+1)$, where p, q, and p_1 are positive integers and $p_1 \geqq 2$, is denoted by $[p, q]$. If p and q are prime each other, it is called a 'prime pair'.

It is obvious that every element of

$C_2 = \{\ldots, [3,1], [2,1], [1,1],[1,2], [1,3],\ldots\}$

is a prime pair. Since C_2 is symmetric, $p > q$ is assumed hereafter. See Table VII.

Next, the element of C_3 which lies between $[p_2,1]$ and $[p_2+1,1]$ takes either of the following two forms:

(a) $[2p_2+1,2] + p_3[p_2,1] = [(p_3+2)p_2+1, p_3+2]$,

(b) $[2p_2+1,2] + p_3[p_2+1,1] = [(p_3+2)p_2+(p_3+1), p_3+2]$

where $p_3 \geqq 0$, and it is easily seen that both of them are prime pairs.

Expression (a) above is rewritten as $[p_3p_2+1, p_3]$ $(p_3 \geqq 2)$, if p_3+2 is replaced by p_3. The element of C_4, which lies between $[p_3p_2 +1, p_3]$ and $[(p_3+1) p_2+1, p_3+1]$, takes either of the following two forms:

C_1	C_2	C_3
2	...	
	[3,1]	
		...
		[11,4]
		[8,3]
		[5,2]
		[7,3]
		[9,4]
		...
	[2,1]	
		...
		[7,4]
		[5,3]
		[3,2]
		[4,3]
		[5,4]
		...
	[1,1]	
	[1,2]	
	[1,3]	
3	...	

Table VII. Alternative expression of Table IV by the use of prime pairs.

(a) $[(2p_3+1)p_2+2,\ 2p_2+1] + p_4\ p_3p_2+1,\ p_3]$

$= [((p_4+2)p_3+1)p_2+(p_4+2),\ (p_4+2)p_3+1],$

(b) $[(2p_3+1)p_2+2,\ 2p_2+1] + p_4[(p_3+1)p_2+1,\ p_3+1]$

$= [((p_4+2)p_3+p_4+1)p_2+(p_4+2),\ (p_4+2)p_3+p_4+1],$

where $p_4 \geqq 0$, and both of them are prime pairs. The same result is obtained for the element of C_3 with expression (b). Similar processes show that positive integers which appear between p_1 and p_1+1 in Table IV are all prime pairs.

Our next task then is to show the fact that when a prime pair $[p,q]$ is given, the position it occupies between p_1 and p_1+1 in Table IV is uniquely determined. If this is shown, it is known that every prime pair appears once and only once between p_1 and p_1+1 in Table IV.

Now, as described above,

$C_2 = \{[p_2,1],\ p_2 \geqq 1\}$.

Next, the general form of the element of C_3, which lies between $[p_2,1]$ and $[p_2+1,1]$ is either of the following two forms:

(a) $[2p_2+1,2] + p_3[p_2,1] = [(p_3+2)p_2+1,p_3+2],$

(b) $[2p_2+1,2] + p_3[p_2+1,1] = [(p_3+2(p_2+1)-1,\ p_3+2],$

where $p_3 \geqq 0$. Thus the general form of C_3 can be expressed as $[p_3p_2' + \varepsilon_3,p_3]$ $(p_3 \geqq 2)$ where $\varepsilon_3 = \pm 1$, and $p_2' = p_2$ if $\varepsilon_3 = +1$, $p_2' = p_2+1$ if $\varepsilon_3 = -1$.

Next, the general form of C_4, which lies between $[p_3p_2'+\varepsilon_3,p_3]$ and $[(p_3+1)p_2' + \varepsilon_3,\ p_3+1]$, takes either of the following two forms:

(a) $[(2p_3+1)p_2'+2\varepsilon_3,\ 2p_3+1] + p_4[p_3p_2' + \varepsilon_3,p_3]$

$= [((p_4+2)p_3+1)p_2'+)p4+2)\varepsilon_3,(p_4+2)p_3+1],$

(b) $[(2p_3+1)p_2' + 2\varepsilon_3,\ 2p_3+1] + p_4[(p_3+1)p_2' + \varepsilon_3,\ p_3+1]$

$= [((p_4+2)(p_3+1)-)p_2'+(p_4+2)\varepsilon_3,\ (p_4+2)(p_3+1)-1],$

where $p_4 \geqq$. Thus the general form of C_4 is given by

$[(p_4p_3' + \varepsilon_4)p_2' + p_4\varepsilon_3,\ p_4p_3' + \varepsilon_4] \qquad (p_4 \geqq 2)$

where $\varepsilon_4 = \pm 1$, $p_3' = p_3$ if $\varepsilon_4 = 1$, $p_3' = p_3+1$ if $\varepsilon_4 = -1$.

Summing up, it is found that the element [p,q] of C_3 is expressed as

$$p = p'_2 q + \varepsilon_3,$$

$$q = p_3,$$

and the element of C_4 as

$$p = p'_2 q + \varepsilon_3 q_3 \qquad (q_3 < q/2)$$

$$q = p'_3 q_3 + \varepsilon_4,$$

$$q_3 = p_4.$$

In like manner, the element [p,q] of C_k can be expressed as follows:

$$p = p'_2 q + \varepsilon_4 q_3 \qquad (q_3 < q/2),$$

$$q = p'_3 q_3 + \varepsilon_4 q_4 \qquad (q_4 < q_3/2),$$

$$q_3 = p'_4 q_4 + \varepsilon_5 q_5 \qquad (q_5 < q_4/2),$$

$$\cdots\cdots$$

$$q_{k-3} = p'_{k-2} q_{k-2} + \varepsilon_{k-1} q_{k-1} \qquad (q_{k-1} < q_{k-2}/2),$$

$$q_{k-2} = p'_{k-1} q_{k-1} + \varepsilon_k,$$

$$q_{k-1} = p_k,$$

where $\varepsilon_i = \pm 1$ for $i = 3,4,\ldots,k$ and $p'_{i-1} = p_{i-1}$ if $\varepsilon_i = +1$, $p'_{i-1} = p_{i-1} + 1$ if $\varepsilon_i = -1$; $p_2 \geqq 1$, $p_i \geqq 2$ for $i = 3,4,\ldots,k$.

The above expression of the prime pair [p,q] indicates the position it occupies between p_1 and p_{1+1} in Table IV. An example is shown in Table VIII where the prime pair [143, 38] is found to belong to C_5.

Thus, it is ascertained that the number of integer $n (n \geqq 5)$ which belongs to C_i's $(i \geqq 2)$ is equal to the number of ways of expressing n by the prime pair [p,q]; namely, the number of ways of expressing n as

$$n = pp_1 + q(p_1 + 1)$$

$$= (p + q)p_1 + q$$

$$= r\,p_1 + q \qquad (r = p + q),$$

$$
\left\{
\begin{array}{lllll}
143 = 4\cdot 38 - 9 & p_2' = 4 & \varepsilon_3 = -1 & p_2 = 3 & q_3 = 9 \\
38 = 4\cdot 9 + 2 & p_3' = 4 & \varepsilon_4 = +1 & p_3 = 4(2) & q_4 = 2 \\
9 = 4\cdot 2 + 1 & p_4' = 4 & \varepsilon_5 = +1 & p_4 = 4(2) & (k = 5) \\
 & & & p_5 = q_4 = 2(0) &
\end{array}
\right.
$$

C_2	C_3	C_4	C_5
$p_2 = 3$	$p_3 = 2$ $\varepsilon_3 = -1$ ↑	$p_4 = 2$ $\varepsilon_4 = +1$ ↓	$p_5 = 0$ $\varepsilon_5 = +1$ ↑
[4,1] (0) [3,1]	[19,5] (2) [15,4] (1) [11,3] (0) [7,2]	(0) [34,9] (1) [49,13] (2) [64,17] [79,21]	(0) [143,38]

Table VIII. This table shows how one discovers position occupied by prime pair [143, 38]. If $\varepsilon_i = \pm 1$, the position is located, as indicated by small arrows, in a half of the set C_i, which consists of smaller/larger prime pairs.

where $p_1 \geqq 2$; p and q are positive integers prime each other; r is prime to q and $r > q \geqq 1$; n is prime to r and $n > rp_1 \geqq 2r$. Hence, $n/2 > r > 1$.

Given a positive integer n, let the positive integers which are less than and prime to n be

$$1 = r_1 < r_2 < r_3 < \cdots < r_s < r_{s+1} < \cdots < r_{\phi-1} < r_\phi = n-1,$$

where ϕ implies Euler's function $\phi(n)$. Since $r_1 + r_\phi = n$, $r_2 + r_{\phi-1} = n, \ldots, r_s + r_{s+1} = n$, it is immediately follows that $s = \phi(n)/2$. Since r is limited to $n/2 > r > 1$, the number of ways of expressing n in the form $n = rp_1+q$ is equal to the number of r, such that

$$r = r_2, r_3, \ldots, r_s,$$

or s-1.

Besides, every positive integer not less than 2 appears once in C_1. Hence, it is concluded that the number of $n(n \geqq 3)$ which appears in Table IV is given by $(s-1) + 1 = \phi(n)/2$.

REFERENCES

(1) L. D. Harmon, "Studies with Artificial Neurons, I: Properties and Functions of an Artificial Neuron," Kybernetik, Vol. 1, pp. 89-101, Dec. 1961.

(2) E. R. Caianiello, "Outline of a Theory of Thought-Processes and Thinking Machines," J. Theoret. Biol., Vol. 1, pp. 204-232, April 1961.

(3) E. R. Caianiello and A. DeLuca, "Decision Equation for Binary Systems, Application to Neuronal Behavior," Kybernetik, Vol. 3, pp. 33-40, Jan. 1966.

(4) G. H. Hardy and E. M. Wright, "An Introduction to the Theory of Numbers," Oxford, Clarendon Press, p. 54, 1960.

(5) E. C. Titchmarsh, "The Theory of Functions," Oxford, Clarendon Press, p. 366, 1968.

(6) J. Nagumo and M. Shimura, "Self-Oscillation in a Transmission Line with a Tunnel Diode," Proc. IRE, Vol. 49, pp. 1281-1291, Aug. 1961.

(7) J. Nagumo, "Some Properties of a Function Which is Derived from Solutions of a Nonlinear Difference Equation," Report of Professional Group on Nonlinear Problems, Inst. Electronics Commun. Eng. Japan, NLP 72-10, Aug. 1972 (in Japanese).

(8) S. Sato, M. Hatta and J. Nagumo, "Response Characteristic of a Neuron Model for a Periodic Input," Report of Professional Group on Medical Electronics and Biological Engineering, Inst.

Electronics Commun. Eng. Japan, MBE 73-21, Oct. 1973 (in Japanese).
(9) J. Nagumo and S. Sato, "On a Response Characteristic of a Mathematical Neuron Model," Kybernetik, Vol. 10, pp. 155-164, Mar. 1972.
(10) S. Sato, "Mathematical Properties of Response of a Neuron Model, A System as a Rational Number Generator," Kybernetik, Vol. 11, pp. 208-216, Nov. 1972.

PERFORMANCE ASPECTS OF STOCHASTIC NONLINEAR SYSTEM CLASSIFICATION BY PATTERN RECOGNITION METHODS

R.F. Hofstadter and G.N. Saridis

Purdue University

West Lafayette, Indiana 47907

ABSTRACT

The feasibility of identifying an unknown nonlinear stochastic system as belonging to a class of such systems by use of pattern recognition methods has recently been demonstrated. This paper examines the performance aspects of the classification technique by means of a theoretical approximation to the Bayes minimum classification error. The error is shown to be strongly dependent upon the pattern vector dimension thus showing that the classification error can be controlled by appropriate choice of pattern dimension. Two detailed classification examples are included, the first of which may be compared to earlier experimental results to substantiate the approximations used.

I. INTRODUCTION

Classification of members of defined classes of stochastic nonlinear systems using pattern recognition techniques has recently been reported by Saridis and Hofstadter [1]. The classification may be done in an on-line or off-line situation and is achieved by means of a pattern vector based only on the measured values of the system input and the system output which is possibly noise corrupted. The systems investigated were of the general form

$$x(j+1) = g(x(j), \zeta(j)) \tag{1-1}$$

$$y(j) = h'x(j) \tag{1-2}$$

$$z(j) = y(j) + \eta(j) \tag{1-3}$$

where $x(j)$ is an n_0-vector, $g(\cdot)$ is a nonlinear n_0-vector, $h' = (1,0,\ldots,0)$ an n_0-vector transposed, and $y(j)$, $z(j)$ are scalars as is $\eta(j)$, the measurement noise sequence. The n_0-vector sequence $\zeta(j)$ is the input disturbance acting on the system. By appropriately specializing the function $g(\cdot)$ various classes of nonlinear stochastic systems may be specified.

In [1] two methods for constructing pattern vectors were proposed and utilized in actual classifications using nonlinear system pattern vectors generated by digital simulation. Because the K-nearest-neighbor (KNN) classification is related to the minimum Bayes error, the KNN algorithm was previously used in actual classification studies. Results of these studies showed that error rates of considerably less than 1% are achievable for six stated classes of low order nonlinear stochastic systems.

The overall classification method is shown in Fig. 1 for single-input single-output systems with the input now specified by the scalar sequence $\xi(j)$. By means of the classification method, an unknown system may be identified as belonging to one of a number of known classes. Knowledge of the class membership of the unknown system may then be subsequently used to establish a parametric model for further system identification and modeling, or it may be used for control purposed by selecting a learning type of feedback controller to asymptotically optimize performance.

In this paper a theoretical achievable minimum classification error will be examined and shown to agree with the experimental results. Also the dependence of classification error on pattern dimension is examined and shown to agree with an intuitive interpretation of the pattern vector. Plots of classification error versus pattern dimension are generated which provide a way of selecting the maximum pattern dimension required for a given achievable error level.

II. STATEMENT OF THE PROBLEM

In the following development a two-class problem will be considered. This is not an essential restriction and the extension to the multiclass problem is conceptually straightforward. Unless otherwise noted, the term "system" will mean a discrete-time nonlinear stochastic system. The mathematical expectation of scalar and vector quantities will be represented by $E[v]$ or by $\bar{v}$. For simplicity only single-input single-output systems will be considered.

In [1] two types of pattern vectors were developed for the classification of systems based only on the input-output measurements. For a system as shown in Fig. 1, the first type of pattern vector is based on the system discrete crosscovariance function which may

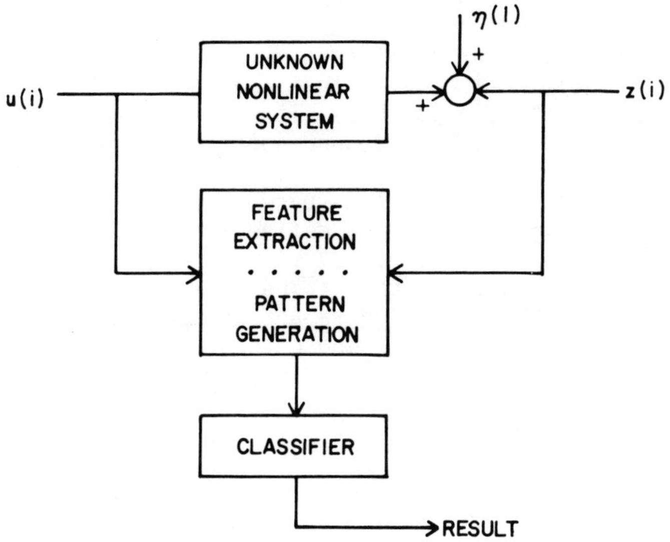

Fig. 1. Nonlinear System Classification Method

be written as

$$C_{\xi y}(k) = E[(\xi(j) - \bar{\xi})(y(j + k) - \bar{y})] \quad (2)$$

for the stationary processes under consideration. By defining $v_k = C_{\xi y}(k)$, the m-dimensional pattern vector may be written as

$$V = (v_1, v_2, \ldots, v_m) \quad (3)$$

with m chosen large enough to provide sufficient information for classification. The suitable choice of m will be discussed below.

Let ω_1 and ω_2 represent two classes of nonlinear systems and for the moment assume the classes contain systems of the same order. A question of primary importance is to determine the theoretical minimum Bayes classification error that can be achieved. Before establishing the error relationships, it will be convenient to first define the probability space and notation to be used.

Thus let R^m denote Euclidean m-space containing all m-tuples of the form $V = (v_1, v_2, \ldots, v_m)$, and define an interval in R^m by the set of all points V such that $\lambda_{k\ell} \leq v_k \leq \lambda_{ku}$. Let Ω^m be the smallest σ-algebra of subsets of R^m containing all intervals. Ω^m is thus an m-dimensional Borel σ-algebra and the pair (R^m, Ω^m) constitute a measurable space. Let μ_k denote the Borel probability measure in the k-th dimension and let $\mu = \prod_{k=1}^{m} \mu_k$ be the probability measure defined over the σ-algebra Ω^m. This defines the triplet (R^m, Ω^m, μ) which constitutes the probability space for the sample patterns. The product measure μ defined over the σ-algebra Ω^m uniquely determines, and is uniquely determined by, the m-variate distribution function $P(V) = P(v_1, v_2, \ldots, v_m)$ [2]. Restrict μ to be absolutely continuous with respect to Lebesgue measure. Then there exists a μ-measurable non-negative function $p(V) = p(v_1, v_2, \ldots, v_m)$, the uniquely determined probability density function of the pattern vectors V.

To obtain the class conditional density functions, let ω_i be the event that a random sample pattern V_j falls in the i'th class and define $Pr[\omega_i]$ to be the a priori probability of event ω_i. Let $p(V|\omega_i)$ denote the class conditional density function conditioned on the event ω_i, and write the mixture probability density as

$$p(V) = \sum_{i=1}^{2} P(V|\omega_i) Pr[\omega_i]$$

In the notation described, the Bayes minimum error of classification associated with the Bayes decision rule is concisely given by the relation

$$\varepsilon^* = \int_{\Gamma_2} p(\omega_1|V)d\mu + \int_{\Gamma_1} p(\omega_2|V)d\mu \tag{4-1}$$

where

$$\Gamma_i = \{V|p(V|\omega_i)Pr[\omega_i] > p(V|\omega_j)Pr[\omega_j]\} \quad i,j = 1,2: i \neq j$$

and $p(\omega_i|V)$ represents the a posteriori densities. By applying Bayes Theorem for densities and noting that $\int_\Gamma d\mu = \int_\Gamma p(V)dV$ for arbitrary regions Γ in Ω^m, the Bayes error may also be expressed as [3]

$$\varepsilon^* = \int_{\Gamma_2} p(V|\omega_1 dV\ Pr[\omega_1] + \int_{\Gamma_1} p(V|\omega_2)dV\ Pr[\omega_2]. \tag{4-2}$$

The integral expressions in (4) appear deceptively simple for it is usually the case in pattern recognition situations that the conditional densities are not available and must be estimated from sample patterns if they are to be used at all. Moreover, when the conditional densities are available the n-fold integrals can become extremely difficult to evaluate when the pattern dimension is even moderately high.

In the following section it will be shown that for the nonlinear system classification problem under consideration, it is possible to arrive at suitable analytical conditional density functions to describe the classes. Following this, methods are developed to evaluate the error expression (4) for patterns of arbitrary dimension. Having accomplished this, the dependence of classification error with pattern dimension is examined and plotted for various values of the parameters involved.

It may be noted before proceeding that the noise free measurement $y(j)$ has been used in formulating the error expression. The fact that the output measurement $z(j)$ in (1-3) can be used just as readily is easily seen from the properties of correlation functions. For example, in the usual case of zero mean, uncorrelated measurement noise $E[\eta(j)] = 0$, $E[\xi(i)\ \eta(j)] = 0$ for all i,j, it is easily shown that $C_{\xi y}(k) = C_{\xi z}(k)$. Also it may be noted that the theoretical correlation expressions have been used thus far. In an actual classification situation however, the correlations have to be estimated from actual system data. This is readily accomplished by means of the following recursive relation.

$$\hat{C}_{\xi z}^{(N+1)}(k) = \frac{1}{N+1}\left[N\hat{C}_{\xi z}^{(N)}(k) + [\xi(N+1-k) - \bar{\xi}][z(N+1) - \bar{z}]\right] \tag{5}$$

When sufficient classification time is available, the estimates $\hat{C}^{(N+1)}_{\xi z}(k)$ may be obtained to a high degree of accuracy. The effects of error that may be present in the estimates will be discussed separately in Section VI.

III. CONDITIONAL CLASS DENSITIES

To define various classes of nonlinear stochastic systems, it will be convenient to consider equations of the form (1-1) which can be written as a nonlinear stochastic difference equation of order n_0 in terms of the system output variable $y(j)$ as

$$\begin{aligned} y(j+n_0) &= a_1^0\, y(j+n_0-1) + a_2^0\, y(j+n_0-2) + \cdots + a_{n_0}^0 y(j) + \\ &\quad f(y(j),\, y(j+1),\cdots,y(j+n_0-1) + \xi(j) \qquad (6) \\ z(j) &= y(j) + \eta(j) \end{aligned}$$

where $f(\cdot)$ is a nonlinear function of the y's, and the superscript o indicates coefficients of the linear part of the system. Some of these coefficients may in general be equal to zero. The input is now $\xi(j) = \zeta_{n_0}(j)$, a white noise sequence with zero mean and unity variance, and $\eta(j)$ will be assumed to have zero mean and variance σ_η^2, with $E[\xi(i)\eta(j)] = 0$ for all i,j. The model of (6) does not admit all possible classes that might be considered, but it is sufficiently general for development of the essential concepts here.

In general the nonlinear function $f(\cdot)$ may consist of a sum of terms such as

$$f(\cdot) = \sum_{k=1}^{N_0} f_k \cdot \sum_{i=1}^{n_0} \gamma_{ki} y(j+n_0-1) \qquad (7)$$

where typically many of the γ_{ki} will be zero in practice. Also, the number of nonlinearities N_0 will typically be small for a given class of systems, usually less than n_0 although this need not be so. For convenience the $f_k(\cdot)$ will also be assumed to be zero mean, single valued, odd functions of their arguments. Other than this the $f_k(\cdot)$ are arbitrary subject of course to the implicit restriction that the overall system of equation (6) be stationary. This definition of the $f_k(\cdot)$ permits a very broad variety of nonlinearities to be utilized including symmetric saturation functions, odd polynomial functions and, for example even powers of the $y(j+n_0-i)$, $i = 1,2,\ldots,n_0$. This is readily done, for example, by defining

$$f_k(y(j+1)) = y^2(j+1) \qquad y(j+1) \geq 0$$
$$= -y^2(j+1) \qquad y(j+1) < 0$$

to obtain an odd function of even powers. The model of equation (6) and the allowable forms for the nonlinearities permit consideration of quite a large number of specific classes of nonlinear systems. More general models and nonlinearities can be considered to enlarge the permissible classes of systems.

To continue the development, consider the $f_k(\cdot)$ replaced by their statistically linearized [8] equivalents δ_k^* chosen to minimize the expected value of the mean square error

$$e^2 = (f_k(y(j)) - \delta_k^* y(j))^2$$

between $f_k(\cdot)$ and $\delta_k^* y(j)$ where δ_k^* is a constant. It is easily shown for example in [8], that the desired value of $\delta_k^* = E(y(j)f_k(y(j)))/E(y^2(j))$ and that in addition to minimizing the e^2, δ_k^* also yields the same crosscorrelation function between $y(j)$ and $\delta_k^* y(j)$ as the nonlinearity itself. Criteria other than the minimization of e^2 are available [8] and result in slightly differing forms for the δ_k^*. In particular it is possible to include a constant bias term d_k^* to result in an equivalent linearization $\delta_k^* y(j) + d_k^*$ which may be used to describe non-zero-mean nonlinearities and thus remove the assumption of zero-mean nonlinearities for the $f_k(\cdot)$, but for brevity this refinement will not be included here.

The above idea is readily extended to cover the classes of systems being considered here. To do so, define the constants γ_{ki} and

$$\sigma_k = \sum_{i=1}^{n_0} \gamma_{ki} y(j+n_0-i) \tag{8}$$

and replace $f_k(\sigma_k)$ by its statistically linearized equivalent δ_k^* to obtain the following equation

$$f(y(j), y(j+1), \cdots, y(j+n_0-1)) = \sum_k f_k(\sigma_k)$$
$$= \sum_k \delta_k^* \sigma_k$$
$$= \sum_k \sum_{i=1}^{n_0} \delta_k^* \gamma_{ki} y(j+n_0-1)$$

$$= \sum_{i=1}^{n_0} \sum_k \delta_k^* \gamma_{ki} y(j+n_0-1)$$

$$= \sum_{i=1}^{n_0} \delta_i \; y(j+n_0-1) \qquad (9)$$

where the variable δ_i has been defined as

$$\delta_i = \sum_k \delta_k \gamma_{ki}. \qquad i = 1,2,\ldots,n_0$$

Note that typically many of the γ_{ki} will be either zero or one, as the input to a specific nonlinearity often consists of just one of the n_0 possible variables $y(j+n_0-i)$. Also, as noted earlier, the summation on k typically involves only a few terms, one for each type of nonlinearity in a member system of a specified class. Thus the general formula simplifies considerably when applied to actual classes of physical nonlinear systems.

It is now possible to combine equations (6) and (9) to obtain

$$y(j+n_0) = \sum_{i=1}^{n_0} a_i^0 y(j+n_0-1) + \sum_{i=1}^{n_0} \delta_i y(j+n_0-1) + \xi(j)$$

$$= \sum_{i=1}^{n_0} a_i y(j+n_0-1) + \xi(j) \qquad (10)$$

where $a_i \triangleq a_i^0 + \delta_i$. It is now apparent that the a_i are random variables of an arbitrary class of nonlinear stochastic systems and that the distribution or occurrence of particular members of a stated class will control the resulting distributions of the random variables a_i. It is important to note that the a_i are not completely arbitrary for it is well known that they must be such that the characteristic equation

$$\lambda^{n_0} - a_1 \lambda^{n_0-1} + \cdots - a_{n_0} = 0$$

has roots less than unity for stationarity of the system in equation (10). Thus the statistical variation of the a_i must be limited to certain finite intervals which in turn implies that only probability distributions with finite interval domains of definition are suitable for establishing the <u>a priori</u> distributions to be associated with the various a_i. This precludes the use of the Gaussian distribution as well as any other permitting arbitrarily large values of the variables to occur. Depending on the amount of a priori information known about a class of systems, various distributions are available which may be assigned to the random variables a_i. For instance, if no a priori information is available, it is convenient to assign a uniform distribution $U(a_{il}, a_{iu})$ to the a_i. Another suitable and quite general distribution to use is the modified Beta distribution with density function $p(t;\alpha,\beta) = K(t-t_1)^\alpha(t-t_2)^\beta$ with compact support $[t_1, t_2]$, α,β rational numbers, and K the constant of normalization as shown in Fig. 2.

It is now possible to proceed with the development of the class conditional density function $p(V|\omega_\nu)$ for an arbitrary class of systems which originated from the general model equation (6). It will be assumed that suitable distributions for the a_i are determined and that all systems of the class ω_ν are of the same order n_0. Let (R,Ω,μ_{a_i}) and $p_{a_i}(a_i|\omega_\nu)$ be respectively the probability space for the variable a_i and its uniquely determined density function conditioned on ω_ν, with compact support $[a_{il}, a_{iu}]$ defined as Γ_{a_i}. For brevity of notation define also the set

$$v^{(k)} = \{v_i | i = 1,2,\ldots,k-1,k\}$$
$$\{\phi | k < 1\}$$

Then by application of the probability chain rule, the conditional class density function for class ω_ν may be written as

$$\begin{aligned} p(V|\omega_\nu) &= p(v_1, v_2, \ldots, v_m) \\ &= p(v_m | v^{(m-1)}, \omega_\nu) p(v_{m-1} | v^{(m-2)}, \omega_\nu) \cdots \\ &\quad \cdots p(v_2 | v^{(1)}, \omega_\nu) p(v_1 | v^{(0)}, \omega_\nu) \\ &= \prod_{i=1}^{m} p(v_i | v^{(i-1)}, \omega_\nu) \end{aligned} \tag{11}$$

defined over the set $\Gamma_{v_1} X \Gamma_{v_2} X \cdots X \Gamma_{v_m} \in \Omega^m$ where X denotes the Cartesian product. Next, multiplying (10) by $\xi(j+n_0-k)$ and taking expected values yields a recursive relation for the theoretical cross-correlations c_k given by

$$c_k = \sum_{i=1}^{n_0} a_i c_{k-1} \qquad k = n_0+1, n_0+2, \ldots \tag{12}$$

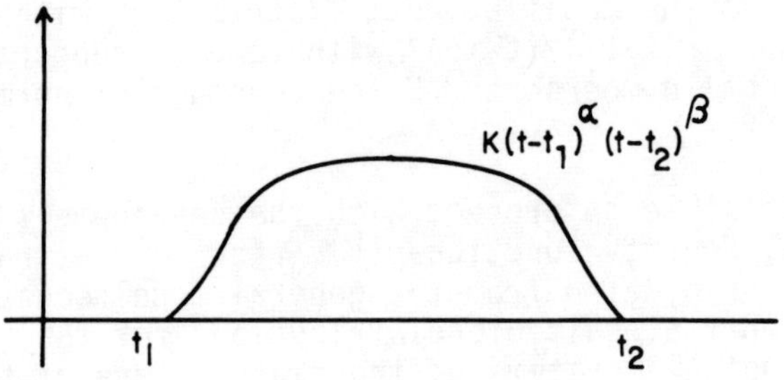

Fig. 2. Representative Density Function $p(t;\alpha,\beta)$ on $[t_1,t_2]$

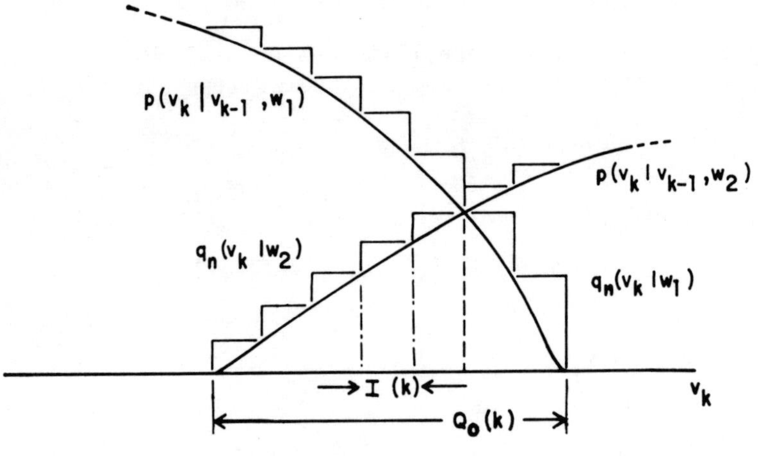

Fig. 3. Class Conditional Density Functions; k'th Dimension.

$$= 1 \qquad\qquad k = n_0.$$

$$= 0 \qquad\qquad k < n_0$$

From consideration of the recursive relation just obtained and the assumption of arbitrarily large N, the transformation of random functions permits writing a general factor of the product in (11) as

$$p(v_i|v^{(i-1)},\omega_\nu) = \frac{1}{|J|}\, p_{a_i}\left(v_i - \sum_{j=1}^{i-1} a_i v_{i-j} \middle| v^{(i-1)},\omega_\nu\right) \quad (13)$$

where $|J|$, the determinant of the Jacobian of the transformation, is equal to unity. Observe now that the product in (11) may actually be written from $i = n_0 + 1$ to $i = m + n_0$ since the conditional densities $p(v_i|v^{(i-1)},\omega_\nu)$ are delta functions $\delta(v_i)$ for $i = 0,1,..,n_0-1$, and $p(v_{n_0}|v^{(n_0-1)},\omega_\nu) = \delta(v_{n_0}-1)$. This follows directly from (12) where it is seen that the first n_0+1 correlations provide no information for classification and thus need not be included in the product (11).

The substitution of the general factor (13) into equation (11) completes the conditional class density description of arbitrary classes of systems based on the model of equation (6). Thus it is possible, for instance, to define two classes of nonlinear stochastic systems based on the model equation (6) and obtain their respective class conditional densities given by the equations above. As noted earlier, these conditional densities along with the a priori probabilities $Pr[\omega_\nu]$, $\nu=1,2$ and the mixture density expression complete the probabalistic description of the two class problem. Consideration of the M-class problem is straightforward, requiring $Pr[\omega_\nu]$, $p(V|\omega_\nu)$ $\nu = 1,2,...M$, and equations (11), (13).

It is perhaps worthwhile to briefly summarize the approximations used in the above development. Up to and including (11), the density expressions are exact although m has not yet been specified. Then (12) utilizes the statistically linearized model (10) to obtain the result (13) to produce the approximated conditional class density expression, which in turn will be used to compute the Bayes error given by (4). Further analytical results have recently been obtained in regard to choice of m and development of a sufficient statistic to represent the available system data, but space does not permit their inclusion here. These will appear in forthcoming papers.

IV. CLASSIFICATION ERROR

In this section a method will be developed for evaluation of the classification error as given by (4) for two classes of systems using joint class densities of the type derived in the previous secion. The effort will be directed towards obtaining a computer algorithm which will permit rapid evaluation of the error for arbitrary classes of systems and for arbitrary dimension of the pattern vector V. This will permit evaluation of the possibility of classification under general conditions.

(1) Evaluation of Error Integrals

The difficulty of evaluating the m-fold integrals in (4) has been noted. A typical approach is to utilize Monte-Carlo integration schemes when the functional forms of the conditional class densities are available as they are here. However, this approach is not suitable when the error expression is to be evaluated repeatedly in general theoretical studies of the effects of pattern dimension and other parameters because the attendant computation time becomes prohibitive. Thus an alternate approach will be devised which conservatively overestimates the error based on the conditional class densities obtained above. This alternate approach is capable of being programmed for solution, for any desired values of parameters and dimension, in a way which requires very little computation time. The solution is based on a theorem to be developed next. A sequence of functions will be constructed that converges to $p(V|\omega_\nu)$ from above and the integration of these functions will provide the desired upper bound for the error integrals in (4).

Let the set of points $Q_0(k) \triangleq \Gamma_1(k) \cup \Gamma_2(k)$ denote the region of overlap of the two conditional class densities $p(v_k|v_{k-1},\omega_1)$ and $p(v_k|v_{k-1},\omega_2)$ in the k'th pattern dimension, and consider a partition of the set $Q_0(k)$ into 2^n cells. Define the cells to be $I_1(k) = [\lambda_{k0},\lambda_{k1}]$, $I_2(k) = [\lambda_{k1},\lambda_{k2}],\ldots,$ $I_{2^n}(k) = [\lambda_{k2^n-1},\lambda_{k2^n}]$ with λ_{k0} the left endpoint of the set $Q_0(k)$, and let $D_r(k) = \lambda_{kr} - \lambda_{k,r-1}$ denote the length of a cell $I_r(k)$. On the interval $Q_0(k)$ construct the simple functions [5] $q_n(v_k|\omega_\nu)$, $\nu = 1,2$ which take on the constant value

$$p_{kr}(n,\omega_i) = \max\left[p(v_k|\omega_\nu)\Big|_{\lambda_{k,r}}, p(v_k|\omega_\nu)\Big|_{\lambda_{k,r-1}} \right] \tag{14}$$

on the r'th cell, for any given interger n. The functions $q_n(v_k|\omega_\nu)$ may be concisely expressed as

$$q_n(v_k|\omega_\nu) = \sum_{r_k=1}^{2^n} \rho_{k,r_k}(n,\omega_\nu)\ \chi_{I_r}(k) \tag{15}$$

where $\chi_{I_r}(k)$ is the characteristic function of the cell interval $I_r(k)$. It is evident from the construction of the functions in (14) that the $q_n(v_k|\omega_\nu)$ are an upper bound for the densities $p(v_k|v_{k-1},\omega_\nu)$, $\nu = 1,2$. The functions are shown in Fig. 3 for the k'th dimension.

Substitution of the $q_n(v_k|\omega_\nu)$ into (11) defines the desired upper bounding function

$$q\ (V|\omega_\nu) = \prod_{k=n_0+1}^{m+n_0} q_n(v_k|\omega_\nu) \qquad \nu = 1,2 \tag{16}$$

for points $V \in Q_0(n_0+1) X Q_0(n_0+2) X \ldots X Q_0(m+n_0) \overset{\Delta}{=} Q$. The product of simple functions (16) is again a simple function in R^m and thus integration of (16) is readily accomplished [5].

<u>Theorem</u>: For Q, $I_r(k)$, n, $q_n(v_k|\omega_\nu)$, $\rho_{k,r_k}(n,\omega_\nu)$ as defined above, and for general conditional class densities $p(V|\omega_\nu)$ based on the linearized model (10), the Bayes minimum error of classification ε is bounded from above by

$$\varepsilon \le \varepsilon(n) = \int_Q \min_\nu \left[q_n(V|\omega_\nu) Pr[\omega_\nu]\right] dV \tag{17}$$

$$= \sum_{r_1=1}^{2^n} \sum_{r_2=1}^{2^n} \cdots \sum_{r_k=1}^{2^n} \min_\nu \left[\prod_{k=n_0+1}^{m+n_0} \rho_{k,r_k}(n,\omega_\nu) D_{r_k}(k) \right] \tag{18}$$

<u>Proof</u>: The proof of the theorem is given in the Appendix.

Although the expression in (18) appears complicated, on inspection it evidently is readily performed by digital computer. Note also that $\varepsilon(n)$ is still a function of n where 2^n is the number of cells in the partitioning of the intervals $Q_0(k)$. As n is increased, the error $\varepsilon(n)$ approaches closer to ε. It can be shown that $\lim \varepsilon(n) = \varepsilon$, but this will not be done here, as only finite sums are involved in the computer evaluation of (18).

(2) Application of the Theorem

The usefulness of the theorem in the present situation is readily seen since the quantities in (18) are computable from *a priori* information about classes of systems. In the event that no information is available, uniform distributions may be used making the expressions in (10) and (18) particularly simple. To study the classification and separability of general classes of systems, a flexible algorithm was programmed to permit evaluation of $\varepsilon(n)$ for arbitrary classes of systems and pattern dimension m. Provision was made for specifying either the modified Beta distribution mentioned above or a uniform distribution to represent *a priori* class information. The algorithm was implemented on a CDC 6500 system and used to obtain the results described in the following section.

V. RESULTS OF CLASSIFICATION ERROR STUDIES

To demonstrate that classification of nonlinear systems is possible with the method proposed, the results of the preceding sections are utilized here to investigate the associated error of classification and its dependence on pattern dimension. All the theoretical error curves in this section are computed by the algorithm described above and represent upper bounds for the theoretical minimum Bayes error ε with respect to the linearized model (7).

(1) A Classification Example

This example deals with two realistic classes of second-order nonlinear stochastic systems defined as follows:

Class ω_1. This class may be referred to as mildly nonlinear since it is represented by an autoregressive process [4] with a small nonlinearity as

$$y(j+2) = a_2 y(j) + \varepsilon f_r(y(j),y(j+1)) + \xi(j)$$

where the subscript r indicates that the general nonlinearity $f(\cdot)$ may consist of positive powers of $y(j)$, $y(j+1)$, and cross terms of the variables. The ε is a small positive parameter that can vary from system to system within the class. This class of systems is fundamental because it represents a first departure from the well-known autoregressive process, and thus it may be used to model suitable processes or systems with small nonlinearities.

Class ω_2. The systems of this class consist of a saturation type of nonlinear element cascaded in front of a linear portion and connected with unity feedback. A general equation for the class may be written as

$$y(j+2) = a_1 y(j+1) + a_2 y(j) + f_r(\xi(j) - y(j))$$

where the subscript r represents nonlinearities that are odd functions with zero mean output for a zero mean input, and approach a limiting constant value for large excursions of the nonlinearity's input. This type of system has been extensively studied for stochastic inputs in, for example, [6,7,8], and may be used to model various physical processes exhibiting saturation nonlinearities. Often the nonlinearities $f_r(\cdot)$ are referred to as simply nonlinear.

Data for these classes of systems was supplied to the algorithm described above which generates the appropriate conditional densities and computes the resultant classification error. The error was computed for pattern vectors of increasing dimension and the results are plotted in Fig. 4 for both uniform and modified Beta types of a priori distributions. As is to be expected, the classification error is consistently smaller when the Beta distributions are used and more a priori information is utilized.

The decrease of error with increasing pattern dimension is in agreement with the heuristic interpretation that classification accuracy should improve as more information about the system is utilized in the pattern vector. The extremely small error probabilities demonstrate that it is possible to identify unknown members of these two classes of nonlinear systems with a very high probability of correct classification.

A comparison of the theoretical error computed by the algorithm and actual error obtained in experimental studies with real patterns [1] is given in the following table for twelve-dimensional pattern vectors.

Classification Error

Theoretical	Experimental
1.1×10^{-11}	1.2×10^{-3}

The very low error probability indicates that the two classes are, for practical purposes, completely separable.

From Fig. 4, the dependence of error probability on pattern dimension shows that these particular classes of systems are classifiable to a high degree of accuracy with only a very small number of features in the pattern vector. It is to be expected that more features may be required to achieve a given error probability for some classes of nonlinear, second-order systems.

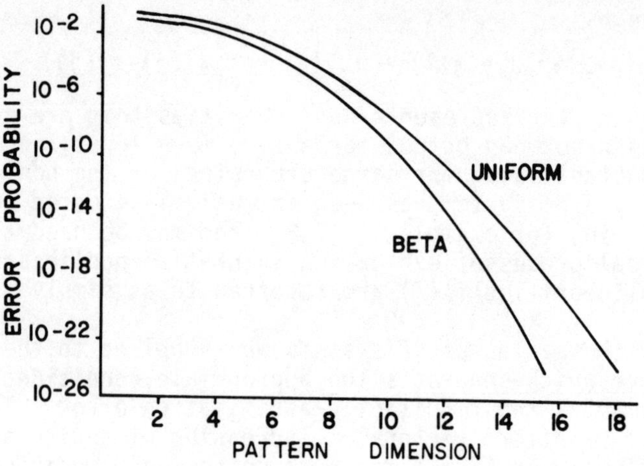

Fig. 4. Classification Error: Example 1.

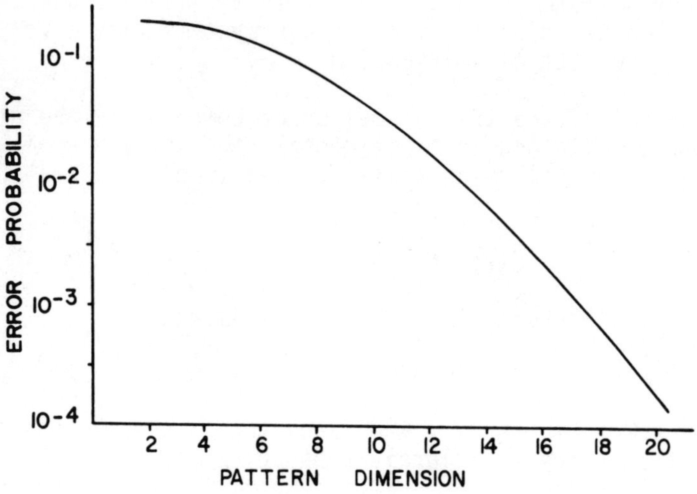

Fig. 5. Classification Error: Example 2.

This is indicated, for instance, in Fig. 5 which shows a classification error for a general hypothetical classification situation for second order nonlinear systems (described in the next section). Comparing Fig. 4 and Fig. 5 it is seen that, for an error of 5×10^{-3}, 5 and 15 pattern features are required respectively to achieve the same error probability. In numerous other studies made with the algorithm, it was found that a maximum of 15 to 17 features were required to achieve very acceptable error levels for second order systems.

(2) Additional Theoretical Studies

Even when the nonlinear system order is specified and the pattern vector dimension is fixed, there remain numerous parameters that enter into a complete specification of the classification problem. To graphically depict the complete behavior of classification error as a function of these parameters is impossible, but it is nevertheless interesting to fix additional parameters and vary the remaining ones to observe the effects on classification error.

From the results of Section III, the parameters δ_1 and δ_2 are the most interesting to examine since they arise directly from the nonlinearities of any given class of systems. The following statistics were used to establish a hypothetical classification situation and obtain the error plots shown in Figs. 6 and 7. For class ω_1, $\bar{a}_1 = 0.25$, $\bar{a}_2 = 0.30$ and for class ω_2, $\bar{a}_1 = 0.35$, $\bar{a}_2 = 0.25$ with the standard deviation of both variables set at 0.02 for both classes. Classification error probabilities were then computed for values of S.D. (δ_1) and S.D. (δ_2) varied over the range 0.0, 0.02, 0.04, 0.06, 0.08, and 0.1 for both classes.

Fig. 6 and Fig. 7 show the results for a pattern vector of 15 features and 20 features respectively. This may be considered a very difficult classification situation and indeed the error levels have risen considerably above the values of example 1, but they are nevertheless acceptably low to achieve practical classification. The appearance of a knee in the curves at a value of $SD(\delta_1) = 2SD(a_1)$ indicates that at this point the pattern densities overlap nearly to the maximum possible for the two classes which results in an error of 5×10^{-3} for a pattern of 20 dimensions. Also, it is interesting to note that the increase from 15 to 20 dimensions has the same error decreasing effect on all members of the classes since the error curves show a nearly uniform decrease for the range of parameters.

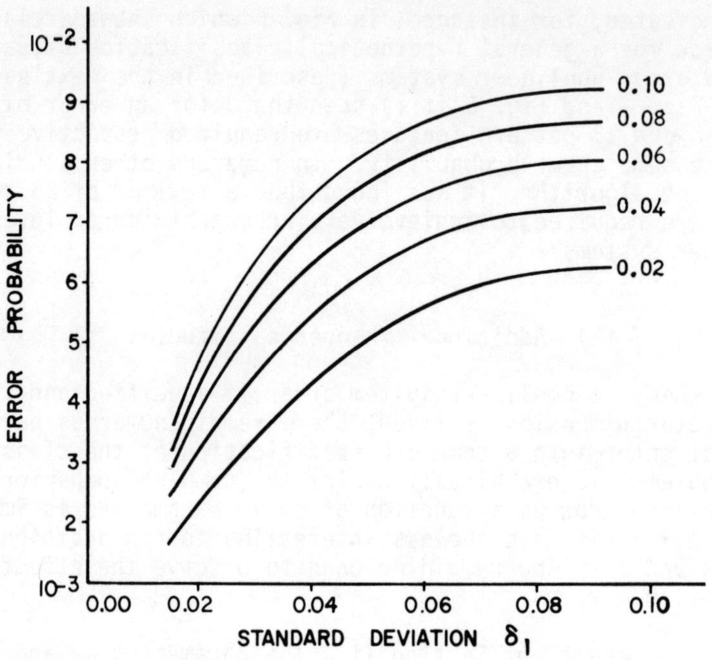

Fig. 6. Classification Error; 15 Dimensions.

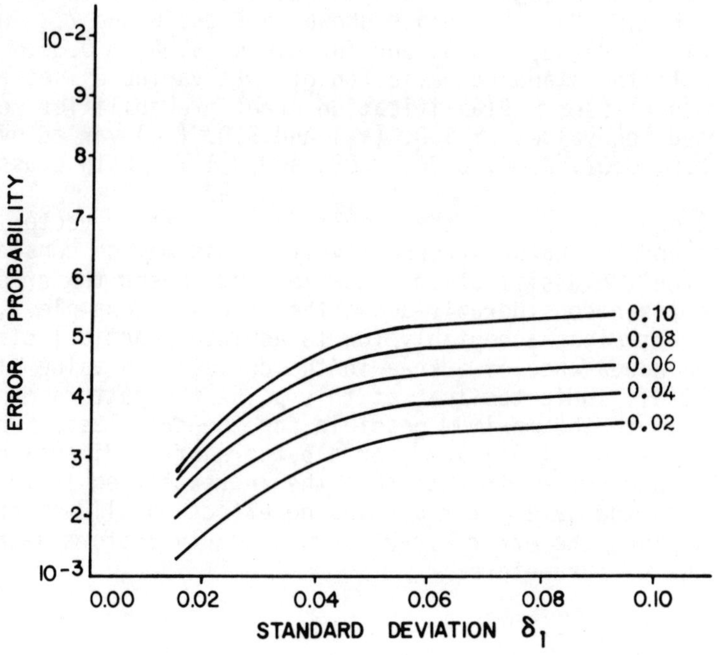

Fig. 7. Classification Error; 20 Dimensions.

VI. SUMMARY AND CONCLUSIONS

The purpose of this paper has been to initiate a theoretical study of the classification of specified classes of nonlinear stochastic systems by means of pattern recognition techniques. This was done by obtaining exact expressions for the conditional class pattern densities, with respect to the linearized model of equation (7), to be used in computing the theoretical Bayes minimum classification error ε. Since this error was shown to be acceptably small, it demonstrates that classes of nonlinear systems can be successfully established and that member systems can be identified as belonging to their respective classes based only on the input-output measurements of an unknown system under examination.

That classification is possible to a high degree of accuracy is shown by the resulting error rates as well as by experimental error rates reported earlier [1] from classification studies using actual patterns. As is to be expected, the theoretical errors are considerably smaller than the experimental values because only a finite number of sample patterns were used in the k-nearest-neighbor experimental studies, and it is well known that the k-nearest-neighbor rule yields accurate error estimates only when the number of nearest neighbors k and the sample patterns become infinite.

With the class conditional densities based on (10) available, a theorem was established for the evaluation of ε to permit a conservative estimate to be made of the classification error by means of a rapid computer algorithm for any desired choice of parameters, pattern dimension, and prior distributions. This resulted in representative plots of classification error versus pattern dimension. Prior distributions other than the ones employed here could be used to advantage where justified and may be expected to result in reduced error rates, but at the cost of additional analytical complexity. It is also important to note that the linearization techniques used in Section III are not used in actual classification systems, but are only a device in the theoretical study of the problem.

Despite the fact that accurate classification has been shown feasible, there are a number of interesting aspects and extensions of the overall problem that remain open for consideration. For instance, the effects of randomness in the estimates (5) could be included to refine the overall error estimates. For use in an on-line classification situation it would be interesting to model the improvement in classification error that results as the estimates in (5) improve with time. In effect, classification is possible before the estimates have completely converged, but how good is the classification under such circumstances?

Applications for the proposed nonlinear stochastic system classification method in Self-Organizing and Learning Control Systems, as well as initial modeling of an unknown system have been detailed in [1]. In [9] the applicability of pattern recognition methods in Learning Control Systems has been especially pointed out, and concrete applications of the present classification method should prove interesting in such systems. Classes of nonlinearities and systems pertinent to biological systems [10,11] are presently being investigated with the classification method.

REFERENCES

1. G.N. SARIDIS and R.F. HOFSTADTER, "A Pattern Recognition Approach to the Classification of Nonlinear Systems From Input-Output Data," IEEE Conference on Systems, Man and Cybernetics, 1973, Preprints.
2. E. WONG, Stochastic Processes in Information and Dynamical Systems, New York, McGraw-Hill, 1971.
3. K. FUKANAGA, Introduction to Statistical Pattern Recognition, New York, Academic Press, 1972.
4. K.J. ASTROM, Stochastic Control Theory, New York, Academic Press, 1970.
5. W. RUDIN, Real and Complex Analysis, New York, McGraw-Hill, 1966.
6. A.T. FULLER, Nonlinear Stochastic Control, London, Taylor and Francis Ltd., 1970.
7. J.F. BARRET, "Applications of Kolmogorov's Equations to Randomly Disturbed Automatic Control Systems," Proc. 1961 IFAC Conference, London, Butterworths, Vol. 2, pp. 724-733.
8. H.W. SMITH, Approximate Analysis of Randomly Excited Nonlinear Controls, MIT Research Monograph No. 34, Cambridge, Mass., MIT Press, 1966.
9. K.S. FU, "Learning Control Systems - Review and Outlook," IEEE Trans. Automatic Control, Vol. AC-15, pp. 210-221, April 1970.
10. B.F. WOMACK, "Identification of Models of Respiratory Sinus Arrhyhmia in Humans: Three Approaches." Proceedings 1972 JACC, pp. 112-116.
11. L. STARK, Y. TAKAHASHI, and G. ZAMES, "Nonlinear Servoanalysis of Human Lens Accommodation," IEEE Transactions on Systems Science and Cybernetics, vol. 1, pp. 75-83. November 1965.

APPENDIX

The theorem stated in the text is proved as follows. Observe first that the integral expressions in (4-2) may be exactly re-written in a more concise form as

$$\varepsilon^* = \int_{\Gamma_1 \cup \Gamma_2} \min_{\nu} \left[p(V|\omega_\nu) Pr[\omega_\nu] \right] dV$$

and that $\Gamma_1 \cup \Gamma_2$ is exactly the region Q defined in the text. This yields

$$\varepsilon^* = \int_Q \min_{\nu} \left[p(V|\omega_\nu) Pr[\omega_\nu] \right] dV \qquad \text{(A-1)}$$

From the method of construction of the values $\rho_{k,r_k}(n,\omega_\nu)$ in (14), it is clear that the simple functions $q_n(v_k|\omega_\nu)$ form an upper bound for the density functions $p(v_k|\, v_{k-1},\omega_\nu)$, $\nu = 1, 2$ on the interval $Q_0(k)$. Similarly the product (16) of simple functions $q_n(v_k|\omega_\nu)$ is again a simple function defined on Q in Ω^m, and since $q_n(v_k|\omega_\nu) \geq p(v_k|v_{k-1},\omega_\nu)$ it is clear that

$$\prod_{k=n_0+1}^{m+n_0} q_n(v_k|\omega_\nu) \geq \prod_{k=n_0+1}^{m+n_0} p(v_k|v_{k-1},\omega_\nu)$$

which together with (14) and (16) yields

$$q_n(V|\omega_\nu) \geq P(V|\omega_\nu) \qquad \text{(A-2)}$$

for all $V \in Q$. Substitution of (A-2) into (A-1) yields

$$\int_Q \min_{\nu} \left[q_n(V|\omega_\nu) Pr[\omega_\nu] \right] dV \geq \int_Q \min_{\nu} \left[p(V|\omega_\nu) Pr[\omega_\nu] \right] dV$$

or equivalently

$$\varepsilon(n) \geq \varepsilon$$

which is the desired result. Note that the * has been dropped since the densities based on the linearized model (10) have been used from (A-1) onward.

*ALGORITHMIC TECHNIQUES FOR MODELING NONLINEAR FUNCTIONS

Lester A. Gerhardt
Rensselaer Polytechnic Inst.
Troy, New York 12181

Takatoshi Miura
Kokushikan University
Tokoyo, Japan

ABSTRACT

Algorithmic Techniques are derived and applied to the problem of modeling nonlinear functions. Two general approaches are considered. The first uses zero order quantizers with adaptable parameters to approximate the given nonlinearity. The second makes use of algorithmic techniques to establish successively better estimates of the coefficients of a Chebyshev polynomial representation of a nonlinear function, with particular emphasis on a hysteresis nonlinearity. Algorithms are derived, convergence considered and results given for each case.

INTRODUCTION

This paper summarizes some of the results of joint research conducted during the 1971 and 1972 academic year by the authors. The paper is divided into two basic Sections.

The first considers the approximation of a nonlinearity by a

*These are selected results obtained by the authors as a result of joint research conducted at Rensselaer, under co-sponsorship by the respective universities. The Rensselaer portion of the work was partially supported by the Air Force Office of Scientific Research (AFSC), Directorate of Mathematical and Information Sciences under the guidance of Lt. Col. Russell Ives and Lt. Col. Thomas J. Wachowski.

zero order adaptive quantizer. It serves to generalize the more restricted methods previously established[1]. Wiener [2,3] showed that any nonlinear system with finite settling time could be characterized by a linear network which characterized the input past, followed by a zero memory nonlinearity. The major objective of the first portion of this study is therefore to explore the nonlinear element modeling capability of an open loop self adaptive nonlinear quantizer. These networks may be applied to single valued continuous nonlinear functions and to a more limited extent nonmonotonic functions.

Algorithms for adjusting the quantizing boundary locations and the levels of an adaptive zero-order quantizer are derived, based on a least mean square error index. The main emphasis has been on developing algorithms which do not require boundary crossings, and on developing in detail solutions for monotonic nonlinear elements. Experimental results are included and indicate that this type of modeler converges in the mean , and to within a prespecified variance solution.

The second Section of the paper considers the identification of nonlinear, single input, single output elements, specifically those having hysteresis characteristics. A mathematical model is used which is described by Chebyshev's orthogonal polynomial expansion, and a self adaptive convergence algorithm is used to estimate its coefficients.

Results are given which show that if the nonlinear element has been modeled by a finite order Chebyshev-Lissajous type function, then the Chebyshev's orthogonal polynomial converges with finite order, (which otherwise would require an infinite series).

A new method called a self adaptive convergence algorithm is introduced to obtain the coefficient of Chebyshev's orthogonal polynomial. This approach avoids usage of the usual correlation integral which has singular points at the extreme values of x.

The hysteresis nonlinearity is considered as a special case, since the representation of a hysteresis type nonlinear characteristics is, in general, not an easy problem. A first attempt was presented in 1955 by Levis [4] and later formulated in 1971 by Clay [5]. Both of their approaches used a Chebyshev orthogonal polynomial expansion. However, each of the coefficients was obtained by the usual correlation integral which has singular points at the extreme values for $x = \pm 1$, (the usual correlation method has singular point to get the coefficient a_n, that is,

$$a_n = \frac{1}{\pi} \int_{1}^{-1} y \frac{T_n(x)}{\sqrt{1-x^2}} \, dx$$

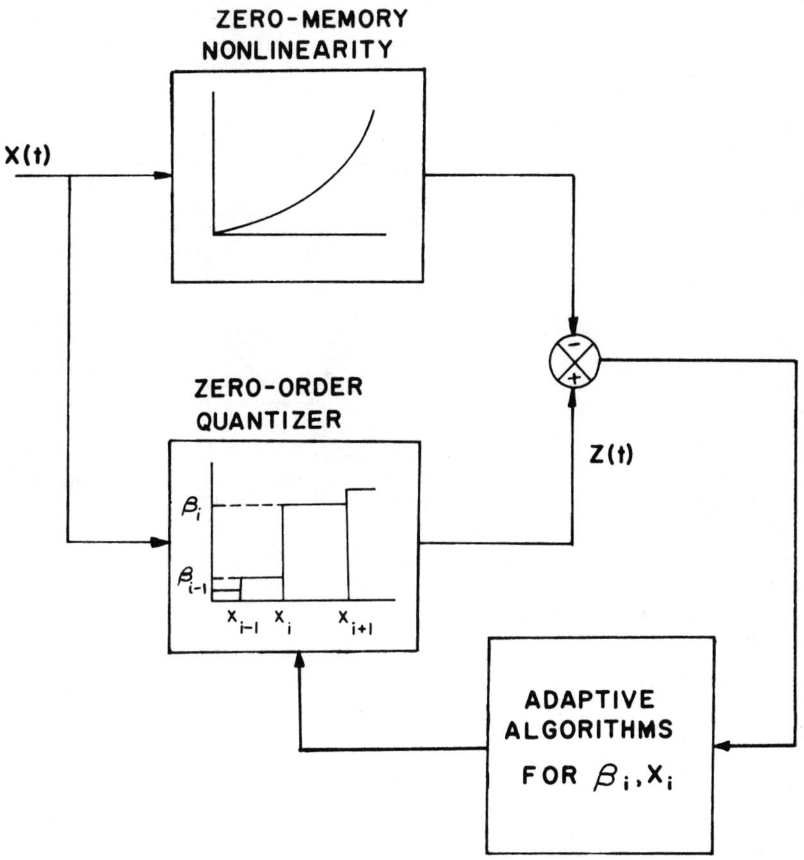

Fig. 1. Zero-Order Quantizer Adaptive System

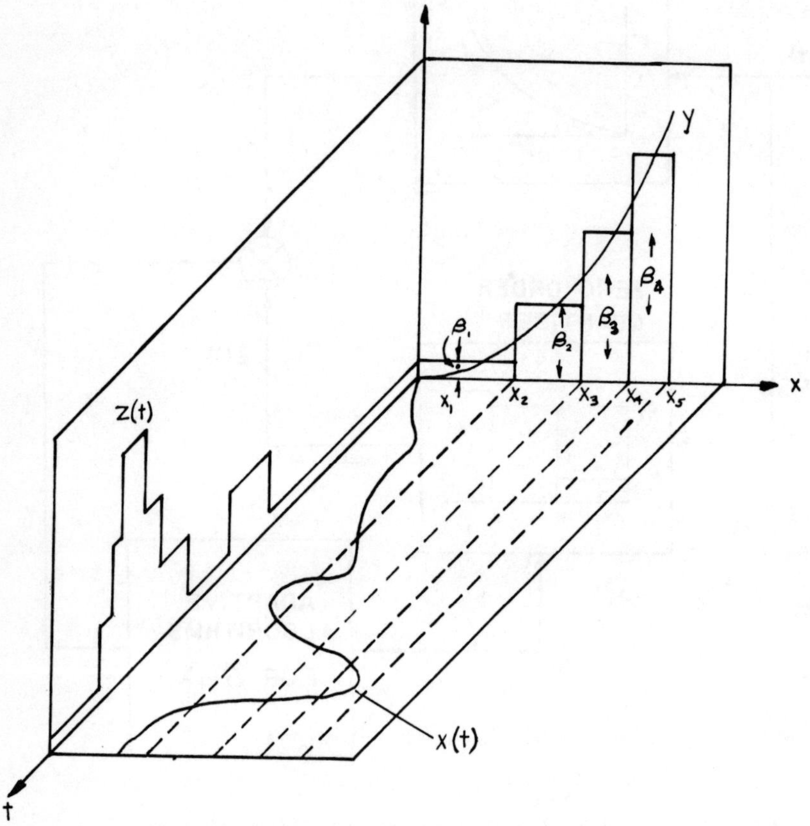

Fig. 2. Zero-Order Quantizer Input-Output Relation With Amplitude and Time

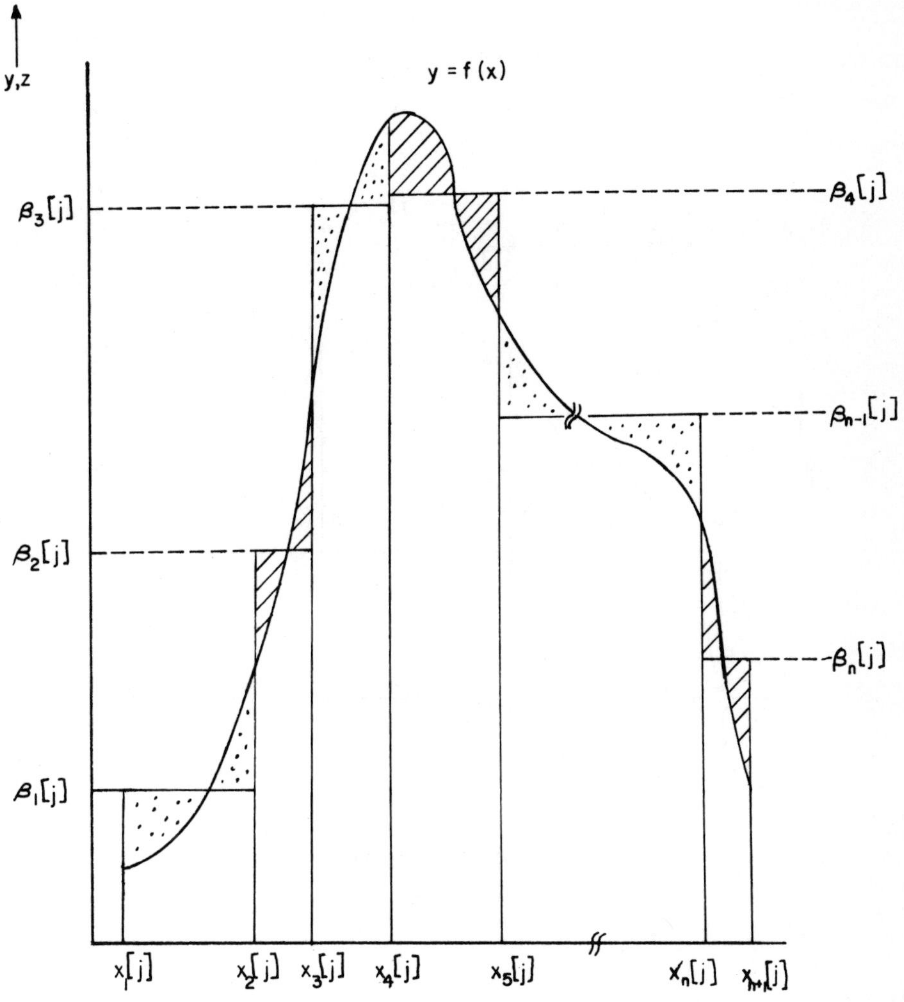

Fig. 3. Zero-Order Quantizer

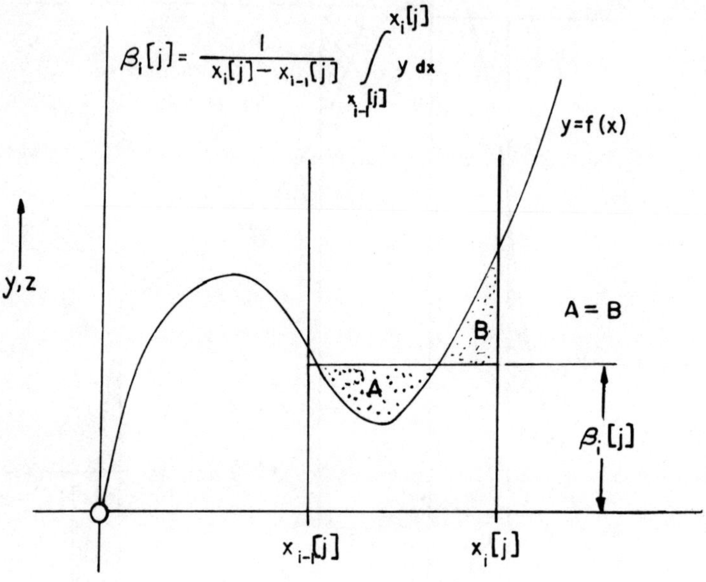

Fig. 4. Optimum Level $\beta_i[j]$

$$a_n = \frac{1}{\pi} \int_{-1}^{1} y \frac{T_n(x)}{\sqrt{1-x^2}} \, dx$$

where $\lim_{|x| \to 1} \frac{1}{\sqrt{1-x^2}} = \infty$.)

This need for performing this calculation introduces the most serious drawback in the application of these prior methods.

This paper (second part) uses an algorithmic technique which does not require the use of this integral directly.

SECTION 1 - MODELING USING A ZERO ORDER QUANTIZER

The zero-order quantizer will be used to approximate nonlinear functions by a function which is composed of simpler N-boundary components where each boundary has level β_i, (i=1,2,...,N). Adaptive algorithms for the boundaries and levels dominate the study using a performance index of least mean square error.

Figure 1, 2 and 3 show a graphical interpretation of the notation and application.

Optimum Level Theorem for a Fixed Single Boundary

For any fixed single boundary $[x_{i-1}[j], x_i[j]]$, (i=1,2,...,N, j=1,2,3,...) there exists a unique optimum level $\beta_i[j]$ which minimizes the squared error for a single valued function y = f(x), as follows, see Figure 4.

$$\beta_i[j] = \frac{1}{x_i[j] - x_{i-1}[j]} \int_{x_{i-1}[j]}^{x_i[j]} y \, dx$$

Proof.

$$\frac{1}{2} \frac{\partial}{\partial \beta_i[j]} \left\{ \int_{x_{i-1}[j]}^{x_i[j]} (\beta_i[j] - y)^2 \, dx \right\}$$

$$= \int_{x_{i-1}[j]}^{x_i[j]} (\beta_i[j] - y) \, dx = 0 \qquad \text{Q.E.D.}$$

Optimum Two Boundary Theorem for Middle Point $x_i(j)$

If $y = f(x)$, a continuous monotonic function ($f'(x) \neq 0$) between the fixed boundaries $[x_{i-1}[j],\ x_{i+1}[j]]$, then the middle boundary point $x_i[j]$ has an optimum value at the point when the following equation is satisfied uniquely.

$$f(x_i[j]) = \frac{\beta_{i-1}[j] + \beta_i[j]}{2} \tag{1}$$

Proof

$$\frac{d\,J_i[j]}{d\,x_i[j]} = (\beta_{i-1}[j] - y)^2 - (\beta_i[j] - y)^2 \Big|_{y=f(x_i[j])} \tag{2}$$

This vanishes for the following condition

$$y = f(x_i[j]) = \frac{\beta_{i-1}[j] + \beta_i[j]}{2}$$

The second derivative of equation (2) is positive definite for a monotonic function $f(x)$. See Figure 5.

Convergence of the Two Boundary Case

Assume $y=f(x)$, a continuous monotonic function over the fixed boundary $[x_{i-1}[j],\ x_{i+1}[j]]$, and the middle point $x_i[j]$, is changed, keeping the relation $x_{i-1}[j] < x_i[j] < x_{i+1}[j]$ without increasing the performance index function $J_i[j]$, where

$$J_i[j] = \int_{x_{i-1}[j]}^{x_i[j]} (\beta_{i-1}[j] - y)^2\,dx + \int_{x_i[j]}^{x_{i+1}[j]} (\beta_i[j] - y)^2\,dx \tag{3}$$

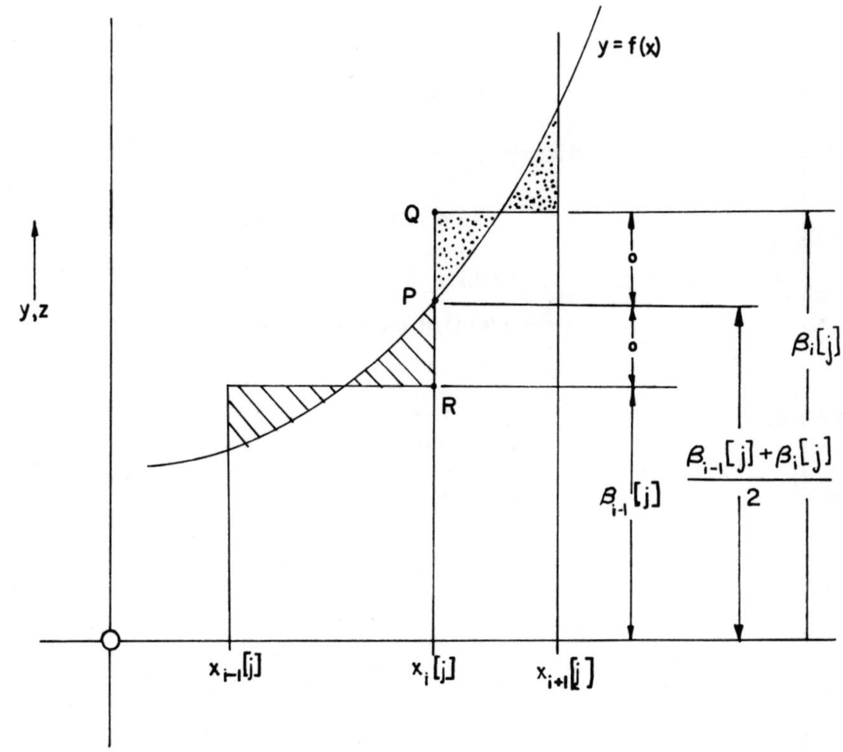

Fig. 5, Optimum Two Boundary Theorem
For Middle Point $x_i[j]$

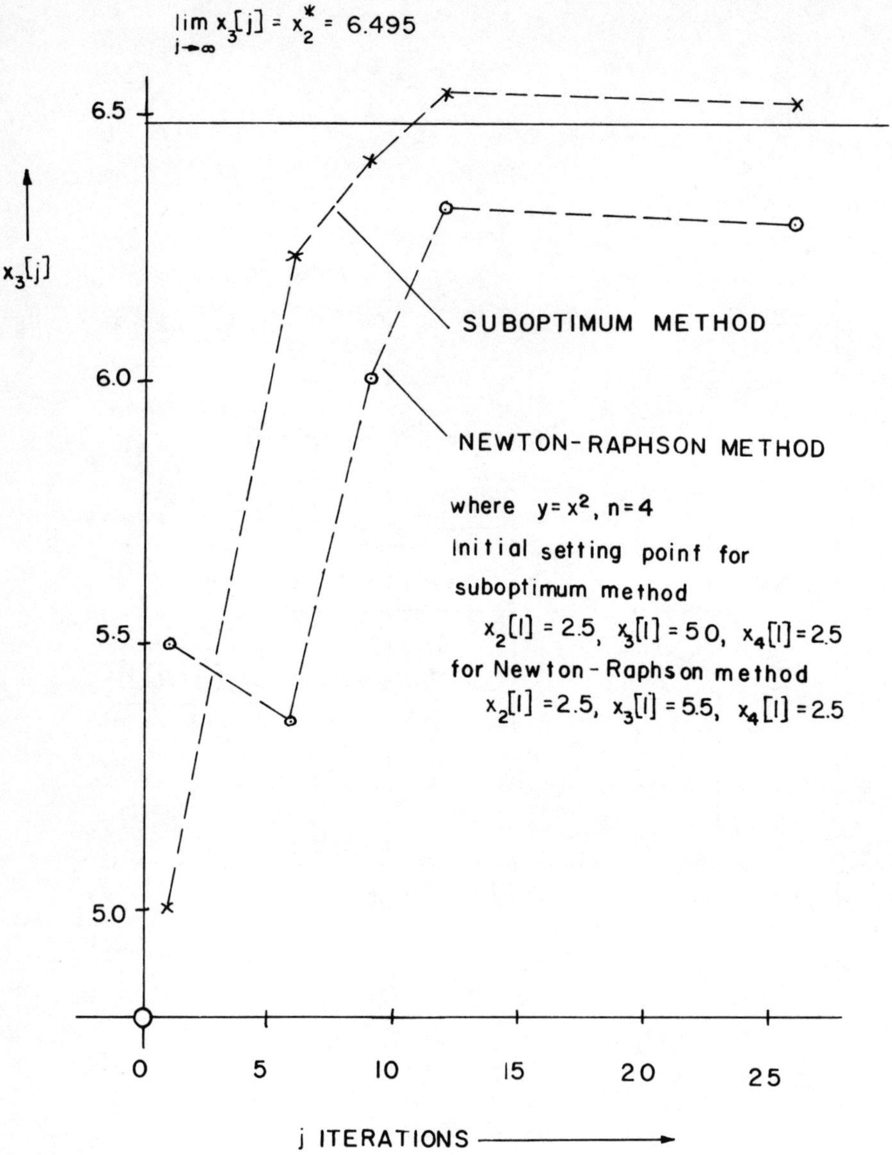

Fig. 5.1. Our Method is Better Than Newton Raphson Method in This Case

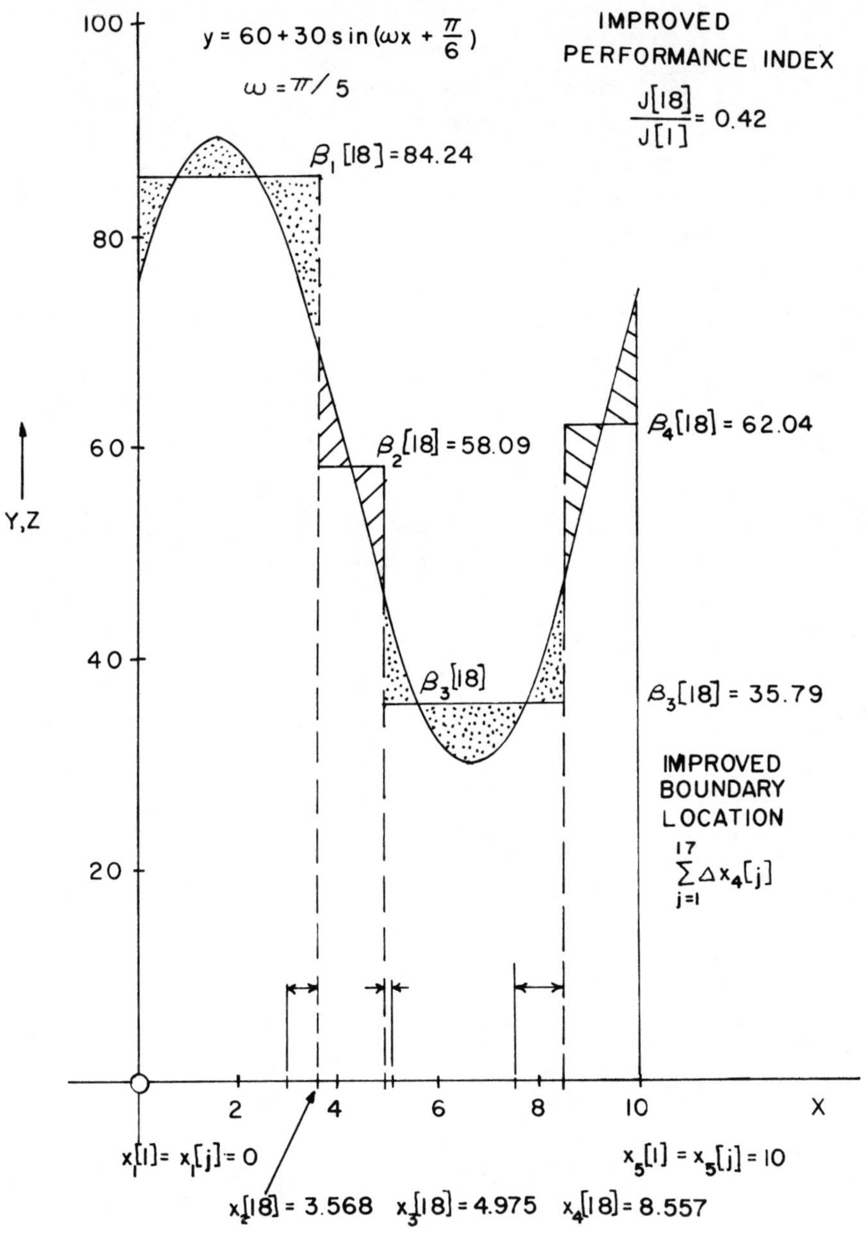

Fig. 5.2. Improvement by Sub-Optimum Method for Nonmonotonic Function

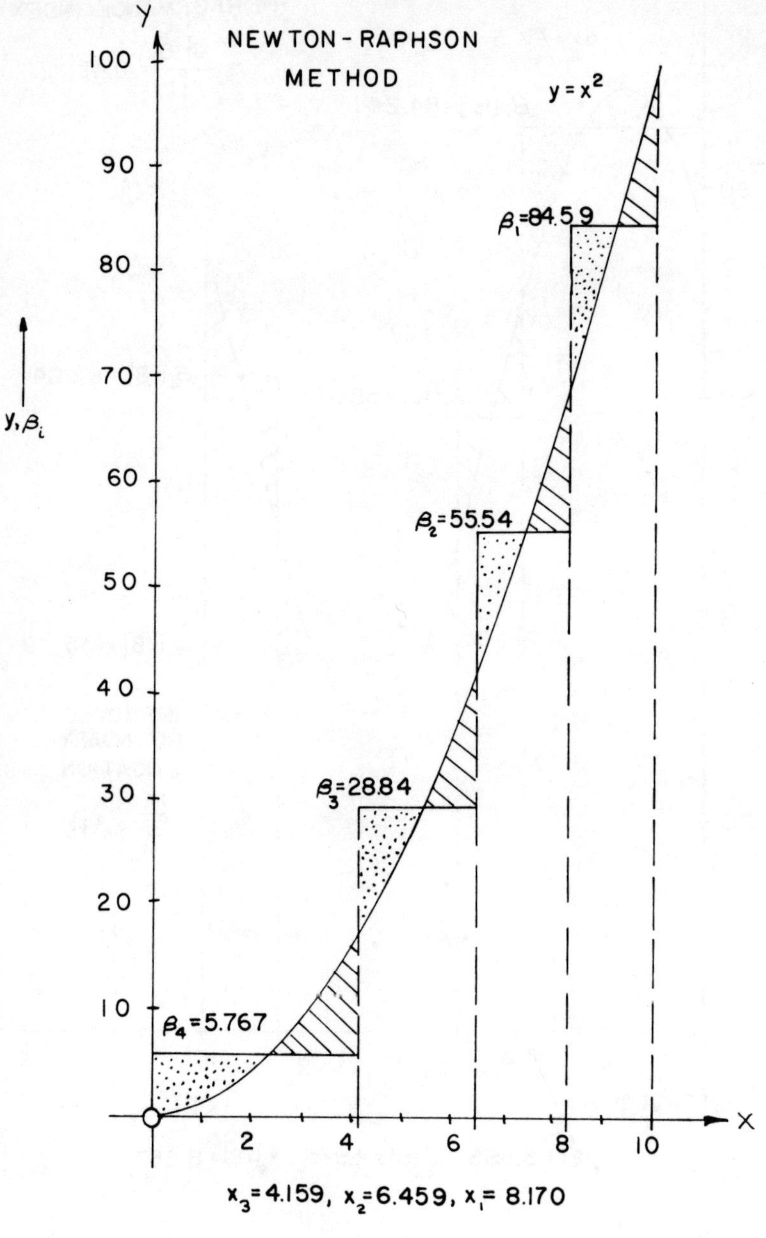

Fig. 5.3. Optimum Boundaries & Levels for $y = x^2$

Define the following auxiliary performance indices $J_i[j+1; +\Delta x]$, and $J_i[j+1; -\Delta x]$ as

$$J_i[j+1;+\Delta x] = \int_{x_{i-1}[j]}^{x_i[j]+ar^{j-1}} \left(\frac{1}{x_i[j]+ar^{j-1}-x_{i-1}[j]}\right)$$

$$\int_{x_{i-1}[j]}^{x_i[j]+ar^{j-1}} y\,dx - y)^2\,dx +$$

$$\int_{x_i[j]+ar^{j-1}}^{x_{i+1}[j]} \left(\frac{1}{x_{i+1}[j]-x_i[j]-ar^{j-1}}\right)$$

$$\int_{x_i[j]+ar^{j-1}}^{x_{i+1}[j]} y\,dx - y)^2\,dx \qquad (4)$$

$$J_i[j+1;-\Delta x] = \int_{x_{i-1}[j]}^{x_i[j]-ar^{j-1}} \left(\frac{1}{x_i[j]-ar^{j-1}-x_{i-1}[j]}\right)$$

$$\int_{x_{i-1}[j]}^{x_i[j]-ar^{j-1}} y\,dx - y)^2\,dx$$

$$+\int_{x_i[j]-ar^{j-1}}^{x_{i+1}[j]} \left(\frac{1}{x_{i+1}[j]-x_i[j]+ar^{j-1}}\right)$$

$$\int_{x_i[j]-ar^{j-1}}^{x_{i+1}[j]} y\,dx - y)^2\,dx \qquad (5)$$

where, a and r are some constants which are discussed later.

A converging sequence may be formulated as follows. ($j = 1,2,\ldots$)

$$x_i[j+1] = x_i[j] + \Delta x_i[j] \qquad (6)$$

$$\Delta x_i[j] = a \cdot r^{j-1} \cdot \text{sgn}(\Delta x) \tag{7}$$

$$\text{sgn}(\Delta x) = \begin{cases} 1 & \text{for } J_i[j+1; \ +\Delta x] < J_i[j] \\ 0 & \text{otherwise} \\ -1 & \text{for } J_i[j+1; \ -\Delta x] < J_i[j] \end{cases} \tag{8}$$

$$\beta_{i-1}[j] = \frac{1}{x_i[j] - x_{i-1}[j]} \int_{x_{i-1}[j]}^{x_i[j]} y \, dx \tag{9}$$

$$\beta_i[j] = \frac{1}{x_{i+1}[j] - x_i[j]} \int_{x_i[j]}^{x_{i+1}[j]} y \, dx \tag{10}$$

where $x_{i-1}[j]$ is the lower fixed boundary point, and $x_{i+1}[j]$ is the upper fixed point. Both of them are, for now, considered fixed although later all boundaries are simultaneously adopted.

$$x_{i-1}[1] = x_{i-1}[2] = \dots x_{i-1}[j] = \dots \tag{11}$$

$$x_{i+1}[1] = x_{i+1}[2] = \dots x_{i+1}[j] = \dots \tag{12}$$

Initial values for the x_i [j] of equation (6) are given as follows

$$x_i[1] = \frac{x_{i-1}[1] + x_{i+1}[1]}{2} \tag{13}$$

for any $x_{i-1}[1]$, $x_{i+1}[1]$.

The initial gain a and decreasing factor r must have some dependence which guarantees that the entire range is covered and boundaries never cross. This relation becomes

$$\frac{a}{1-r} = x_{i+1}[1] - x_i[1] = x_i[1] - x_{i-1}[1] \tag{14}$$

$$0 < a < x_{i+1}[1] - x_i[1] = x_i[1] - x_{i-1}[1] \tag{15}$$

This initial gain a has the greatest effect on convergence speed; a bigger initial gain yields faster convergence, and smaller gains result in slower convergence. The convergence speed will be discussed later in more detail.

Then $x_i[j] \to x_i^*$, as $j \to \infty$ and the performance index $J_i[j] \to J_i^*$, as $j \to \infty$. At each step from $j \to j+1$, the performance indices never increase, so this $x_i[j]$ sequence converges uniformly to the convergence point x_i^* for monotonic function $y = f(x)$. The levels $\beta_{i=1}[j]$ also converge to their optimal point β_{i-1}^*, β_i^*.

Sub-Optimum Method, Each Boundary Contributed Equal Square Error

Assume $y = f(x)$ is a continuous monotonic function over the fixed boundary $[x_{i-1}[j], \; x_{i+1}[j]]$. Move the middle point $x_i[j]$, keeping the relation $x_{i-1}[j] < x_i[j] < x_{i+1}[j]$ decreasing the index function J_i until it equals the sub-performance index $J_{sub}[i,j]$. There exists a convergent middle point x_i^*, and some adaptive convergence sequence $x_i[j+1] = x_i[j] + \Delta x_i[j]$.

Define the following sub-performance index $J_{sub}[i,j]$

$$J_{sub}[i,j] = \int_{x_i[j]}^{x_{i+1}[j]} (\beta_i[j] - y)^2 \, dx \qquad (16)$$

The performance index function over the two boundaries $[x_{i-1}[j], x_i[j]]$, $[x_i[j], x_{i+1}[j]]$ is also defined as

$$J_i[j] = J[i-1; j] + J[i,j] \qquad (17)$$

Then the adaptive convergence algorithm is

$$x_i[j+1] = x_i[j] + \Delta x_i[j] \qquad (18)$$

$$\Delta x_i[j] = a \cdot r^{j-1} \cdot \operatorname{sgn}(\Delta J_i[j]; \; S)$$

where

$$\operatorname{sgn}(\Delta J_i[j]; \; S) = \begin{cases} 1 & \text{for } S > 0 \\ -1 & \text{for } S < 0 \\ 0 & \text{for } J_i[j+1] \geq J_i[j] \end{cases}$$

$$S = J[i;j] > J[i-1; \; j]$$

a is the initial gain, which will be discussed later in detail. r is the decreasing factor and related to a, and initial boundary length $[x_{i-1}[1], x_{i+1}[1]]$ for the two boundary problem.

$$\frac{a}{1-r} = x_{i+1}[1] - x_i[1] = x_i[1] - x_{i-1}[1] \qquad (19)$$

The Rapid Convergence Theorem

If $y = f(x)$, and its derivative y' is a continuous monotonic function over the fixed boundary $[x_{i-1}[j], x_{i+1}[j]]$, and the middle point $x_i[j]$ is changed, along with the mean value of both levels $\beta_{i-1}[j]$ and $\beta_i[j]$, then $x_i[j]$ converges uniformly in the mean to the $x_i^*[j]$ as $j \to \infty$ with convergence speed $x_i[j+1]/x_i[j] = 1 + \{((\beta_{i-1}[j] + \beta_i[j])/2) - f(x_i[j])\} / (x_i^*[j] f'(x_i[j]))$ with a decreasing performance index $J_i[j]$. It remains at the convergence point x_i^*.

The basic idea of this convergence theorem is shown in Fig. 6. If $x_i[j]$ are not on the convergence point x_i^*, there exists a non-zero level difference $\Delta D_i[j]$.

$$\Delta D_i[j] = \frac{\beta_{i-1}[j] + \beta_i[j]}{2} - f(x_i[j]) \qquad (20)$$

Decreasing $\Delta D_i[j]$ decreases the performance index $J_i[j]$. If we find a sequence which monotonically decreases $\Delta D_i[j]$ in each step until it reaches $\Delta D_i[j] = 0$ as $j = \infty$, independently from initial point $x_i[1]$, then this sequence must converge to x_i^* uniquely. Because on half the boundary the $\Delta D_i[j]$ becomes positive and from continuity of $f(x)$, the other part of the boundary $D_i[j]$ becomes negative, so there exists $\Delta D_i[j] = 0$ from the continuity of $f(x)$.

n Boundary Problems

Previously, we have discussed the two-boundary problem of a zero order quantizer. Now we discuss the general n-boundary problems.. Many characteristics of these problems will be the same as before. The only difference for the boundary $x_i[j]$, is that it will never cross another as $j \to \infty$.

We can keep following boundary inequalities by the influence of 'never cross theorem'.

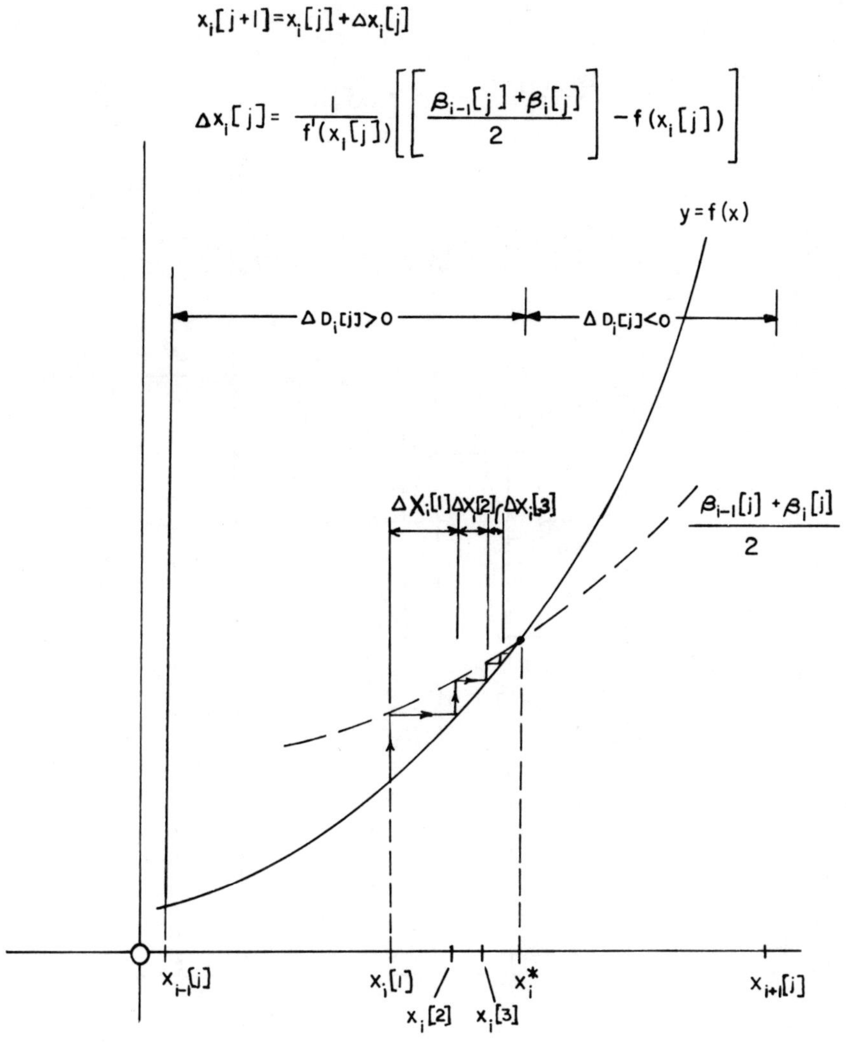

$$x_i[j+1] = x_i[j] + \Delta x_i[j]$$

$$\Delta x_i[j] = \frac{1}{f'(x_i[j])}\left[\left[\frac{\beta_{i-1}[j] + \beta_i[j]}{2}\right] - f(x_i[j])\right]$$

Fig. 6. The Rapid Convergence Theorem

$$\Delta x_i[j] \cong \frac{\Delta D[j]}{f'(x_i[j])}$$

$$\frac{\Delta D[j]}{\Delta x_i[j]} \cong f'(x_i[j])$$

$$\Delta D_i[j] = \frac{\beta_{i-1}[j] + \beta_i[j]}{2} - f(x_i[j])$$

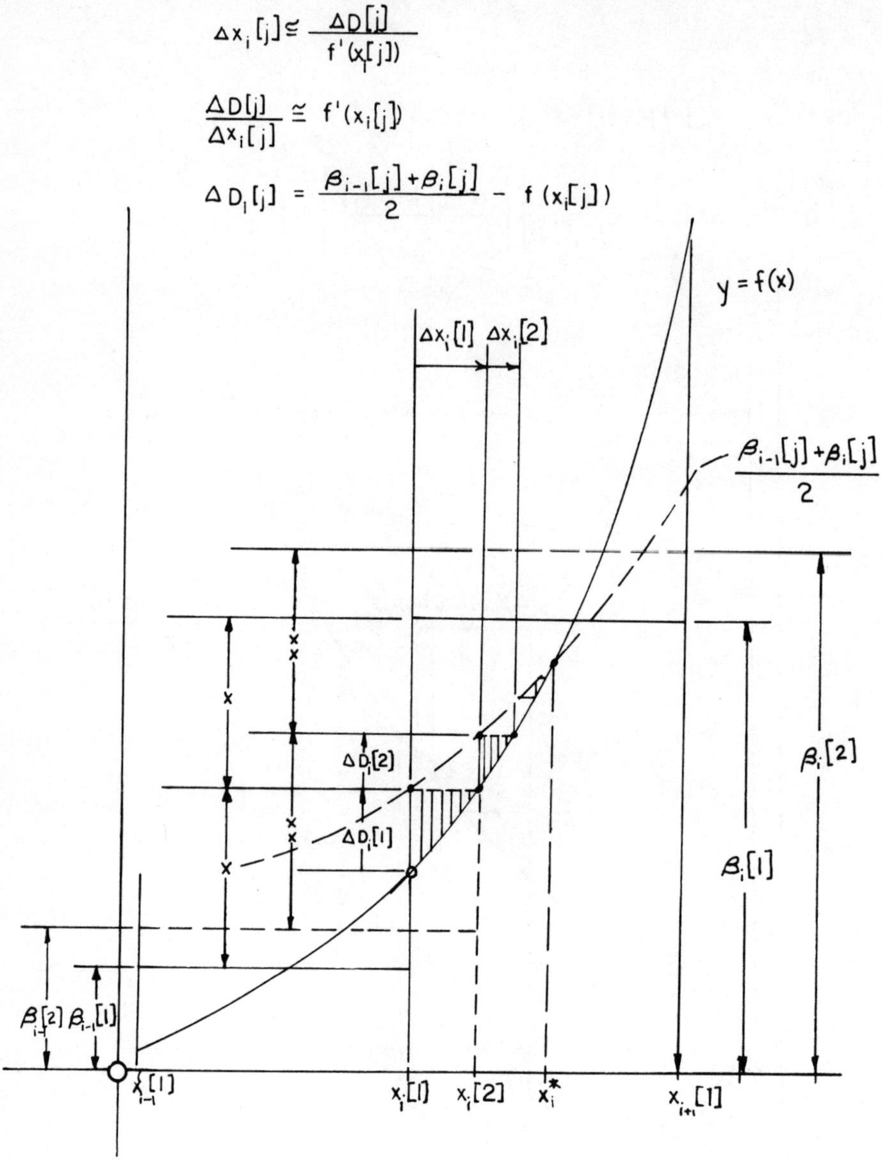

Fig. 7.   The Rapid Convergence Algorithm
$$x_i[j + 1] = x_i[j] + \Delta x_i[j]$$

$$x_i[j] < x_{i+1}[j] < x_{i+2}[j] \tag{21}$$

Let the initial conditions be given as fixed end points $x_1[1]$, and $x_{n+1}[1]$ for the n-boundary problems, and each of the initial boundary points are set as follows, (refer to Figure 8).

$$x_{i+1}[1] = x_i[1] + h_0 \tag{22}$$

where

$$h_0 = \frac{x_{n+1}[1] - x_1[1]}{n}, \qquad i = 1,2,\ldots,n \tag{23}$$

n: the number of divided boundaries on the x-axis.

The distance between from the boundary point $x_i[j]$ to the $x_{i+1}[j]$ is defined as follows by $h_i[j]$.

$$h_i[j] = x_{i+1}[j] - x_i[j] \tag{24}$$

where

$i = 1,2,\ldots,n-1$

$j = 1,2,\ldots..$

Also, the increment of $h_i[j]$ is defined over the fixed two-boundary point $x_i[j]$ and $x_{i+2}[j]$ as follows

$$\Delta h_i[j] = \Delta x_{i+1}[j] \tag{25}$$

$$h_i[j+1] = h_i[j] + \Delta h_i[j] \tag{26}$$

where

$i = 1,2,\ldots,n-1$

$j = 1,2,\ldots..$

Then, in any $n \geq 3$, n-boundary problem, for any i-th boundary ($i=1,2,\ldots,n-1$) and for any j-th iteration, suppose the increment $\Delta x_i[j]$ for fixed two-boundary point $x_i[j+1]$ and $x_{i+2}[j]$ has the following restriction.

$$\Delta x_{i+1}[j] = \Delta h_i[j] < \min(h_i'[j], h_{i+1}[j]) \tag{27}$$

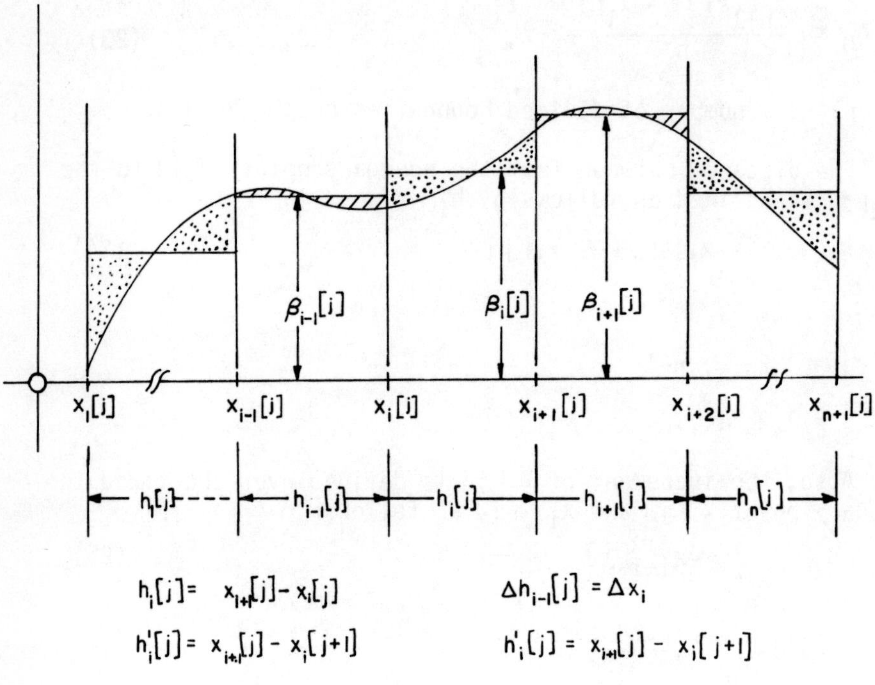

Fig. 8. Never Cross Theorem

$$h_i'[j] = x_{i+1}[j] - x_i[j+1] \tag{28}$$

Then $x_{i+1}[j+1]$ remains on the boundary between $x_i[j+1]$ and $x_{i+2}[j]$. That is the following inequality is maintained

$$x_i[j+1] < x_{i+1}[j+1] < x_{i+2}[j] \tag{29}$$

The proof is evident from the restriction equation (27), and $x_i[j] = x_1[j+1]$.

Coverability

It is possible that in some **cases the i-th boundary region** $h_i[j]$ will converge to the entire boundary region $[x_i[1], x_{n+1}[1]]$ as $j \to \infty$. And another boundary region may vanish entirely. In this case, every boundary point $x_{i+1}[j]$, $(i=1,2,\ldots,n-1)$ will have to remain on the entire boundary $[x_1[1], x_{n+1}[1]]$.

Consider one of the boundary region $h_i[j+1]$.

$$\begin{aligned} h_i[j+1] &= h_i[j] + \Delta h_i[j] \\ &= h_i[1] + \sum_{k=1}^{j} \Delta h_i[k] \end{aligned} \tag{30}$$

where

$$h_i[1] = h_0$$

Consider

$$\begin{aligned} \lim_{j \to \infty} \max_{y} h_i[j+1] &= x_{n+1}[1] - x_1[1] \\ &= n\, h_0 \end{aligned} \tag{31}$$

then

$$h_i[1] + \sum_{j=1}^{\infty} \Delta h_i[j] = n\, h_0$$

$$\max_{y} \sum_{j=1}^{\infty} \frac{\Delta h_i[j]}{h_0} = n-1 \tag{32}$$

The above equation is very useful to establish a relation between initial gain a and the decreasing factor r in our adaptive convergence algorithms.

Initial Gain a, Decreasing Factor r

From the discussion of coverability, we must now obtain the relation between the initial gain a and the decreasing factor r for the never cross theorem and with the equation (27).

It is sufficient to set the initial gain constant a considering the never cross restriction (equation 27).

$$\frac{a}{h_0} < \frac{1}{2} \tag{33}$$

And from the coverability equation (32), and by the similar convergence algorithm equation (7),

$$\sum_{j=1}^{\infty} \frac{\Delta h_i[j]}{h_0} = \frac{a}{h_0(1-r)} = n-1 \tag{34}$$

the relation between initial gain a, and the decreasing factor r then becomes

$$r = 1 - \frac{1}{n-1} \cdot \frac{a}{h_0} \tag{35}$$

where

$$\frac{a}{h_0} < \frac{1}{2}$$

This relation is shown in Table 1.

Multiple Boundary Convergence Theorem

Assume $y = f(x)$, a continuous monotonic function over the entire boundary region $[x_1[1], x_{n+1}[1]]$, and the boundaries are divided into n-regions, keeping the relation $x_i[j] < x_{i+1}[j] < x_{i+2}[j]$, $(i=1,2,..,n-1, j=1,2,....)$ without increasing the performance index function $\sum_{i=1}^{n-1} J_i[j]$ as $j \to \infty$, then there exist convergence boundary points x_{i+1}^* $(i=1,2,...,n-1)$ which minimize

Table 1. Never cross initial gain a and decreasing ratio r by n.

$\frac{a}{h_0}$	$r = 1 - \frac{1}{(n-1)} - \frac{a}{h_0}$				$\frac{a}{h_0} < \frac{1}{2}$		
	n = 2	n = 4	n = 8	n = 16	n = 32	n = 64	n = 128
$\frac{1}{2}$	$\frac{1}{2}$	$\frac{5}{6}$	$\frac{13}{14}$	$\frac{29}{30}$	$\frac{61}{62}$	$\frac{125}{126}$	$\frac{253}{254}$
$\frac{1}{4}$	$\frac{3}{4}$	$\frac{11}{12}$	$\frac{27}{28}$	$\frac{59}{60}$	$\frac{123}{124}$	$\frac{255}{256}$	$\frac{507}{508}$
$\frac{1}{8}$	$\frac{7}{8}$	$\frac{23}{24}$	$\frac{55}{56}$	$\frac{119}{120}$	$\frac{247}{248}$	$\frac{511}{512}$	$\frac{1015}{1016}$
$\frac{1}{16}$	$\frac{15}{16}$	$\frac{47}{48}$	$\frac{111}{112}$	$\frac{239}{240}$	$\frac{495}{496}$	$\frac{1023}{1024}$	$\frac{2031}{2032}$
$\frac{1}{32}$	$\frac{31}{32}$	$\frac{95}{96}$	$\frac{223}{224}$	$\frac{479}{480}$	$\frac{991}{992}$	$\frac{2047}{2048}$	$\frac{4063}{4064}$

the performance index function, and $x_{i+1}[j] \to x_{i+1}^*$ as $j \to \infty$, from given initial boundary point $x_{i+1}[j]$, for $j = 1$.

The proof is evident from the characteristics of the two-boundary case.

n Boundary Convergence Algorithms

The same convergence algorithms apply in the n boundary problems as in the two boundary problems, for continuous monotonic or non-monotonic function $y = f(x)$.

$$x_{i+1}[j+1] = x_{i+1}[j] + \Delta x_{i+1}[j] \tag{36}$$

$$\Delta x_{i+1}[j] = a \cdot r^{j-1} \cdot \mathrm{sgn}(\Delta x) \tag{37}$$

$$\mathrm{sgn}(\Delta x) = \begin{cases} 1 & \text{for } J_i[j+1;\ +\Delta x] < J_i[j] \\ -1 & \text{for } J_i[j+1;\ -\Delta x] < J_i[j] \\ 0 & \text{for otherwise} \end{cases} \tag{38}$$

$$a < \frac{h_0}{2} \tag{39}$$

$$h_0 = \frac{x_{n+1}[1] - x_1[1]}{n} \tag{40}$$

$$r = 1 - \frac{1}{n-1} \cdot \frac{a}{h_0} \tag{41}$$

where the initial values $x_{i+1}[1] = x_i[1] + h_0$ for $(i=1,2,\ldots,n-1)$
$x_1[1]$: is a given lower fixed boundary end point.

SECTION 2 - NONLINEAR MODELING USING CHEBYSHEV POLYNOMIALS

Problem Description

If a single output function y of a nonlinear element which driven by an input $x = \cos\Theta = \cos\omega t$ is a square integrable function, and has a fundamental period, then the function y can be expanded with Chebyshev's orthogonal polynomials as in equation (42).

$$y = \frac{\alpha_0}{2} T_0(x) + \sum_{n=1}^{\infty} (\alpha_n T_n(x) + \beta_n U_n(x)) \tag{42}$$

where

$|x| < 1$

$T_n(x)$: 1st kind of Chebyshev's function

$U_n(x)$: 2nd kind of Chebyshev's function

$$\alpha_n = \frac{1}{\pi} \int_{-1}^{1} y \frac{T_n(x)}{\sqrt{1-x^2}} dx \tag{43}$$

$$\beta_n = \frac{1}{2\pi} \left(\int_{-1}^{1} y \frac{U_n(x)}{\sqrt{1-x^2}} dx - \int_{1}^{-1} y \frac{U_n(x)}{\sqrt{1-x^2}} dx \right)$$

However, the major difficulty in using this expansion is obtaining of the coefficient α_n because this calculation in equation (43) involves the singular point at x = ± 1.

In 1955 Lewis[4] had shown an approximate method for computing the coefficient α_1. Clay[5] in 1971 introduced the expansion to represent the hysteresis characteristics of a nonlinear element.

We propose here a new method to get these coefficients α_n, β_n by model reference techniques.

The new method depends on a recurrence equation which is very convenient for digital implementation.

Modelling Function

Define the recurrence formula of the modelling function $f_n[k,j]$, where k is the k-th coefficient index, and j is the j-th recurrence index, as

$$f_n[k,j] = f_n[k,1] + \begin{cases} \frac{a_0[j]}{2} T_0(x) & , k = 0 \\ a_k[j] T_k(x) & , 1 \le k \le n \\ a_k[j] U_{k-n}(x) & , n+1 \le k \le 2n \end{cases} \tag{44}$$

where

$$k = 0,1,\ldots,2n \qquad j = 1,2,\ldots,\infty$$

$$\lim_{j \to \infty} f_n[k,j] = f_n[k+1,1] \tag{45}$$

the initial value at k=0, j=1 is

$$f_n[0,1] = 0 \tag{46}$$

Also, define the recurrence formula of the modelling performance index at the step j of the k-th coefficient as

$$J_n[k,j] = \frac{1}{2\pi} \int_1^{-1} (f_n[k,j] - y)^2 \, dx + \frac{1}{2\pi} \int_{-1}^{1} (f_n[k,j] - y^2)dx \tag{47}$$

from equation (45)

$$\lim_{j \to \infty} J_n[k,j] = J_n[k+1,1] \tag{48}$$

Define the sign function sgn[j]at the j-th step of the k-th coefficient

$$\mathrm{sgn}[j] = \begin{cases} \mathrm{sgn}[j-1] & \text{for } J_n[k,j] < J_n[k,j-1] \\ -\mathrm{sgn}[j-1] & \text{for } J_n[k,j] > J_n[k,j-1] \end{cases} \tag{49}$$

The initial value is at j=1

$$\mathrm{sgn}[1] = 1 \tag{50}$$

The recurrence equation for the coefficient $a_k[j]$ may be shown to be

$$a_k[j+1] = a_k[j] + \mathrm{sgn}[j] \cdot \Delta a_k[j] \tag{51}$$

where

$$\Delta a_k[j] = \Delta a_k[1] \cdot r^j \tag{52}$$

$$\Delta a_k[1] = \begin{cases} 2\sqrt{J_n[0,1]} & \text{for } k=0 \\ \sqrt{2\, J_n[k,1]} & \text{for } k\neq 0 \end{cases} \tag{53}$$

$$r = \frac{\sqrt{2}}{2} \tag{54}$$

the initial value

$$a_k[1] = 0 \qquad \text{for all } k \tag{55}$$

These coefficients $a_k[j]$ constitute a sequence which converges to the values a_k as $j \to \infty$. The a_k's are unique, and this sequence has the property that inequality (15)

$$|a_k[j] - a_k| \leq \frac{r^j}{1-r} \Delta a_k[1] \tag{56}$$

Let

$$j = 2N \tag{57}$$

then

$$\frac{1}{\Delta a_k[1]} \left| a_k[2N] - a_k \right| \leq \frac{2+\sqrt{2}}{2^N} \tag{58}$$

These inequalities govern the convergence speed which is seen to be logarithmic for any step j, and k. Therefore,

$$\lim_{j\to\infty} a_k[j] = a_k \tag{59}$$

$$a_k = \begin{cases} \alpha_k & \text{for} \quad 0 \leq k \leq n \\ \beta_{k-n} & \text{for} \quad n+1 \leq k \leq 2n \end{cases} \tag{60}$$

The modelling function $f_n[k,j]$ which was started at k=0, j=1 becomes sequentially up to the final coefficient of k=2n, j = ∞,

$$\lim_{j\to\infty} f_n[2n,j] = f_n(x) \tag{61}$$

$$f_n(x) = \frac{a_0}{2} T_0(x) + \sum_{k=1}^{n} (a_k T_k(x) + a_{k+n} U_k(x)) \tag{62}$$

Reasonable Value of $\Delta a_k[1]$ and r

The value $\Delta a_k[1]$ in equation (53) will now be explained. The $\Delta a_k[1]$ takes a maximum value of $|a_k|$ with respect to y.

Then for $k \geq 1$

$$\Delta a_k[1] = \max_y |a_k| \tag{63}$$

$$\Delta a_k[1] = \sqrt{2 J_n[k,1]} - \min_y \sqrt{2 J_n[k+1,1]} \tag{64}$$

where

$$\min_y \sqrt{2 J_n[k+1,1]} = 0 \tag{65}$$

therefore

$$\Delta a_k[1] = \sqrt{2 J_n[k,1]} \tag{66}$$

also for k = 0

$$\Delta a_k[1] = 2\sqrt{J_n[0,1]} \tag{67}$$

The constant r is selected by considering the convergence speed which requires selecting the smallest value. However, the convergence characteristics for the entire boundary require satisfying the following inequality,

$$r - \sum_{k=2}^{\infty} r^k \leq -1 \tag{68}$$

This means that

$$\frac{\sqrt{2}}{2} \leq r < 1 \tag{69}$$

But the convergence speed requires the smallest value

$$\left| \frac{a_k[j+2] - a_k}{a_k[j] - a_k} \right| = r^2 \ll 1 \tag{70}$$

so from the equation (69), (70) r becomes $\frac{\sqrt{2}}{2}$.

Application to a Chebyshev Lissajous Type Function

Let's define Chebyshev Lissajous type function $g_m(x)$ as in equation (71)

$$g_m(x) = \sum_{k=0}^{m} A_k x^k - \sqrt{1-x^2}\ \text{sgn}(x) \left(\sum_{\ell=1}^{m} B_\ell x^{\ell-1} \right) \quad (71)$$

where

$|x| \leq 1$

$$\text{sng}(x) = \begin{cases} 1 & \text{for} \quad x = \frac{dx}{dt} > 0 \\ -1 & \text{for} \quad x = \frac{dx}{dt} \leq 0 \end{cases}$$

A_k, B_ℓ constant coefficient.

This function $g_m(x)$ is the same as the m order of Chebyshev's orthogonal polynomial $f_m(x)$.

So, it is clear that this proposed method gives us the 100% modelling for the Chebyshev-Lissajous type nonlinear function $g_m(x)$ if and only if $m \leq n$ where n is the order of modelling function $f_n(x)$. Otherwise it gives us the best approximation in the sense of least mean square error approximation.

A detailed analysis has been performed and may be made available upon request from the authors.

CONCLUSIONS

This paper describes algorithmic procedures for adjusting the parameters of two general classes of modelers which have proven useful in modelling both monotonic and non-monotonic nonlinear functions. The first class of modellers is that of n order quantizers. The same algorithms presented here can be used for an arbitrary modeler of this type, although only the zeroth order quantizer is shown here. The other class of modelers uses orthonormal expansions to estimate the nonlinear function. Although only Chebyshev polynomials are considered in detail, the algorithms have been applied to several other types of orthonormal expansions. So far all problems to which these two classes of modelers have been applied have proven operational.

The key remaining problems are (a) the selection of a suitable index of performance and (b) the selection of the type of modeler class to use, e.g., quantizer or orthonormal expansion. Further, the specific kind of modeler within the class chosen (zeroth order, 1st order, or which orthonormal expansion to use) remains generally unknown. It is obviously a function of the class of problems to be solved. It is recommended that similar research directed in this area be formulated along these lines so as more general conditions can be established.

REFERENCES

1. "Learning the Shape of a Class of pdf's" Lester A. Gerhardt and Kenneth W. Drake, IEEE Transactions on SMC, July 1972.
2. "Nonlinear Problems in Random Theory," N. Wiener, The Technology Press of M.I.T., Cambridge, Mass., and John Wiley and Sons, New York, 1949.
3. "System Identification and Pattern Recognition," Rob Roy and James Sherman, IFAC, Prague, 1967.
4. "Harmonic Analysis for Nonlinear Characteristics," Laurel J. Lewis, Transaction of the AIEE, Vol. 74, 1955, pp. 693-700.
5. "Nonlinear Network and Systems," Richard Clay, Wiley-Interscience, 1971, pp. 126-129.

A SURVEY OF HEURISTIC SEARCH METHOD OF MULTIMODAL OPTIMUM POINT

Moriya Oda, Kahei Nakamura and B.F. Womack

Nagoya University
Nagoya, Japan

The University of Texas
Austin, Texas 78712,U.S.A

This is a survey of a theoretical and experimental study of a heuristic method of searching the globally optimum point of a multimodal, two-dimensional, nonlinear, and unknown criterion function by using the sighted, blindfolded, and blind subjects. First, the paper defines a heuristic search, establishes an experimental method for such a heuristic search, and develops a series of heuristic search experiments. Second, the paper shows some examples of the heuristic search behavior. Five kinds of the heuristic search model are extracted through five kinds of data processing based on a mode analysis. Some extracted models are proven to be appropriate and average. Third, the process of the heuristic search is studied from a viewpoint of concept formation by using a new idea of concept matrix.

1. INTRODUCTION

A subproblem of optimization is commonly contained not only in a control problem but also widely in any decision problem. The criterion function $f(\underline{X})$ to be optimized (maximized) in this paper is generalized to be unknown, two-dimensional, multimodal, and nonlinear. The authors have studied theoretically and experimentally a heuristic approach to find an optimum point $\underline{X}^0$ of control variables $\underline{X} = (x_1, x_2)$ so as to maximize the function $f(\underline{X})$ by using human subjects in three kinds of sighted, blindfolded, and totally blind states, [1] ~ [5].

The heuristic search behavior evolved by the human subjects seems very difficult to explain by any other conventional search methods developed so far for such an optimization problem, and needs a new approach to it. There are several important papers by other authors which also concern the heuristic search: J. Opačić [6] uses two typical kinds of heuristic search rules in her search method, a reflection rule and an extension rule: the former concerns the searching direction and step width, and the latter the search span. Two M. S. Theses by S. Harinasuta [7] and R. Graham [8] have shown new interesting phases of heuristic search by human subjects under the direction of Prof. B.F. Womack and assisted by Visiting Prof. M. Oda.

This paper surveys the heuristic phases of the human searcher's behavior. First, the paper defines a heuristic search, formulates the problem of heuristic search, clarifies important assumptions embedded in the heuristic search, establishes an experimental method for such a heuristic search, and develops a series of heuristic search experiments. Second, the paper surveys some examples of the heuristic search behavior, deviations of the trial number of the search, and the degree of heuristic search performance. In addition, five kinds of the heuristic search model are extracted through the data processing on a mode analysis. By the model-checking test, the extracted models V1, T1, and T2 are proved to be appropriate and average.

Third, the process of the heuristic search (especially including the subject's contour formation based on an adaptive clustering of trial points) is studied from a viewpoint of concept formation. That is, new ideas of matrix expression (structural expression) of concept formation process and five hierarchical levels of concept formation in the heuristic search behavior and their example are seen in the paper.

2. DEFINITION AND FORMULATION OF HEURISTIC SEARCH PROBLEM

2.1 Definition of Heuristic Search

In this section we will define a heuristic search on the same base as the definition of heuristic method, [9] - [11].

<u>Definition 1</u> A <u>heuristic search</u> is a search of a solution which satisfies some <u>given conditions</u> of a problem through a series of trials by using promising clues (<u>heuristic clues</u>) under a special promising priority (<u>heuristic priority</u>) among them.

2.2 Formulation of the Heuristic Search Problem

Our problem of heuristic search is formulated as follows:

Problem: Clarify the method of heuristic search which is evolved by a human searcher in a process of searching the optimum (highest) point on an unknown non-linear two-dimensional multimodal hill.

2.3. Important Assumptions in Heuristic Search

The heuristic search defined and formulated above includes the following assumptions implicitly in this paper:

A. The global representative-point search (global search made at representative points, simply called global RP search) is tried only at some of (not all) the possible representative points.

B. There are possibly several maxima in the vicinity of each trial point by the global RP search.

C. The value of the criterion function at the trial point in a global RP search is not enough to represent the local surface of the function.

These three assumptions, which have not been pointed out explicitly in any other previous papers, will lead to a new approach to the multimodal optimum-seeking problem, [2] and [3].

3. EXPERIMENTAL METHOD

3.1 Scene and Procedure of Search Test

Fig. 3.1 shows the scene of the heuristic search test to solve the problem in 2.2. A subject (sighted, blindfolded or blind subject) and an experimenter sit at the sides of a table. There is a search board in front of the subject. The board is equally divided into 22x18=396 cells, and each center of the cell has a hole which helps the sighted subject to recognize the center visually or does the blindfolded or blind subject tactually with his fingers and also is used for putting a stick showing the height of the test hill at this point (cell) when the subject chooses the point in his search process. There is a test hill in front of the experimenter*1. When the subject selects an i th trial point $\underline{X}_i = (x_{1i}, x_{2i})$, the experimenter tells him the

---

*1 There is a masking screen between the sighted subject and the experimenter, which prevents the subject from seeing the test hill in front of the experimenter.

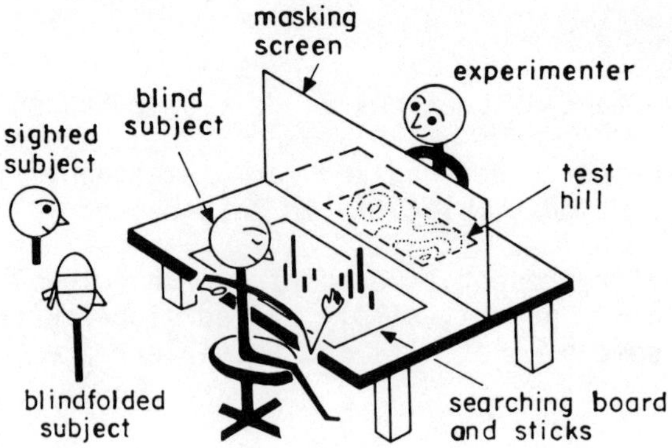

Fig. 3.1. The scene of search test.

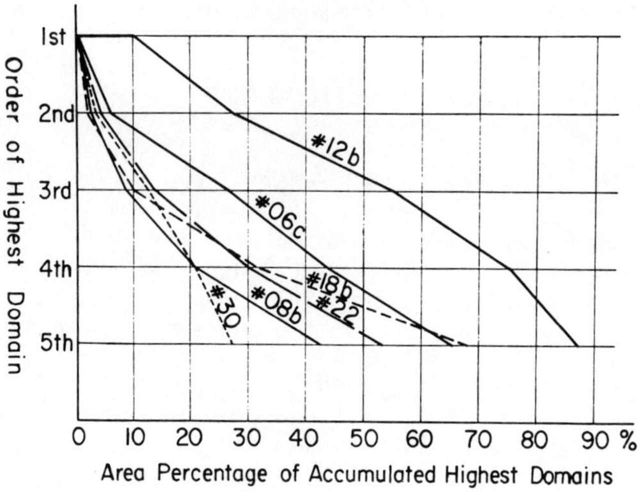

Fig. 3.2. Equivalent unimodal test hills.

corresponding height $f(\underline{X}_i) = y_i$ of the chosen point, and places a stick at the chosen point on a search board*2, where the height of the stick is proportional to y_i, and the subject selects a next trial point $\underline{X}_{i+1}$ after recognizing visually or tactually the height(s) tried previously. Repeating the above procedure, the subject seeks a globally optimum point on the hill.

After the completion of the subject's search trials, the experimenter asks the subject his thinking process during his search and also draws the contours of the searched hill which the subject imagined at the final state of his search.

3.2. Goals and Conditions of Search Test

Goals of Search Test: The main goal is to search for the globally optimum point [the value $f(\underline{X}^m)$ and the coordinates $\underline{X}^m = (x_1^m, x_2^m)$] on the test hill.

An incidental goal is to keep the value at every trial point as big as possible*3.

Conditions of Search Test: There is generally no restriction for a subject to select a next step width $\Delta\underline{X}_i = \underline{X}_i - \underline{X}_{i-1}$, but a weak restriction of each trial mode exists in some experiments.

There is generally a weak (implicit) restriction for the total number of admitted trials such as "free but as few as possible", but a severe restriction of trial number (for instance, twenty-two) in some experiments.

The number of peaks hidden in the test hill is also unknown to the subject.

---

*2 Instead of placing a stick on the board, recording the height y_i on a search sheet is adopted sometimes for the sighted subject.

*3 This incidental goal concerns strongly the optimization problem in control theory, and is understood by some subjects as (a) not to search a point being expected low, (b) not to search a point being expected lower than the present height, or (c) to search the optimum point with as few trials as possible.

3.3 Test Hills

There are total thirty kinds of test hills covering widely different contour variations. In each of them there are hidden some regularities (rules) related to the number, arrangement, and shape (sharpness) of the peaks, which will be some heuristic clues for a subject to discover about the globally highest peak.

The average number of the local peaks embedded in the six test hills (which are seen as thick solid lines in Figs. 4.1 ~ 4.4) used in the experiments by blind subjects is 2.5, the numbers of their peaks ranging from one to five.

The relative shapes of the six test hills can be seen in Fig. 3.2 showing their corresponding equivalent unimodal test hills.

Definition 2: An equivalent unimodal test hill is a unimodally modified hill which has the same area-percentage of the accumulated $\underline{i}$ th ($i = 1,2,...$) highest and higher domain as its corresponding original multimodal test hill.

Using this concept, we will develop our idea on the degree of search success in 4.3.

3.4 Subjects

The subjects who participated in the experiments are Japanese male undergraduate students, graduate students, the staffs of the laboratory at Nagoya University, Nagoya, Japan, totally blind American high school students at the Texas School for the Blind, Austin, Texas, including six boys and four girls, and Thai graduate students at The University of Texas at Austin, Austin, Texas.

When a subject joins the experiment for more than two tests, he is allowed to have plenty of time between two adjacent tests (generally several days in order to suppress the learning effect in the subject).

4. RESULTS AND DISCUSSION OF HEURISTIC SEARCH BEHAVIOR

In this chapter, we will investigate examples and main results of the heuristic search behavior.

4.1 Some Examples of Heuristic Search

Here we will see four examples of heuristic search made by

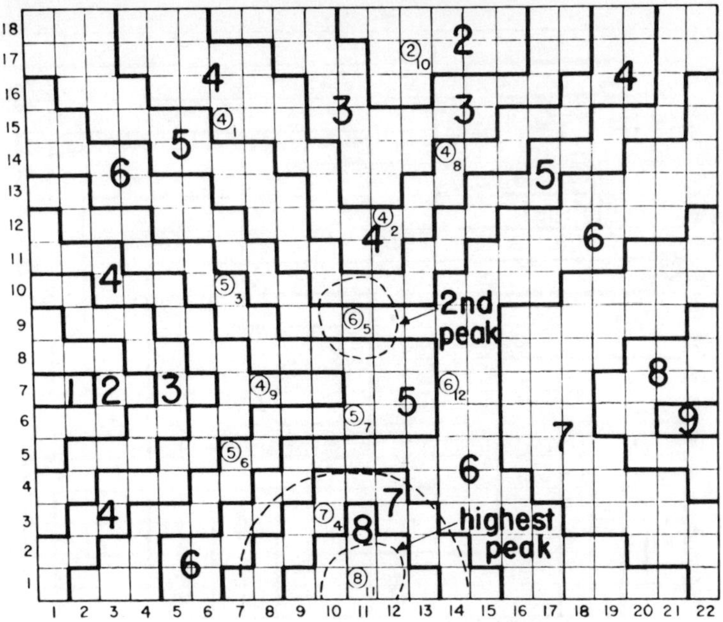

Fig. 4.1. End stage example of heuristic search (Subject T.B., Test Hill 06c, 12 trials).

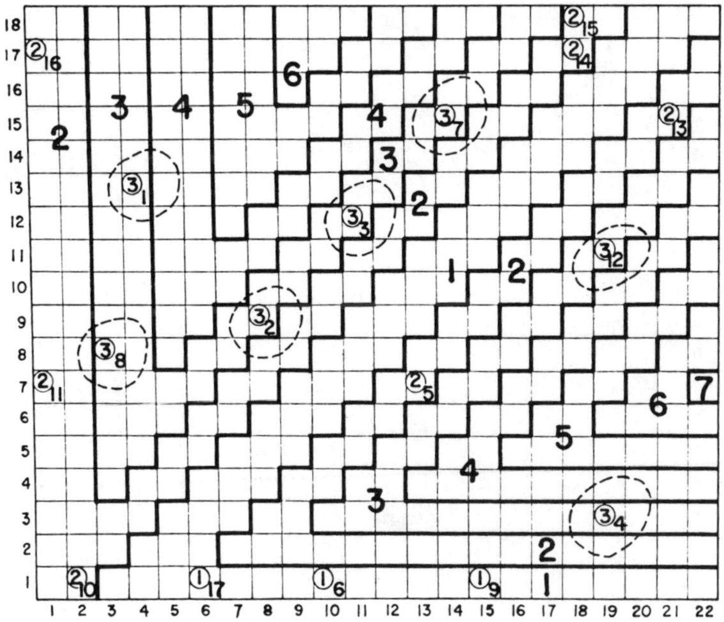

Fig. 4.2. Example of heuristic search (Subject E.B., Test Hill 22, 17 trials).

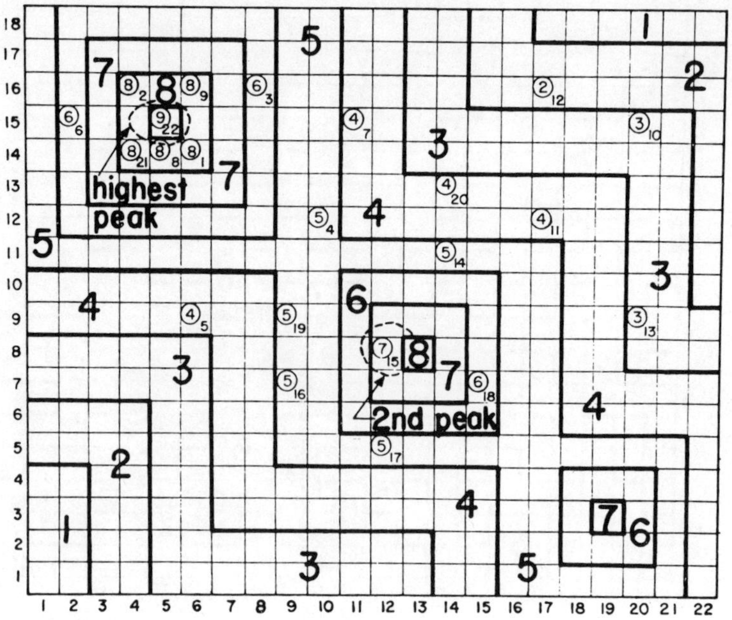

Fig. 4.3. Example of heuristic search (Subject R.B., Test Hill 08 b, 22 trials).

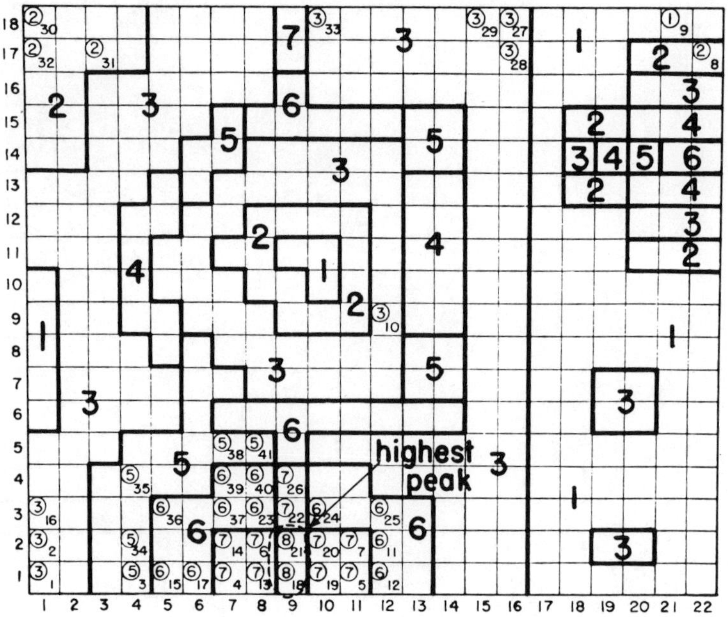

Fig. 4.4. Example of heuristic search (Subject C.S., Test Hill 30, 41 trials).

blind subjects*4

Example 4.1.1 Fig. 4.1 shows an example of heuristic search by Subject T.B. for Test Hill 06c. In the figure, a circle shows a searched trial, where the numeral in the circle shows the value of the criterion function (the height of the test hill) at the trial point and another numeral outside at the right side of the circle shows the trial order.

In the figure, the first four trials are made as a global search in the early stage of the search behavior. The next four trials made in the middle stage seem to be local search trials compared to the first four. The last four trials are made as a local search in the end stage. In this figure, the dotted line is the contour which the subject imagined and sketched after the completion of his search with twelve trials, while the thick solid line shows the contour of the criterion function (test hill) which is unknown to the subject.

This is an example where a small number of trials was used to find the second highest domain.

Example 4.1.2. Similarly, Fig. 4.2 shows another example at the final state of search made by Subject E.B. for Test Hill 22. This is an example with 17 trials which failed to locate the real highest domain and found unluckily seven of the fifth highest domains.

Example 4.1.3. Fig. 4.3 shows another example at the final state of search made by Subject R.B. for Test Hill 08b. This is an example with 22 trials which succeeded in finding the highest domain.

Example 4.1.4. Fig. 4.4 shows another example at the final state of search made by Subject C.S. for Test Hill 30. The subject made 41 trials and successfully located the highest domain.

4.2 Deviation of Trial Number of Search

4.2.1. Deviation Due to Sighted and Blindfolded Subjects

Fig. 4.5 shows the trial numbers searched in the experiment of twelve Japanese subjects and eleven test hills. In Fig. 4.5, any of the twenty-seven points shows the trial numbers of the sighted and blindfolded searches made by the same subject on the same test hill. Fig.4.5 shows that the blindfolded search needs in general a somewhat smaller trial number than the sighted search. This fact is the first reason for us to conclude that the blindfolded search is made differently from the sighted one.

*4 Other search examples are seen in Refs. [1]~[3].

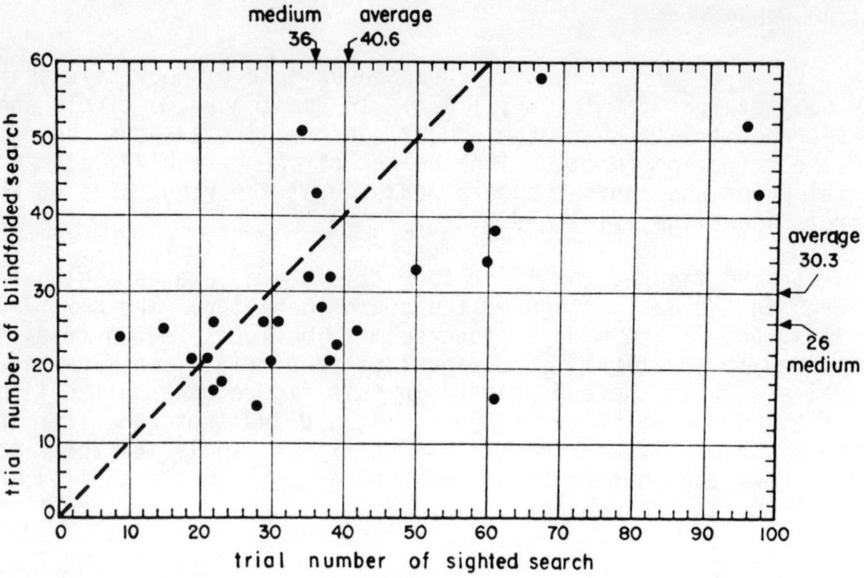

Fig. 4.5. Correlation between trial numbers of sighted and blindfolded searches by twelve subjects.

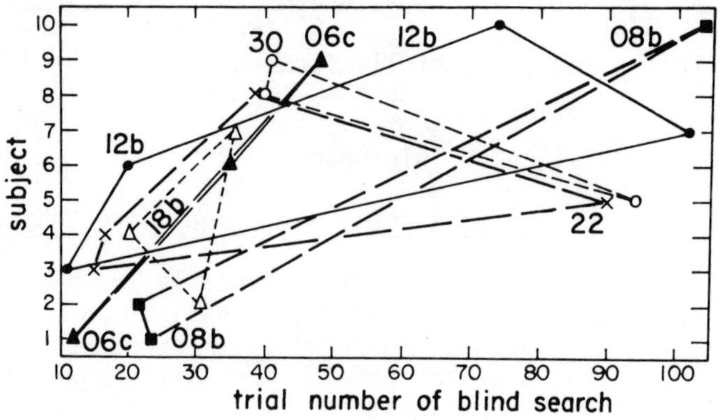

Fig. 4.6. Deviation to trial number due to subject and test hill in the experiment by blind subjects.

4.2.2 Deviation in Blind Search

Fig. 4.6 shows the deviation of trial numbers due to blind subject and test hill (two tests for each ten blind subjects of an American high school; six test hills in total).

4.2.3 Deviation Due to Nationality of Subjects

We had experiments by subjects of different nationality of Japan, the U.S., and Thailand. As a result of the experiments, we can say there is almost no difference in the deviation of trial number of search due to the nationality of subjects.

4.3 Degree of Success in Heuristic Search Performance

In Fig. 4.7, we can see a typical example of the degree of success p(i) of heuristic search performance by ten blind subjects, where p(i) means the accumulated probability of such a success that ith highest and higher domain of the test hill was found by a subject. The average degree of success $\overline{p(i)}$ are as follows: $\overline{p(1)} = 42.5\%$, $\overline{p(2)} = 72.5\%$, and $\overline{p(3)} = 97.5\%$.

4.4 Speciality of Global Search

Almost all the trials of global search are generally tried at an early stage of the heuristic search process, while some others are done at middle or even late stage of the process. Some of the global trials are supposed to be base points B_i of local search around them. The base point of a local area is apt to shift to another better base point B_i when a higher point B_i' is found in the local area, i.e. $f(B_i) < f(B_i')$, where $f(X)$ is the height of a point X. In general, the higher the $f(B_i)$ is, the bigger is the local area to be searched around the base point B_i, [1].

5. HEURISTIC SEARCH MODELS AND DISCUSSION

Although the deviation of trial numbers of heuristic search is not so small due to subject and test hill, as described in the previous chapter, we can extract five kinds of gross but essential models of heuristic search. First, we will define a distance and trial mode in preparation for a mode analysis of heuristic search trial.

Definition 3 The distance $d(\underline{X}_i, \underline{X}_j)$ between two points, $\underline{X}_i = (x_{1i}, x_{2i})$ and $\underline{X}_j = (x_{1j}, x_{2j})$ is defined by

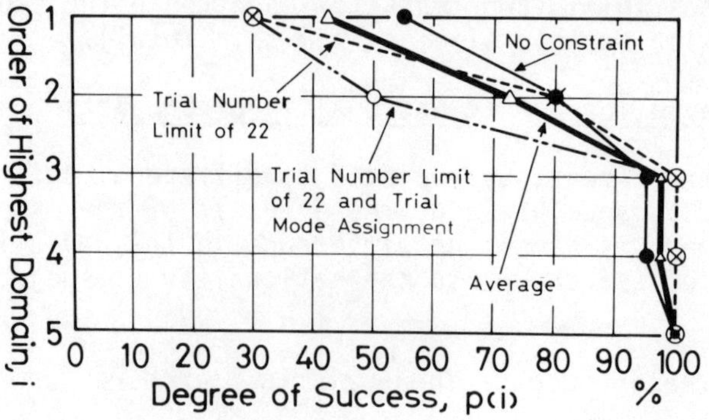

Fig. 4.7. Degree of success in heuristic search by blind subjects.

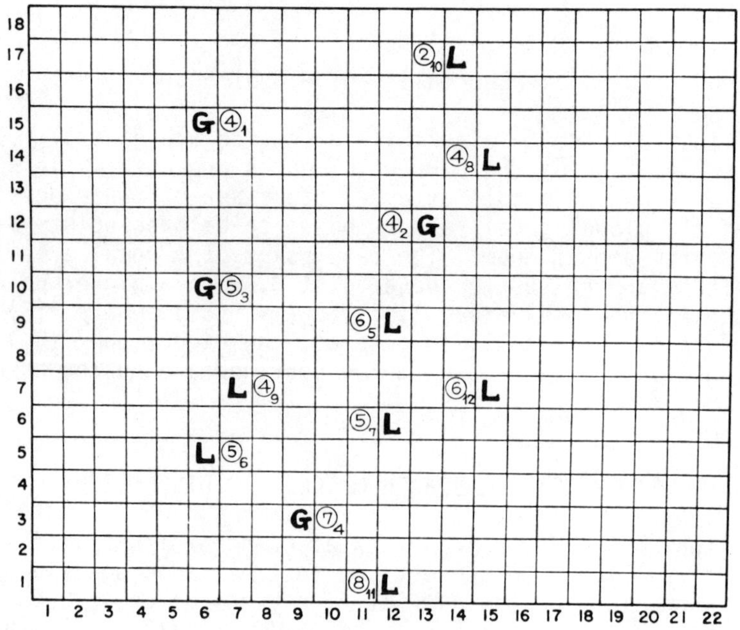

Fig. 5.1. Example of mode analysis (Subject T.B., Test Hill 06 c, 12 trials).

$$d(\underline{X}_i,\underline{X}_j) = \max\{|\Delta x_1|,|\Delta x_2|\} \tag{5.1}$$

where $\Delta x_1 = x_{1i} - x_{1j}$ and $\Delta x_2 = x_{2i} - x_{2j}$.

Next, we define the mode $\underline{M}(\underline{X}_i)$ of the ith search trial $\underline{X}_i$ by using the minimum $d(\underline{X}_i,\underline{X}_j)$, where $j = 1,2,\ldots,i-1$.

Definition 4

$$\underline{M}(\underline{X}_i) = \begin{cases} \text{G-mode (Global mode) if } \min_j d(\underline{X}_i,\underline{X}_j) \geqq 4, \\ \text{L-mode (Local mode) if } 2 \leq \min_j d(\underline{X}_i,\underline{X}_j) \leq 3 \\ \text{C-mode (Convergent mode) if } \min_j d(\underline{X}_i,\underline{X}_j) = 1. \end{cases} \tag{5.2}$$

where $j - 1,2,\ldots,i-1$, and $i \geqq 2$. $\underline{M}(\underline{X}_i)$ = G-mode when $i = 1$.

5.1 Examples of Mode Analysis of Search Behavior

Example 5.1.1. For instance, the trial modes of Example 4.1.1 (cf. Fig. 4.1) are analyzed as shown in Fig. 5.1 and written in individual mode expression

$$G\ G\ G\ G\ L\ L\ L\ L\ L\ L\ L\ L \tag{5.3}$$

or simply in clustered mode expression

$$G^4L^8. \tag{5.4}$$ *5

Example 5.1.2. Fig. 5.2 shows the trial modes of Example 4.1.3 (cf. Fig. 4.3), which are written

$$G\ L\ L\ G\ G\ L\ L\ C\ L\ G\ L\ L\ L\ L\ L\ L\ L\ L\ L\ L\ C\ C \tag{5.5}$$

or

$$G^1L^2G^2L^2C^1L^1G^1L^{10}C^2. \tag{5.6}$$ *6

5.2 Extraction of Heuristic Search Models

The extracting procedure of heuristic search models TiB (i = 1,2,...,5) is explained for the experiment by ten blind subjects under no experimental constraints.

*5,*6 X^m, where X=G, L or C, means that m trials were made continually by X-mode search.

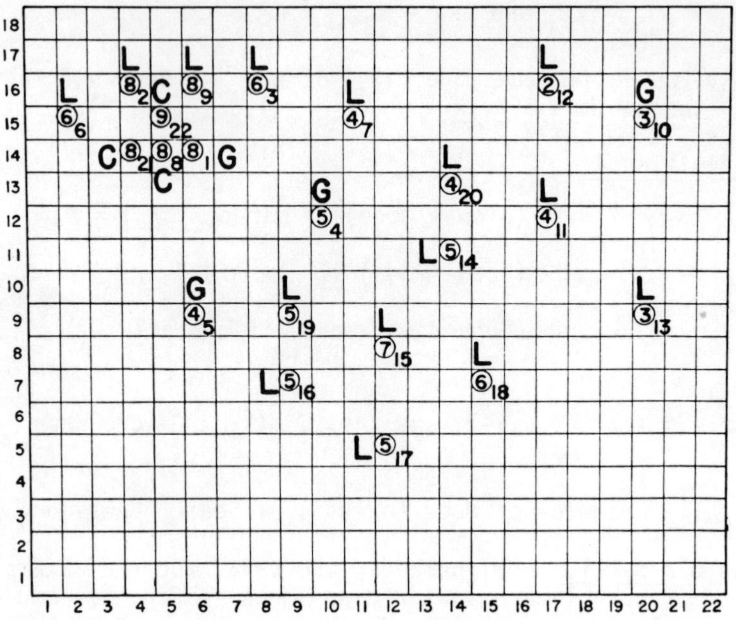

Fig. 5.2. Example of mode analysis
(Subject R.B., Test Hill 08 b, 22 trials),

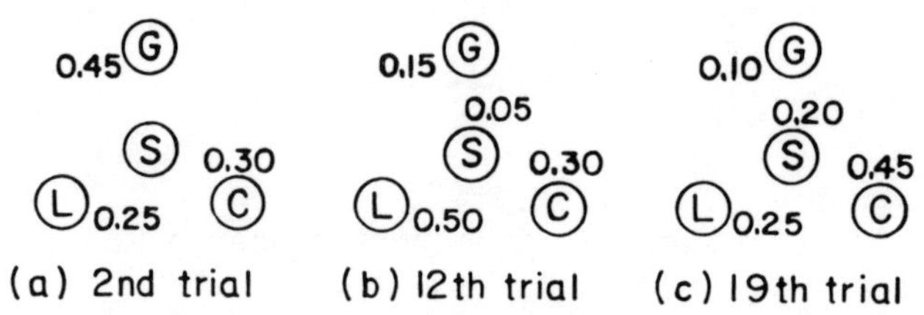

Fig. 5.3. Examples of occurrence probability of trial mode in the experiment by blind subjects under no constraints.

5.2.1 First Model

We can compute the occurrence probability of each trial mode at each trial order. Some of the probabilities are shown in Fig. 5.3, where S means S-mode (Stopping mode) or no trial made.

Selecting the mode with the maximum occurrence probability at each trial order as the representative mode (for instance, G at 2nd trial, L at 12th, and C at 19th in Fig. 5.3), we can extract the first model of heuristic search T1B as shown below:

$$\text{Model T1B}=G^2L^1 \begin{pmatrix} G^1 \\ L^1 \end{pmatrix}\begin{pmatrix} L^1 \\ C^1 \end{pmatrix}\begin{pmatrix} L^1 \\ C^1 \end{pmatrix} L^6 \begin{pmatrix} G^1 \\ L^1 \\ C^1 \end{pmatrix} L^1C^1 \begin{pmatrix} L^1 \\ C^1 \end{pmatrix} L^2C^1L^1C^{10} \qquad (5.7)^{*7}$$

5.2.2 Second Model

We get all the transition probabilities from each ith (i = 1,2,...) trial of G-,L-, and C-mode to each(i + 1)th trial of G-,L-,C-, and S-mode, and also from each ith trial of S-mode to each (i + 1)th trial of S-mode. Some of the probabilities are shown in Fig. 5.4.

Selecting the mode with the maximum transition probability from ith trial to (i + 1)th trial as the representative mode at (i + 1)th trial (for instance, L at 2nd trial, L at 14th, and C at 25th in Fig. 5.4), we can extract the heuristic search model T2B as shown below:

$$\text{Model T2B}=G^1L^3G^1C^2L^1 \begin{pmatrix} L^1 \\ C^1 \end{pmatrix} L^3 \begin{pmatrix} L^1 \\ C^1 \end{pmatrix} L^1 \begin{pmatrix} L^1 \\ C^1 \end{pmatrix} C^1L^2C^8 \qquad (5.8)$$

5.2.3. Third Model

We will focus our point of view on the ith (i = 1,2,...) clustered modes $G_i^{g(i)}$, $L_i^{l(i)}$, $C_i^{c(i)}$, where $g(i)$, $l(i)$, and $c(i)$ are the trial numbers of ith clustered modes G, L, and C respectively.

---

*7 $\begin{pmatrix} X^1 \\ Y^1 \end{pmatrix}$, where X=G or L, Y=L, or C, means that one trial is made by X- or Y-mode search. Similarly $\begin{pmatrix} G^1 \\ L^1 \\ C^1 \end{pmatrix}$ means one trial is made by one of three modes.

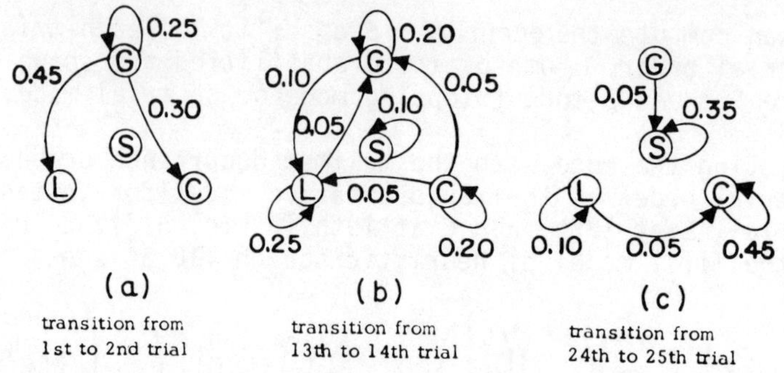

Fig. 5.4. Examples of transition probability of trial mode in the experiment by blind subjects under no constraints.

Subject		Test Hill: 06c	08b	12b	18b	22	30
1	T.B.			difficult 22:enough			
2	R.B.					not hard 22: enough	
3	M.D.						not difficult 22: enough
4	E.B.			not difficult 22: enough			
5	F.B.	difficult 22: enough					
6	G.W.				difficult 22: enough		
7	J.S.					15 more trials	
8	J. H.		not difficult 22: enough				
9	C.S.		⋆				
10	R.R.				22: enough		

⋆Necessary questions were forgotten.

Table 5.1. Subjects' answers in model checking test with the constraint of trial number limit of 22.

For instance, Subject R.B.'s search behavior for Test Hill 18b is written as

$$G_1^{\,1} L_1^{\,3} C_1^{\,4} G_2^{\,1} L_2^{\,1} C_2^{\,1} L_3^{\,1} C_3^{\,4} G_3^{\,1} L_4^{\,4} G_4^{\,1} L_5^{\,5} C_4^{\,2} L_6^{\,1} C_5^{\,1}$$

where the arrows trace a sequence of subscripts.

Next, we will compute the average numbers $\overline{g(i)}$, $\overline{l(i)}$, and $\overline{c(i)}$, of g(i), l(i), and c(i) respectively only when the number of G_i's, L_i's, and C_i's counted is equal to or more than half of total test number. Here, we can define the heuristic search model T3B= $[\{\overline{g(i)}\}/\{\overline{l(i)}\}/\{\overline{c(i)}\}]$ or $[\{G_i^{\overline{g(i)}}\}/\{L_i^{\overline{l(i)}}\}/\{C_i^{\overline{c(i)}}\}]$ as shown below:

$$\text{Model T3B} = \begin{cases} [2,2,2,1/2,3,2,2,2,2/4,2,4,8,3,2] \\ \text{or} \\ [G_1^2,G_2^2,G_3^2,G_4^1/L_1^2,L_2^3,L_3^2,L_4^2,L_5^2,L_6^2/\ C_1^4,C_2^2,C_3^4,C_4^8\ C_5^3,C_6^2] \end{cases} \quad (5.9)$$

We should call this kind of Third Model imcomplete, since only the trial number of each clustered mode is determined and the connection*8 (order) of the clustered modes is not determined in Model T3B.

5.2.4 Fourth Model

We can compute the occurrence probabilities of the clustered modes of G. L, C, and S at the ith (i = 1,2,...) order of clustered mode. Selecting the mode with the maximum occurrence probability at the ith order as the representative mode, we can extract the framework of the heuristic search model T4B as shown below:

$$\begin{matrix}\text{Framework of} \\ \text{Model T4B}\end{matrix} = G\ L\ G\ L\ \binom{G}{C}\ L\ C\ L\ C\ G \qquad (5.10)$$

On the other hand, we can get the average number $\overline{g(i)}$, $\overline{l(i)}$, or $\overline{c(i)}$ for trials for the clustered mode with the maximum occurrence probability at each order of clustered mode; i.e.

$$\begin{matrix}\text{Trial Number of} \\ \text{T4B}\end{matrix} = \begin{matrix}[\{\overline{g(i)}\}/\{\overline{l(i)}\}/\{\overline{c(i)}\}] \\ [2,2,(1),2/2,3,2,3/(4),4,9]\end{matrix} \qquad (5.11)^{*9}$$

*8 We can have 64 possible variations of Model T3B, [7]

*9 (1)=(g(3)) and (4)=(c(1)) in Eq. 5.11 correspond to $\binom{G}{C}$ in Eq. (5.10).

Combining Eqs. (5.10) and (5.11), we can get Model T4B as follows:

$$\text{Model T4B} = G^2L^2G^2L^3\binom{G^1}{C^4}L^2C^4L^3C^9G^2 \qquad (5.12)$$

We should notice the $\overline{g(i)}$, $\overline{l(i)}$, and $\overline{c(i)}$ in Fourth Model are different from those in Third Model respectively due to their different averaging processes.

5.2.5 Fifth Model

We will try to produce Model T5B by combining Model T3B and T4B: We combine the framework of Model T4B and the average number of trials for each clustered mode of Model T3B, and get a better*10 Model T5B.

However unfortunately, we cannot succeed in combining Models T3B and T4B to produce T5B, since the numbers of clustered modes are not the same in both cases.

We should notice there is a successful example of producing a fifth model T5C in the experiment by ten blind subjects under the trial number limit of 22:

$$\text{Model T5C} = G^4L^2G^3L^2\begin{pmatrix} G^1C^2 \\ C_2G_1 \end{pmatrix} L^2C^2L^2C^2 \qquad (5.13)$$

5.3 Other Examples of Heuristic Models

We can extract other examples of heuristic search models from other experiments. For instance, from the experiment by twelve normal subjects (twenty-seven tests for both sighted and blindfolded states; cf. Fig. 4.5) we can extract first and second models (Models V1 and V2 corresponding to the sighted or visual states, and Models T1 and T2 corresponding to the blindfolded or tactual states)*11. They are shown as follows:

$$\text{Model V1} = G^6\binom{G^1}{L^1}\binom{G^1}{L^1}L^6G^1C^3L^3C^1L^3C^1\binom{L^1}{C^1}C^3$$

*10 $\overline{g(i)}$, $\overline{l(i)}$, and $\overline{c(i)}$ in Model T3B seem to be more reasonable than those in Model T4B respectively due to the difference of their averaging processes.

*11 Other examples of the heuristic search model and a discussion on them are seen in Ref. [3].

Subject		TEST HILL 06c	08b	12b	18b	22	30
1	T.B.				2 more trials		
2	R.B.	3 more trials good model					
3	M.D.				22: enough good model		
4	E.B.						22: enough average m.
5	F.B.		not difficult fine model				
6	G.W.		22: enough not wrong				
7	J.S.						easy 20 more trials good model
8	J.H.	not difficult 22: enough average m.					
9	C.S.			difficult 22: enough good model			
10	R.R.		pretty easy 22: enough good model				

Table 5.2. Subjects' answers in the test of checking models Tl and Tw.

$$\text{Model V2} = G^7 \begin{pmatrix} G^1 \\ L^1 \end{pmatrix} L^1 \begin{pmatrix} G^1 \\ L^1 \end{pmatrix} L^4 \begin{pmatrix} L^1 \\ C^1 \end{pmatrix}\begin{pmatrix} G^1 \\ C^1 \end{pmatrix} C^3L^1 \begin{pmatrix} L^1 \\ C^1 \end{pmatrix} C^1L^3C^4$$

$$\text{Model T1} = G^6L^3G^1L^2C^1L^9 \tag{5.14}$$

$$\text{Model T2} = G^8L^4C^2L^1 \begin{pmatrix} L^1 \\ C^1 \end{pmatrix} L^5 \begin{pmatrix} G^1 \\ L^1 \\ C^1 \end{pmatrix}$$

5.4 Appropriateness of Heuristic Models

We had two kinds of tests by ten blind subjects for checking the appropriateness of the extracted heuristic search models T1 and T2.

5.4.1 Experimental Method

In the first experiment of checking the appropriateness of the extracted models, a constraint to the subject is added of a trial number limit of 22. In the second experiment of checking the models, two constraints are added of trial number limit of 22 and trial mode assignment defined by Models T1 and T2. In other words, a subject in the second experiment has to search the optimum point with 22 trials according to each trial mode of Model T1 or T2 (6 subjects checked Model T1 and the other 4 subjects Model T2.)

At the end of the model checking test the subject is asked (1) if it was or was not difficult for him to search with the constraint(s), (2) if 22 trials were enough, or how many more trials he needed and (3)*12 if the extracted model is good, wrong, or average.

5.4.2 Subjects' Judgement of Models

Table 5.1 shows the subjects' answers at the end of the model checking test in the first experiment and Table 5.2 shows those in the second experiment. These tables indicate the extracted models T1 and T2 consisting of 22 trials are good (average) models.

In a similar experiment by normal subjects, the appropriateness of heuristic models V2 and T2 are also proved, [3].

*12 Question (3) is made only in the second experiment.

					X_1	X_2	X_3
E_1	D_1	C_1	B_1	A_1	x_{11}	x_{21}	x_{31}
				A_2	x_{12}	x_{22}	x_{32}
			B_2	A_3	x_{13}	x_{23}	x_{33}
				...	...	...	...
	⋮		⋮	A_i	x_{1i}	x_{2i}	x_{3i}
...	⋮	...	⋮	...	...	...	...
E_n Global-Surface Concept	D_m Local-Surface Concept	C_l Plane-Segment Concept	B_k Line-Segment Concept	A_i Point Concept	Horizontal Axis	Vertical Axis	Height Axis

Primitive Information of Trial Point (X_1, X_2, X_3 columns)

← Development of Concept

Table 6.1. Matrix expression of concept formation

6. CONCEPT FORMATION IN HEURISTIC SEARCH

In this chapter, we will investigate the heuristic search process from a viewpoint of concept formation, [5] and [12]. Some subjects participated in another experiment for the analysis of concept formation process in the heuristic search. The subject was asked to express all his ideas concerning the unknown test hill before and/or after each trial of his heuristic search.

6.1 Matrix Expression of Concept Formation

Suppose a point concept A_i consists of an ith trial point $\underline{X}_i = (x_{1i}, x_{2i})$ and its height x_{3i}, and define its relation as follows:

$$A_i = f_i^A(x_{1i}, x_{2i}, x_{3i}) \qquad (6.1)^{*13}$$

An association of two points A_i and $A_{i'}$ produces a one-level higher concept of line-segment concept B_k as follows:

$$B_k = f_k^B(A_i, A_{i'}) \qquad (6.2)$$

Similarly, a plane-segment concept C_l, a local-surface concept D_m, and a global-surface concept E_n are formed by its corresponding one-level lower concepts as follows.

$$C_l = f_l^C(B_k, B_{k'}, \ldots) \qquad (6.3)$$

$$D_m = f_m^D(C_l, C_{l'}, \ldots) \qquad (6.4)$$

$$E_n = f_n^E(D_m, D_{m'}, \ldots) \qquad (6.5)$$

The above hierarchical mapping process of five-level concept-formation is shown by a matrix expression as in table 6.1.

6.2 Example of the Process of Concept Formation

Fig. 6.1 adds two thick solid arrows to Fig. 4.3. The arrows are examples of line-segment concept B_k. The arrow shows a line-segmental descent, whose lower side is shown by the arrowhead. In the figure, the line-segment concept shown by the arrow between the first and 7th points is formed after the 7th trial, and another

*13 In this case, the dimensions of the concept A_i consists of the horizontal, vertical, and height axes, and the value of the dimension is the coordinate of its corresponding axis.

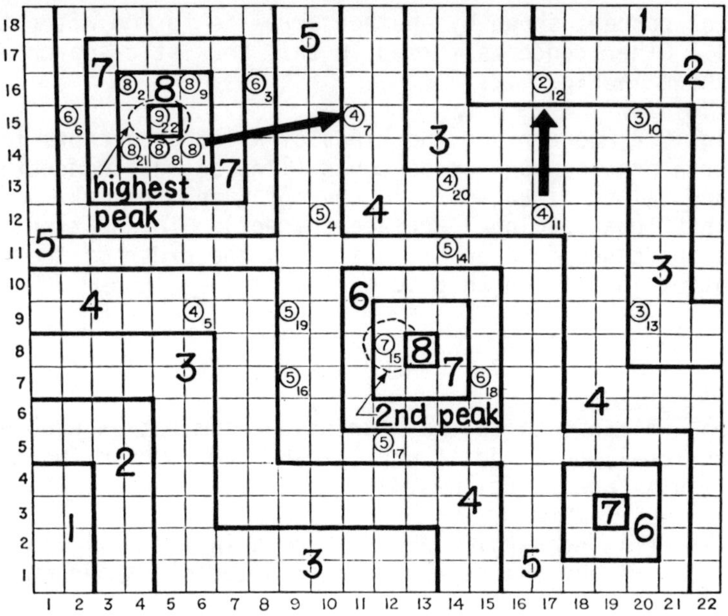

Fig. 6.1. Example of concept formation in heuristic search by subject R.B.

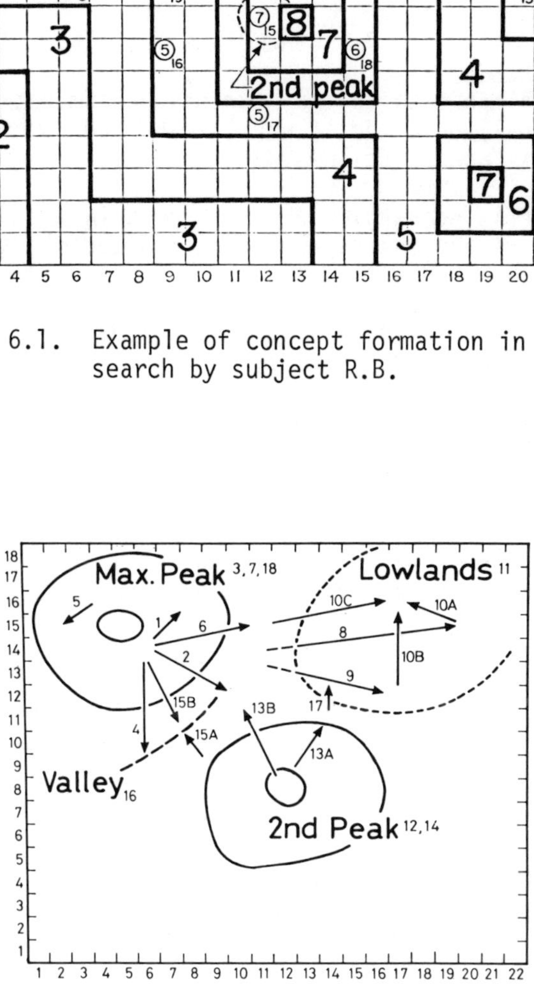

Fig. 6.2. Process of concept formation in heuristic search by subject R.B.

line-segment concept shown by another arrow is formed after the 12th trial. Other concepts formed in this heuristic search process are shown in Fig. 6.2.

In Fig. 6.2, we can see the kinds of all the concepts formed in the search process such as descents (line-segment concept), lowlands and valley (local-surface concept), and the maximum and second highest peaks (global-surface concept). In addition, the numeral on the concept in the figure shows the order of the concept formation in the search process.

7. CONCLUSION

We took an approach to the heuristic optimization problem of a two-dimensional, multimodal, non-linear, and unknown function. Our main results are as follows:

(1) There is no particular difference in the deviation of the trial numbers and the degree of search success among American, Japanese, and Thai subjects.

(2) Five kinds of heuristic search models are extracted through the mode analysis of search trial.

(3) By the model-checking test, the extracted Models T1, T2, and V2 are verified to be good or average. In addition, the number of 22 trials (only 5.5% of total possible trials) is verified to be almost enough to search.

(4) The process of concept formation in the heuristic search behavior is studied from a viewpoint of concept matrix (concept structure).

The studies in this paper are hoped to be helpful in the near future in making a simulation program for heuristic search.

ACKNOWLEDGMENT

The authors would like to thank Mr. T. Nagaoka, Automatic Control Lab., Nagoya Univ., Nagoya, Japan, and Mr. S. Harinasuta, Univ. of Texas at Austin, Austin, Texas, for their helpful devotions in the experiments, and also the many subjects who participated in the experiment.

REFERENCES

(1a)* M. Oda and K. Kakamura; "Heuristic Method of Searching an Optimum Point of a Two-Dimensional Multimodal Criterion Function", J. of Society of Instrument and Control Engineers of Japan, 7-12, 16/22 (Dec. 1968).

(1b) Also available in English, Res. Rept. of Automatic Control Lab., Nagoya Univ., Nagoya, Japan, 16,1/10 (Aprl. 1969).

(2) M. Oda and B.F. Womack; "Experimental Study of Heuristic Search of Multimodal Optimum Point by Human Subjects", Proceedings of 3rd Houston Conf. on Computer and System Sciences, Houston, Texas, 689/700 (Apr. 26-27, 1971).

(3) M. Oda and B.F. Womack; "Experimental Study of Heuristic Search Method for Two-Dimensional Multimodal Optimum Point", Res. Rept. of Automatic Control Lab., Nagoya Univ., Nagoya, Japan, 19, 15/34 (June 1972).

(4) M. Oda, B.F. Womack, and K. Nakamura; "Heuristic Search Behavior by Blind Subject", Conf. of Automatic Control, Inst. of Ele. Engrs. of Japan (may 30, 1972).

(5)* M. Oda, T. Nagaoka, and K. Nakamura; "Heuristic Search and Concept Formation", 12th Annual Conf. of Soc. of Instrument and Control Engrs., Japan, 367/374 (Aug. 1973).

(6) J. Opačić; "A Heuristic Method for Finding Most Extrema of a Non-linear Function", IEEE Trans., SMC-3-1, 102/107 (Jan. 1973).

(7) S. Harinasuta; "Heuristics for Search Techniques", M.S. Thesis, Dept. of Ele. Eng., Univ. of Texas at Austin, Austin, Texas, U.S.A. (Dec. 1971).

(8) R.H. Graham; "Basic Pattern in Heuristic Searches", M.S. Thesis, Dept. of Ele. Eng., Univ. of Texas at Austin, Austin, Texas, U.S.A. (Feb. 1972).

(9) M. Oda and B.F. Womack; "A Unified Discussion on Heuristics in Artificial Intelligence", Proceedings of Third Asilomar Conf., 387/400 (Dec. 1969).

(10) M. Oda and B.F. Womack; " A Detailed Investigation of Various Heuristic Approaches in Artificial Intelligence," 1970 SWIEEECO Record, 125/129 (Apr. 1970).

(11) M. Oda and B.F. Womack; "New Definition and General Discussion on Heuristics", invited paper at 1971 JACC, St. Louis, Mo., 3/5 (Aug. 1971).

(12) M. Oda and K. Nakamura; "Concept Formation in Heuristic Search Behavior", 1st International Joint Conf. on Pattern Recognition, Washington, D.C. (Oct. 30 - Nov. 1, 1973).

* Written in Japanese.

BASIC SEARCH PATTERNS IN HEURISTIC SEARCH

Baxter F. Womack

The University of Texas at Austin

Austin, Texas 78712

ABSTRACT

A general discussion of heuristics is followed by a discussion of a particular type of problem to be solved by a heuristic method. The problem, a common one in control theory, is the searching of the highest point on a multi-dimensional, multi-modal surface. An experiment was conducted in which human subjects performed heuristic searches to find the peak of a two-dimensional, multi-modal surface. Results of this experiment indicate a subject searches a hill according to a predetermined pattern called the basic search pattern. Using the concept of a basic search pattern a theoretical procedure for a heuristic search is postulated.

The search made of a trial point is only one of the attributes of a trial point, and not an independent one at that. A trial point has several important attributes: its location on the map, height, area of maximum relevance, and position with respect to all other trial points and its boundaries on the map. These attributes and other possible information are inputs to the subject's search algorithm. The output of the algorithm is the location of the next trial point. This search algorithm is the core of the heuristic search process. The interaction of global and local trials by means of the algorithm leads to a pattern described as the basic search pattern.

Experiments are summarized in the paper which illustrate the existence of a basic search pattern.

INTRODUCTION

The increase in research in artificial intelligence in recent years has resulted in an increased interest in the area of heuristics. The word heuristic is often used to mean a science, method, or technique of discovery. The importance of heuristics lies in the fact that heuristic solutions are necessary in problems where the conventional sequential or analytical methods are regarded as unable, or very difficult in a practical sense, to produce a correct solution. The area of heuristics will be of increasing importance as engineers and scientists more and more attempt to understand and model complicated, large-scale systems.

This paper presents the results of an experiment performed in the area of heuristics. The purpose of the experiment was to gain insight into a particular type of heuristic method in the hope that such insight may eventually lead to a model of the heuristic method. The heuristic method investigated was the heuristic searching process used by a human subject to search the highest point on a two-dimensional, multi-modal hill [1 - 4]. The results of the experiment indicate that the search is performed according to a definite pattern determined by the subject before the search begins.

GENERAL DISCUSSION OF HEURISTICS

Heuristics have long been associated with education. In fact, the adjective heuristic is defined in Webster's New World Dictionary as follows: "helping to discover or learn; sometimes used to designate a method of education in which the pupil is trained to find out things for himself." The "things" it is hoped the pupil will find out for himself are usually laws or principles. This type of discovery could be termed discovery in its "pure" sense, since, although the pupil is guided somewhat along his path to discovery, the subject matter of his discovery is not constrained, and the student is free to discover whatever he might. What is discovered is not as important as the act of discovery itself, the underlying philosophy of this approach to education being that it is important for the pupil to learn to think for himself in an age of rapid progress. In a world where the truths of today might not necessarily be the truths of tomorrow, rather than teach the pupil truths, it is better to teach him how to discover truths for himself at any time. In the heuristic method of education, therefore, the method of discovery is not a means to an end, discovery, but is, in fact, the end itself.

The same statement cannot be made about the application of heuristics to the fields of science and engineering. Here the emphasis is on the discovery, which is a discovery in the narrow sense of the word, in that the discovery is constrained to be of a certain type. In particular, heuristics are used to arrive at a solution of a problem. As opposed to the area of education where a heuristic method has intrinsic value independent of the discovery arrived at, in the area of science and engineering a heuristic method has value only in relation to its ability to arrive at a "good" solution to a particular problem.

The types of problem for which heuristic methods are used are those in which, during the process of arriving at a solution, certain decisions must be made from among a large number of alternatives. If the number of alternatives is large enough, a complete enumeration of the consequences of each alternative would be, if not impossible, then certainly impractical. An example of this type problem occurs in the writing of a computer program to play chess, where a strategy of considering one by one all possible consequences of all possible moves would simply be infeasible.

When heuristic methods are applied to problems in these areas, the number of alternatives available at each stage of the solution process is reduced by omitting from consideration all those alternatives which seem unlikely, according to some criterion, to produce a satisfactory solution to the problem. Since not all possible cases of solution are considered when using heuristic methods, the solution arrived at is only an approximate solution. The value of a heuristic method is dependent upon the "goodness" of the approximate solution it produces relative to the true solution to the problem.

Heuristic processes are decision-making processes in which the decision arrived at has no mathematical proof. The heuristic process arrives at a decision by using a heuristic method. A heuristic method, then, increases computational efficiency by utilizing all available knowledge to eliminate unlikely prospects from the domain of possible solutions.

This heuristic process could contain within it other heuristic processes, which, in turn, contain other heuristic processes, and so on _ad infinitum_. This is simply a statement of the obvious fact that all decisions are based on previous decisions, and, as such, is not very significant. The point to be made is that when discussing heuristic processes as with any other decision-making processes it is important to delimit the scope of the process under discussion.

The problem faced by a man buying a house is a simple example of a heuristic process which illustrates the point made in the preceding paragraph and also unifies some of the concepts presented in this secion [4]. It was stated that the decision arrived at by a heuristic process had no mathematical proof. Certainly the decision of which house to buy cannot be subjected to mathematical proof. It is arrived at by use of a heuristic method. If we work backwards in stages, the final decision is undoubtedly made from several houses that the man had under final consideration, the houses under final consideration came from a larger number of houses that the man had looked at, and these houses had come from still a larger number of houses whose advertisements he had seen in the newspapers. At each stage the number of possible solutions was decreased. This is the essence of the heuristic method.

If we look at a particular stage such as that of deciding which houses to look at from those whose ads he saw in the newspaper, we see that the man uses heuristic clues (price, location, size, etc.), according to some heuristic priority (size more important than location, but not as important as price) to make his decision. This means that at each stage a heuristic method is utilized to decrease the number of alternatives. Therefore, the individual stages could be considered separate heuristic processes or parts of a larger heuristic process. If we decide that the result of the heuristic process with which we are concerned is the final decision of which house to buy, we have only specified the end of the heuristic process. We must also specify a beginning, the first decision that is a part of the heuristic process. In the present example, the process of deciding which house to buy could be said to begin with the final decision or with the decision made at any of the previous stages.

Obviously, the determination of the beginning point will at times be quite arbitrary. For instance, it would be impossible to specify all decisions leading to the final decision of which house to buy. However, it is important that a beginning point be designated, however arbitrarily, because this beginning point will determine what constitutes the input to the heuristic process. In the present example, if the heuristic process were said to begin with the final decision, then the input to the process would consist of only those houses under final consideration. On the other hand, if the process was said to begin with the decision of which houses to look at, then the input to the process would be all the houses which were advertised in the newspaper.

EXPERIMENTAL STUDY OF HEURISTIC SEARCHES

A. Purpose of the Study

Nothing has been said about how heuristic methods can be developed. Before we turn to this question it is necessary to make a distinction between formal heuristic methods and human heuristic methods.

Formal heuristic methods are those in which the decision rules are well defined and not subject to interpretation. In most cases, the use of the same formal heuristic method by different persons will result in the same solution to a given problem.

Human heuristic methods are those methods used by a human to solve a problem in the absence of formalized methods. The decision rules in a human heuristic method are not well-defined and exist only in the mind of the person solving the problem. In fact, the person may not even be aware of the existence of these decision rules and may well confuse decisions with "guesses". A human heuristic method is peculiar to the particular person using it, and the identical method cannot be applied by another person. Moreover, since the decision rules are not well defined, it is highly unlikely that a person will ever twice solve a given problem by the identical heuristic method.

Given the distinction between formal and human heuristic methods, it is clear that decisions can be made by machines only through the use of formal heuristic methods.

The experiment described in this chapter was undertaken in the hopes of clarifying a particular human heuristic method. The purpose was to provide insight, not a complete model, although that would hopefully be the eventual outcome of work of this type.

B. Description of the Problem

The problem that the experiment is concerned with is that of searching an optimum point $\underline{X}^0$ of variables $\underline{X} = (x_1, x_2, \ldots)$, so as to optimize (in this case maximize) a certain kind of criterion function f $(\underline{X})$. This problem is very common, occurring as a subproblem in almost all control problems. It can be generalized as "to search an optimum (maximum) point on a multidimentional hill formed by a nonlinear criterion function".

Several methods have been developed to solve this particular problem. These methods can be divided into three main categories: (1) pure random methods, (2) combined random method and local method, and (3) approximate equation method.

C. Basic Description of Heuristic Search

The three methods of solving the optimization problem given in the last section all involve heuristic searches. The type of heuristic search that we are particularly interested in is that performed by a human. Figure 1 shows an example of a heuristic search performed by a subject in the course of the experiment

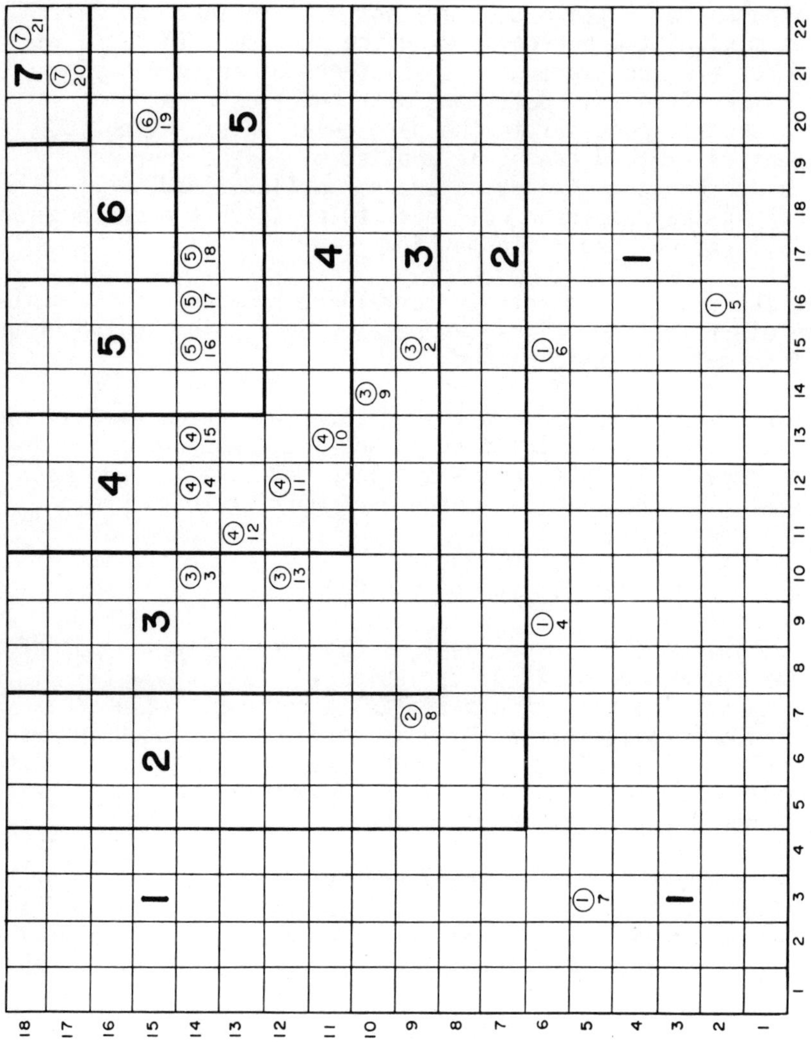

Fig. 1. Example of a Heuristic Search

described in the next section. The search domain consists of 18 x 22 = 396 cells. The contours of the criterion function (unknown to the subject) are shown by the thick lines. The subject's trial points are indicated by two numbers in a cell. The first number, the one in a circle, indicates the value of the criterion function (the height of the test hill) at that particular point. The other number shows the trial order of that particular trial point. So in Figure 1 the numbers in cell (11, 13) (the first number refers to the cell numbers along the left hand side of the map while the second number refers to the cell numbers along the bottom of the map) indicate that that point was selected at the tenth trial and the value of the criterion function at that point is four.

The total number of trials needed for the heuristic search shown in Figure 1 was twenty-one, or only 5.3% of all the searchable cells. Although the test hill shown in this particular example is relatively simple, the ease with which the subject found the maximum value of the criterion function indicates the reason for the studying of human heuristic searching techniques.

BASIC SEARCH PATTERNS

Although the models presented in [1,2] are useful in describing the searching behavior of an "average" subject, they provide limited insight into the actual search process used by a subject. A subject conducting a search guided by one of the models would select the ith trial point in the following manner: First, he would determine the mode of the ith trial from the model, and then he would pick a trial point which would be of that mode. The position of the trial point is not specified by the model except to the extent that the trial point must be of a certain mode. The modes of all trial points of the model are determined before the search begins. Therefore, information gained during the search cannot influence the choice of modes.

Although a search performed using one of the models would be considered a heuristic search in the sense that the subject would utilize all the information gained during the search to pick the ith trial point from among all possible trial points having the correct mode for the ith trial, it could not be considered an efficient heuristic search. Since the subject conforming to a model is not at liberty to pick which ever trial point he feels is dictated by all the available information, he is in effect denied full use of the information. Therefore, the models lack adaptability, an essential element of an efficient heuristic search.

The problem with the models is that the analysis leading to them assumes that trial points are completely specified by their modes. The mode of the ith trial point, and hence, the trial point itself, is determined by the mode of previous trial points, if it is determinable at all. However, the mode of a trial point is only one of the attributes of a trial point, and not an independent one at that.

A trial point has several important attributes: its location on the map, height, mode, area of maximum relevance, and position with respect to all other trial points and the boundaries of the map. Obviously, some of the attributes are not independent of each other. In fact, the location is the only independent attribute; all others are dependent on it. Strictly speaking, a trial point is completely specified by its location. However, since the criterion function which relates the height of the trial point to its location is unknown to the subject, it is preferable to think of a trial point as being completely specified by its location and height.

In selecting a trial point, a subject specifies the location of the trial point. The location of the point, along with the locations of all previous trial points, determines the mode of the trial point. Therefore, in unconstrained search the mode is not selected first and then the location as the models suggest. In addition, the subject considers more than just the mode of the previous trial points in making his selection. He considers all of the attributes of the previous trial points. These attributes and other possible information are inputs to the subject's search algorithm. The output of the algorithm is the location of the next trial point. This search algorithm is the core of the heuristic search process.

Some insight into the search algorithm can be gained from an investigation of the global trials of a subject. When a subject begins a test run he has no information at all about the test hill. Therefore, his first goal is to obtain as much information as possible as quickly as possible. The best way to do this is with a series of global trials. Although the subject is unlikely to pick the highest point on the hill in a series of global trials, he does gain a considerable amount of information about the general shape of the hill. Using this information he can concentrate on those parts of the hill which are most likely to contain the highest point.

But how does the subject pick the location of these global trial points? In particular, how does he pick the location of the first trial point? It was stated that information provided by previous trial points is the input for the search algorithm. But what if there are no previous trial points and, hence, no input? One possibility is that the subject picks the location of the first trial at random. However, the results of the experiment indicate that the choice of the location of the first trial is not random.

Instead, this point seems to be part of a pattern of global trials. This pattern, which can be called the basic search pattern of the subject, exists in the mind of the subject at the start of the trial run and is an essential part of the search algorithm.

Before describing the basic search pattern, it should be pointed out that the writer does not mean to infer that the basic search pattern exists in the subject's mind as a series of coordinates. Instead, the basic search pattern consists of a number of small, hazily-defined areas of the hill. When a subject selects a point in his basic search pattern, he is not trying to select any one particular point but simply a point that lies within a particular one of these areas. (As a practical matter, even if the subject did want to choose a particular point he would have a hard time doing it since the only reference points he has are the boundaries of the search board.) It should be understood that when reference is made to the points in a basic search pattern the writer is in fact referring to a group of points each one coming from one of the small areas of which the basic search pattern actually consists.

The basic search pattern can be viewed in two ways. First, the pattern consists of those points which the subject feels will give him the most information about the hill in the shortest amount of time (fewest trials). Using this information he can search the parts of the hill most likely to have the highest point. Second, the basic search pattern can be viewed from the standpoint of the termination of the search. The subject will not make a decision concerning the highest point until he has the information provided by all the points in the basic search pattern, regardless of what other information he may have. This means the subject will never terminate the search until all the points of the basic search pattern have been selected.

In a sense, the information provided by the points in the pattern can be considered the minimum amount of information on which a subject is willing to make his decision. In practice, the subject will almost invariably not make his decision until he has much more information than just the information provided by the points in the pattern. Since a series of global trials is unlikely to contain the highest point, the subject will have to select points of L and C mode to find the highest point. The information provided by the points in the pattern can be considered the minimum required for a decision in the sense that it is this information that assures the subject that no other higher point exists on the hill.

A subject will not normally select all of the points in his basic search pattern before he selects any other points. The basic search pattern is simply the sequence of points that would be selected by the search algorithm if it were not provided with any information. Once points are selected, the algorithm is provided with information, and the sequence of points will be altered to take advantage of the information. It is in this manner that while maintaining the same basic search pattern, the search process can adapt to different types of hills.

Using the concept of a basic search pattern, it is possible to theorize on the nature of the search process used by a subject. The first trial point selected by a subject will always be a point in his basic search pattern. Normally, the subject will continue selecting points in the pattern, until the subject feels the information provided by the selected points is sufficient to decide what part of the hill the highest point will be in. The subject will abandon his basic search pattern and choose trial points of L and C modes. Eventually the subject will find a point which he feels on the basis of all the available information is the highest point on the hill. If the subject has already completed his basic search pattern, then he will terminate the search. However, if he has not completed the basic search pattern, the search cannot be terminated. Therefore, the subject will resume selecting points from his basic pattern. If the subject completes his pattern still believing he has found the highest point, then the search is terminated. However it is possible that one of the points selected will shake the subject's belief that he has found the highest point. If so, the subject may again abandon his basic pattern for L and C mode trial points. This process is repeated until the subject finally completes his basic search pattern, at which point the search is terminated.

While the evidence seems to support the existence of a basic search pattern it is far from conclusive. There are two major problems in determining a basic search pattern. First, while the basic search pattern is defined as consisting of all global trials, the distinction between a G-mode trial and an L-mode trial is really quite arbitrary. A subject may have chosen a trial point as part of his basic search pattern, but if it is only a distance of three from the nearest previous trial instead of a distance of four, it won't be included in the pattern. Ideally, the criterion for distinguishing between L and G-mode trials would depend on the size of the hazily-defined areas of which the basic search pattern really exists. If these areas are approximately three by three then, any two points within a distance of three could not be considered as two distinct points in the pattern. In such a case a cutoff distance for three for L-mode trials would be appropriate. However, if the areas were two by two a cutoff distance of two would be correct. Since it is impossible to tell what size the areas are, it must be assumed that they are approximately three by three. This is not an unreasonable assumption since a subject probably could not locate a specific point within a smaller area a high percentage of the time.

Table 1
Global and Total Trials for Different Test Runs

Subject	Test Hill										
	1	2	3	4	5	6	7	8	9	10	11
A			11/30	10/42			9/17	10/37		11/16	
B	6/12										
C		5/31		10/33							
D		10/22				6/12					
E				4/19					4/31		
F			7/20								5/21
G			1/17						7/31		
H					5/12	5/21					
I							9/37				
J							12/38		5/33		7/44

number of global trials/total number of trials

Table 2
Areas of Maximum Relevance for Subject H, Test Hill 6.

Trial Point	Height	Unshared Area of Maximum Relevance	Total Area of Maximum Relevance
1	3	27	36
2	3	9	20
3	3	14	56
4	1	33	49
5	1	36	48
6	1	16	31
7	1	43	54
8	2	44	62
9	3	1	8
10	4	1	8
11	4	1	10
12	4	1	5
13	3	5	37
14	4	1	28
15	4	1	26
16	5	1	30
17	5	1	27
18	5	4	26
19	6	11	19
20	7	5	12
21	7	1	3

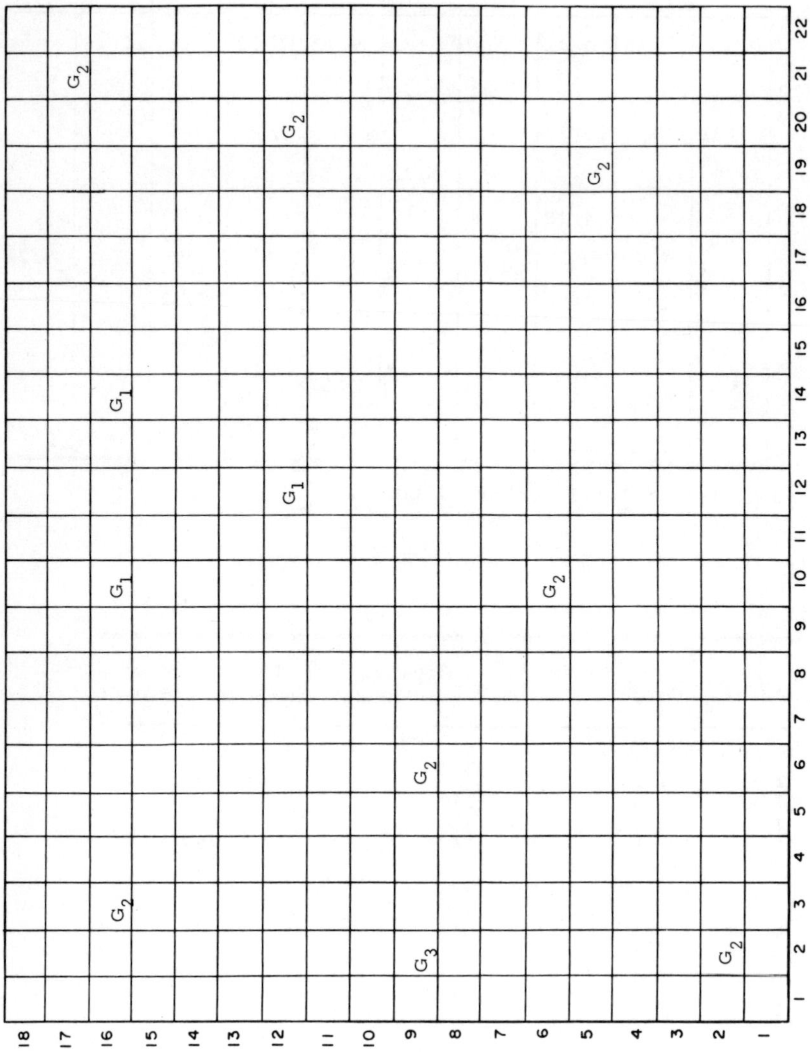

Fig. 2. Global Trials - Subject A, Test Hill 3

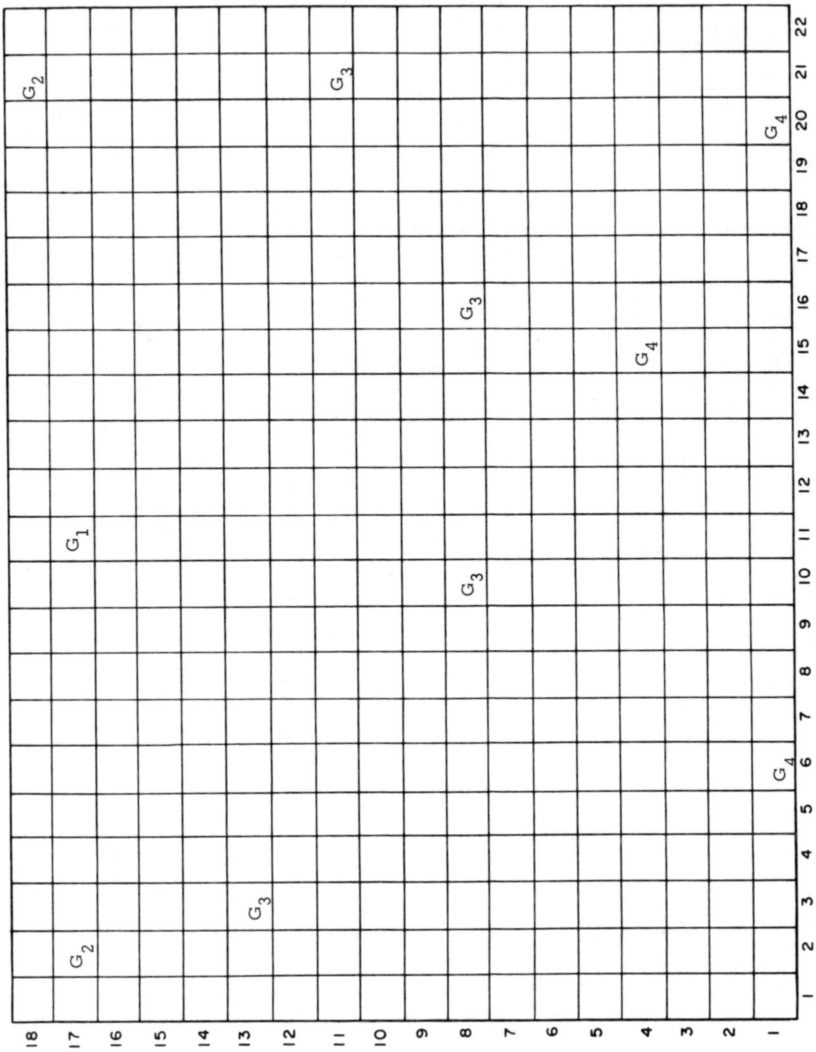

Fig. 3. Global Trials - Subject A, Test Hill 4.

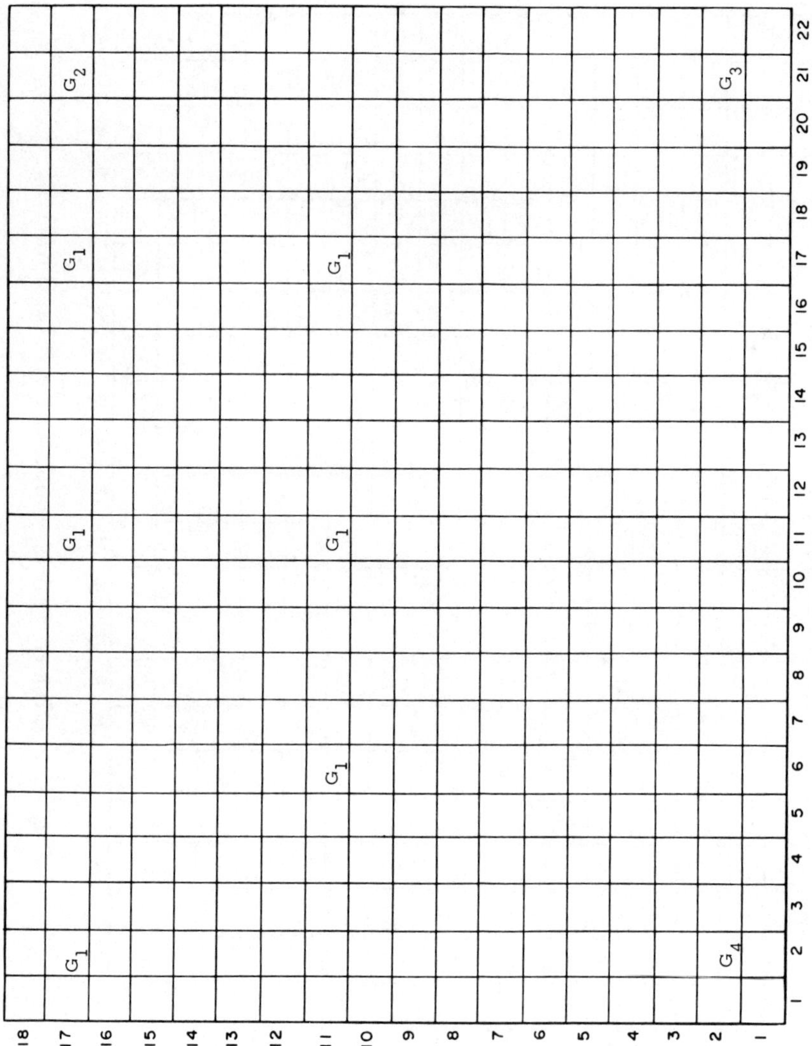

Fig. 4. Global Trials - Subject A, Test Hill 7.

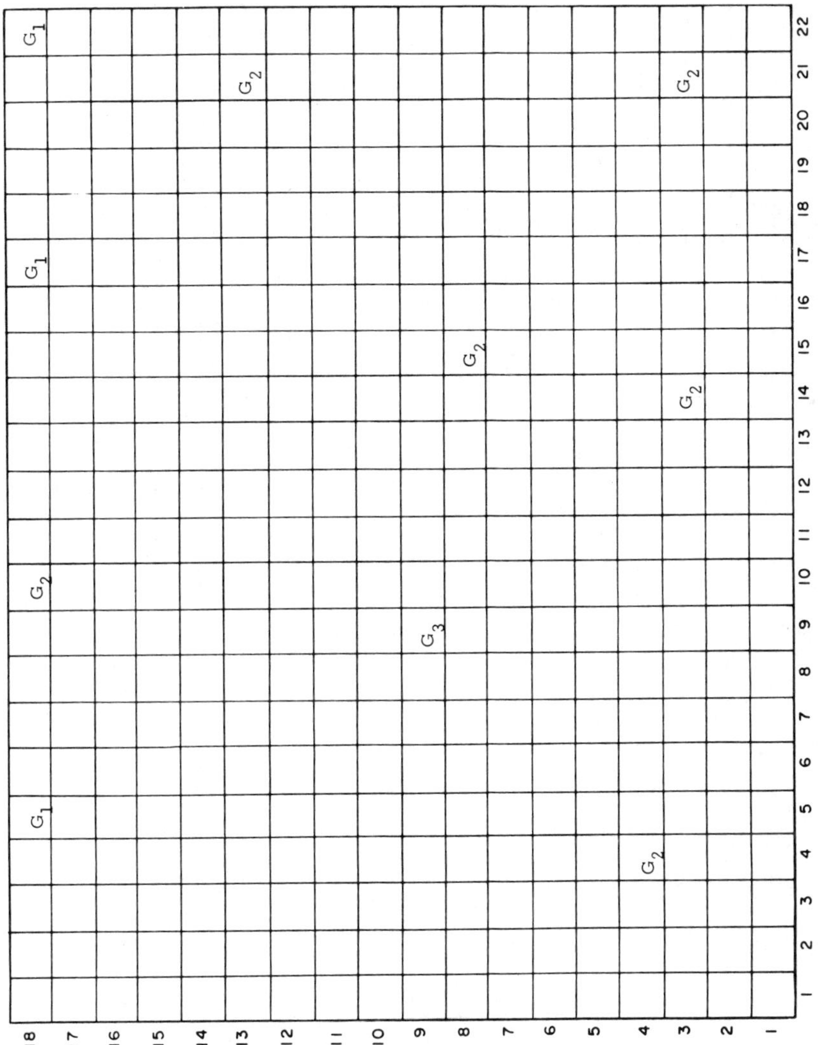

Fig. 5. Global Trials - Subject A, Test Hill 8.

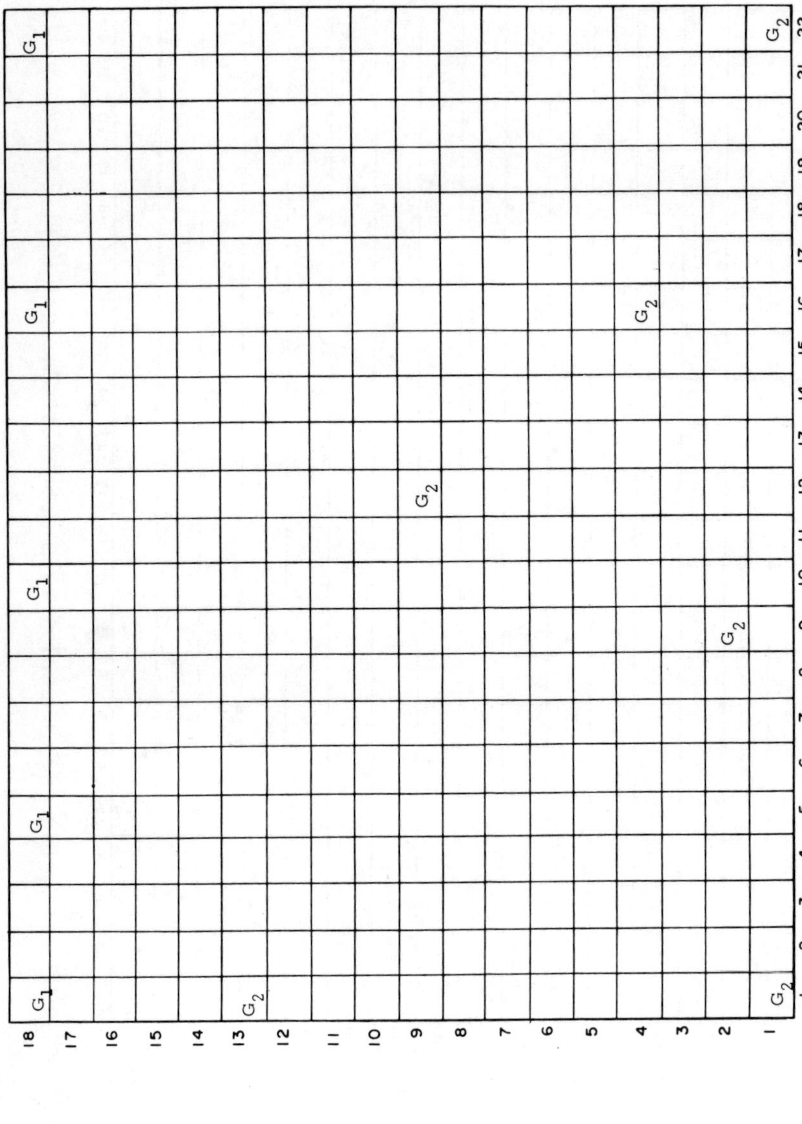

Fig. 6. Global Trials - Subject A, Test Hill 10.

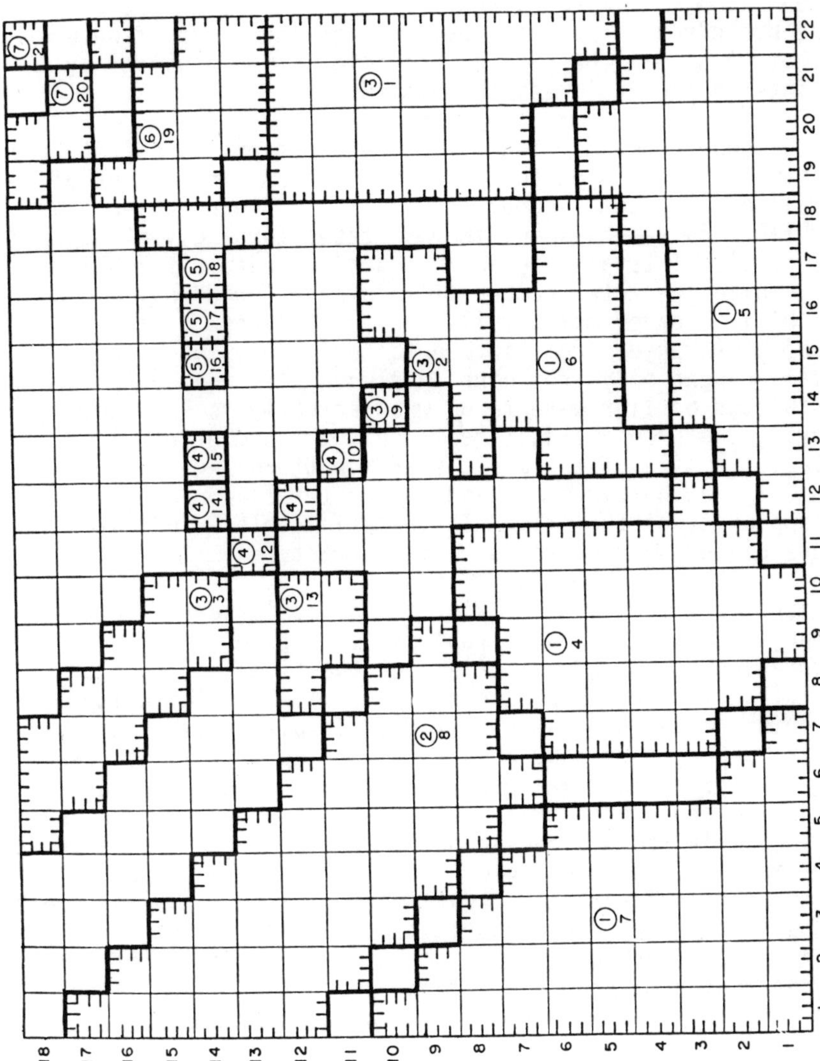

Fig. 7. Unshared Areas of Maximum Relevance for Subject H, Text Hill 6.

The second problem occurs when a subject is using L and C-mode trials to try to find the highest point in a particular area. Through constantly shifting base points an L-mode search may proceed away from the original base point (a point in the basic search pattern) and eventually come within another area in the basic search pattern. In this case the point chosen in the area will be an L-trial instead of a G-trial and, hence, will not be recorded as part of the basic search pattern, even though the information provided by that point will be part of the information provided by all the points in the basic search pattern.

Despite these problems, the evidence does seem to indicate the existence of basic search patterns. Table 1 shows the number of global trials and the total number of trials for each subject and each test hill. As can be seen, in most cases for a given subject the number of global trials was approximately the same for all test hills, while the total number of trials varied widely depending on the hill. For example, while the total number of trials used by subject A ranged from sixteen to forty-two, the number of global trials ranged only from nine to eleven. If the search procedure postulated earlier were to be what actually took place, then, ignoring the problems just mentioned, the number of global trials (i.e., the size of the basic search pattern) would remain the same. Only the number of C- and L-mode trials would change with the complexity of the hill. The data in Table 1 indicates that the postulated search procedure is correct.

Additional evidence of the existence of basic search patterns can be found in the test runs of subject A. Figures 2 through 6 show the location of the global trials of the five test runs of subject A. (The subscripts are mode numbers.) The similarity in the positioning of the points is obvious. In all cases a point was chosen in the upper right-hand corner of the map. In three cases this point was the third trial point, and in the other two cases it was the sixth and eighth trial points. In two cases there were no L or C-mode trial points chosen before this point while in three cases there were. This choosing of the same point in the basic search pattern at different times during the search is in agreement with the postulated search procedure in which the subject may abandon the basic search pattern at any time to conduct L or C-mode searches, but must complete the pattern before the search is terminated.

The positioning of the other subjects' global trials for different test hills does not show as much similarity as did subject A's. This would indicate that the subject changed his basic search pattern between test runs, possibly as a result of learning gained from the previous test runs. However, the number of points in the basic search pattern does not seem to change much, because the number of points in a subject's basic search pattern depends on how much risk the subject is willing to take in choosing the highest point.

A subject can never be sure he has correctly selected the highest point on a test hill unless he has searched every point on the hill. (Actually, because of the step-wise continuity of the criterion function, it would not be necessary to search every point.) Whenever there are some unsearched points, there exists the possibility that the selected highest point will be incorrect.

Since a search will not be terminated until all the points in the basic search pattern have been selected, the number and location of these points determines the maximum relevance existing when the search is terminated. Although the areas actually existing at the time the search is terminated will invariably be smaller than these maximum areas, the maximum areas do give an indication of how much risk the subject is willing to assume in his selection of the highest point. Therefore, the number of points in the basic search pattern will depend to a large extent on how certain the subject wants to be of his selection. For example, subject H had five points in his basic search pattern for both his trial runs. This indicates that subject H is willing to take a greater risk in his selection than is subject A whose basic search pattern always had at least nine points. Figure 7 shows the unshared areas of maximum relevance existing at the termination of a search by subject H, while Table 2 gives the sizes of both the total and unshared areas of maximum relevance. The willingness of subject H to accept a large amount of risk is evidenced by the fact that the largest area of maximum relevance is greater than 15% of the total area of the hill. As might be expected, subject H once failed to correctly identify the highest point, while subject A (in similar experiments) never failed to find the highest point.

CONCLUSION AND RECOMMENDATIONS

This paper has presented the results of a heuristic searching experiment. On the basis of these results the existence of basic search patterns has been postulated and supporting evidence has been presented. It is recommended that further experiments be conducted with the purpose of verifying or refuting the existence of these patterns in various types of heuristic search.

REFERENCES

1. "A Survey of Heuristic Search Method of Multimodal Optimum Point", M. ODA, K. NAKAMURA and B.F. WOMACK, this conference: U.S. - Japan Seminar on Learning Control and Intelligent Control, Oct. 22-26, 1973, Gainesville, Florida.

2. "Experimental Study of Heuristic Search Method for Two-Dimensional Multimodal Optimum Point", M. ODA and B.F. WOMACK, T.R. No. 113, Electronics Research Center, The University of Texas at Austin, Sept. 15, 1971, 80 pp.
3. "Heuristics for Search Techniques", S. HARINASUTA, M.S. thesis, The University of Texas at Austin, Dec., 1971, 89 pp.
4. "Basic Patterns in Heuristic Searches," R.H. GRAHAM, M.S. thesis, The University of Texas at Austin, May, 1972, 72 pp.

MULTI-MODAL SYSTEM IDENTIFICATIONS BY A LEARNING PROCEDURE

Setsuzo Tsuji, Kousuke Kumamaru, Naotoshi Maeda and
Katsuji Tsuruda
Kyushu University

Fukuoka, Japan

I. INTRODUCTION

The multi-modal search is one of the most important and interesting problems in the field of learning control because this might offer the basic guide to solving many scientific or industrial problems accompanying the optimization. Among many approaches to the multi-modal search, [(1)-(4)] the method due to the global function learning is considered as the fundamental one since this can obtain the total aspect of performance ranging over the significant domain, in which the information about local extremal points is also included. The straightforward method of function learning will be the estimation of parameters included in the assumptive function representing that multi-modal system.

In the Seminar on the Learning Process in Control Systems in 1970,[(5)] authors represented that learning identifications are possible for unknown linear systems or multi-modal systems through parameter searching by applying an augmented state estimation based upon a nonlinear filter. In that case, however, we also indicated that the adequate selection of *a priori* values will be the important problem to be solved hereafter because, in order to get a reasonable convergence of estimation, it is necessary that the proper *a priori* values have to be settled at the beginning of that search.

This paper represents a consistent learning procedure for multi-modal systems which can be expressed by Gaussian functions, accompanying the preliminary determination of *a priori* values from a standpoint of global function learning through parameter estimation. Our learning procedure is composed of three successive processes, i.e. preliminary search by a function learning using Kalman filter, parameter a priori value searching by a nonlinear programming and search by a second order filter. In each process, a stochastic estimation is performed due to the noisy observation of the multi-modal function value. The data obtained at the first and second processes are used as the initial a priori values of the second and third processes respectively. By this consistent way, the reasonable a priori covariance can be obtained for a second order filter in our learning procedure in addition to a priori value of estimate. The proper selection of initial covariance value has been the key to secure the desirable convergence of nonlinear estimation, especially, for the multi-dimensional system. Consequently, the convergence of our estimate may be improved remarkably. Numerical example shows the satisfactory results to a two-dimensional, tri-modal system.

II. STATEMENT OF THE PROBLEM

We will consider an unknown gain function to be searched for the system optimization. In most cases, such a gain function may be multi-modal and represented by a linear combination of Gaussian functions, i.e.

$$L(\underline{x},\underline{a},\underline{r},\underline{p}) = \sum_{i=1}^{q} a_i \exp(-\tfrac{1}{2}\|\underline{x}-\underline{r}_i\|^2_{P_i}) \qquad (1)$$

where $\underline{x}$ is the n-dimensional state in the space $\Omega(\underline{x})$, and $\underline{a}$, $\underline{r}$, and $\underline{p}$ are unknown parameters, indicating the shape of each Gaussian function, defined as follows,

$\underline{a} = (a_1 a_2 \dots a_q)'$; qx1 vector

$\underline{r} = (\underline{r}'_1 \underline{r}'_2 \dots \underline{r}'_q)'$; qnx1 vector

$P_i = (p_i^{jk})$; nxn positive definite symmetric matrix

$1 \leqq j,k \leqq n, \quad i=1,2,\dots,q$

$\underline{p}_i = (p_i^{11}\ p_i^{12} \dots p_i^{1n}\ p_i^{22} \dots p_i^{nn})'$; $\frac{n(n+1)}{2}$x1 vector

$\underline{p} = (\underline{p}_1'\ \underline{p}_2' \dots \underline{p}_q')'$; $\frac{qn(n+1)}{2}$x1 vector

Let's assume that the actual gain value corresponding to a state $\underline{x}_k$ at the k-th step is sequentially observed with the additive noise as below,

$$y_k = L(\underline{x}_k,\underline{a},\underline{r},\underline{p}) + v_k \qquad (2)$$

where v_k is a white Gaussian noise with zero mean and known variance R_k.

In order to identify the multi-modal system given by eq.(1), it is necessary to estimate the unknown parameters with higher nonlinearity in the observation.

Now define the augmented state vector $\underline{X}$ composed of the unknown parameters, then the following discrete-time augmented system is obtained from eq. (2),

$$\underline{X}_{k+1} = \underline{X}_k = \underline{X} \tag{3}$$

$$y_k = f(\underline{x}_k, \underline{X}_k) + v_k \tag{4}$$

where $\underline{X} = (\underline{a}'\underline{r}'\underline{p}')'$; Nx1 vector, $N = q(n+1)(\frac{n}{2}+1)$

$$f(\underline{x}_k, \underline{X}_k) = L(\underline{x}_k, \underline{a}, \underline{r}, \underline{p})$$

As a result, the identification problem of the multi-modal system is transformed to the nonlinear parameter estimation problem. Hitherto, many researches concerning nonlinear filter synthesis(5),(6),(11)-(16) have been represented for such nonlinear estimations. Among them,some nonlinear filters, such as a second order filter (5),(6), may be directly applied to estimate the augmented state vector $\underline{X}$, if the adequate a priori values are given which guarantee the convergence of the estimation.

In the general case, however, such satisfactory a priori values are initially unknown. Therefore, it is necessary to develop a consistent learning procedure for the multi-modal system identification, accompanying the pre-determination of the a priori values. This learning procedure is composed of the following three processes: The first process is the global function learning for the approximate determination of parameters, the second one is the maximum likelihood estimation for the improvement of the estimated parameters combining the selection of a priori covariance, and the final process is the second order filtering for the search of exact parameter values. The first two processes play their peculiar roles in the learning of the adequate a priori values for the nonlinear filter. In the first process, a proper approximate value of unknown parameters can be obtained in spite of the initial vague information, but the error covariance of parameters is still obscure. Following it, the second process may give the adequate a priori error covariance for the final process with no initial covariance information besides the improvement of the estimate. As a result, the simple second order filtering is pursued satisfactorily after the previous two processes.

In the following, those learning processes are investigated in detail.

III. LEARNING PROCEDURES

3-1 Preliminary Search by a Function Learning Using Kalman Filter

This process is to obtain the approximate value of $\underline{X}$ by a global function learning using the Kalman filter.[6]

In the first, the unknown multi-modal function $L(\underline{x},\underline{a},\underline{r},\underline{p})$ is approximated by the following linear combination of m Gaussian functions which are uniformly distributed on the state space $\Omega(\underline{x})$,

$$\sum_{i=1}^{m} h_i \exp\left(-\frac{1}{2}\|\underline{x}-\underline{s}_i\|_A^2\right) \tag{5}$$

where $\underline{s}_i$ and A are known parameters and h_i are unknown coefficients $(i=1,2,\ldots,m)$. In order to estimate h_i, we set up the following discrete-time system for the augmented parameter by using eq. (2),

$$\underline{h}_{k+1} = \underline{h}_k = \underline{h} \tag{6}$$

$$y_k = H_k \underline{h}_k + v_k + e_k \tag{7}$$

where k indicates the observation step, and

$$\underline{h} = (h_1\ h_2\ \ldots\ h_m)'$$

$$H_k^i = \exp\left(-\frac{1}{2}\|\underline{x}_k-\underline{s}_i\|_A^2\right)$$

$$H_k = (H_k^1\ H_k^2\ \ldots\ H_k^m)$$

In the above, e_k is regarded as an equivalent white Gaussian noise with zero mean and variance E_k, corresponding to the approximation error in formula (5). By applying the Kalman filter to the above system, the optimal estimate $\hat{\underline{h}}_{k/k}$ and estimation error covariance $S_{k/k}$ of the unknown coefficients $\underline{h}_k$ are obtained as follows,

$$\underline{\hat{h}}_{k/k} = \underline{\hat{h}}_{k-1/k-1} + G_k(y_k - H_k\underline{\hat{h}}_{k-1/k-1}) \tag{8}$$

$$G_k = S_{k-1/k-1}H_k'(H_kS_{k-1/k-1}H_k' + R_k + E_k)^{-1} \tag{9}$$

$$S_{k/k} = S_{k-1/k-1} - G_kH_kS_{k-1/k-1} \tag{10}$$

This estimation process is a linear one and may converge for the ordinary selection of an observation state because it is easy to obtain the observability condition.

Next, the approximate value of $\underline{X}$, that is $\bar{\underline{X}}$, can be obtained by aggregating m Gaussian functions with estimated coefficients into q Gaussian functions in the form of eq. (1) (see Appendix I).

3-2 Parameter *A Priori* Value Searching by a Nonlinear Programming

In this process, we will calculate the *a priori* values for the nonlinear estimation of $\underline{X}$ by using Cox's maximum likelihood estimation method(7) with nonlinear programming (NLP).

Let $p(\underline{X}/Y^k)$ be a conditional probability density function of $\underline{X}$ on $Y^k=\{y_1,y_2,\ldots,y_k\}$. By Bayes' rule and mutual independence of measurement noises $\{v_k\}$, $p(\underline{X}/Y^k)$ may be rewritten as

$$p(\underline{X}/Y^k) = p(y_1/\underline{X})p(y_2/\underline{X}) \ldots p(y_k/\underline{X})p(\underline{X})/p(Y^k) \tag{11}$$

By considering the large uncertainty of the initial information about $\underline{X}$, eq. (11) becomes from eq. (4),

$$p(\underline{X}/Y^k) = \text{Const.}\exp(-\frac{1}{2}\sum_{i=1}^{k} \|y_i - f(\underline{x}_i,\underline{X})\|^2_{R_i^{-1}}) \tag{12}$$

And we define the cost function $J(\underline{X})$ for the Cox's maximum likelihood estimation by

$$J(\underline{X}) = \sum_{i=1}^{k} \|y_i - f(\underline{x}_i,\underline{X})\|^2_{R_i^{-1}} \tag{13}$$

Suppose that $J(\underline{X})$ is approximated by the second order Taylor expansion about $\underline{X}^*$ which minimizes $J(\underline{X})$, as follows,

$$J(\underline{X}) = J(\underline{X}^*) + \frac{1}{2}\|\underline{X}-\underline{X}^*\|^2_{J_{XX}(\underline{X}^*)} \tag{14}$$

where $J_{xx}(\underline{X}^*)^{ij} = \left.\frac{\partial^2 J(\underline{X})}{\partial X^i \partial X^j}\right|_{\underline{X} = \underline{X}^*}$, $1 \leq i,j \leq N$

Substituting eqs. (13),(14) into eq. (12) yields

$$p(\underline{X}/Y^k) = (2\pi)^{-N/2}(\det C^*)^{-1/2} \quad (-\frac{1}{2}\|\underline{X}-\underline{X}^*\|^2_{C^{*-1}}) \qquad (15)$$

where $C^* = (\frac{1}{2}J_{xx}(\underline{X}^*))^{-1}$

From eq. (15), we may regard approximately the distribution of $\underline{X}$ as the Gaussian distribution with mean $\underline{X}^*$ and covariance matrix C^*. NLP may be applied to estimate $\underline{X}^*$ and C^* by the following way. By minimizing $J(\underline{X})$ with Fletcher-Powell's method,[8] the estimate of $\underline{X}^*$ can be obtained in the recurrence formula using $\bar{\underline{X}}$ searched in the previous process as the initial value, and at the same time the approximate value of $J_{xx}(\underline{X}^*)^{-1}$ is attained as the converged result of H_i (see Appendix II). Thus C^* in eq. (15) can be determined without any covariance information.

When there is a constraint of the positive definite on P_i that is a matrix composed of the partial elements of $\underline{X}$, we should use the following modified cost function by SUMT,[9],[10] instead of $J(\underline{X})$,

$$I(\underline{X},c_j) = J(\underline{X}) + c_j \sum_{i=1}^{q} \sum_{t=1}^{n} \frac{1}{\det P_i(t)} \qquad (16)$$

where $c_j > 0$ for any j, $c_1 > c_2 > c_3 > \ldots,$ $c_j \to 0$ as $j \to \infty$,

and $\det P_i(t)$ is a principal minor of order t in P_i.

We will use the estimate of $\underline{X}^*$ and the diagonal matrix composed of diagonal elements of the estimated C^* as the a priori values $\hat{\underline{X}}_{0/0}$ and $C_{0/0}$ respectively for the next filtering process.

3-3 Search by Second Order Filter

In this final process, a second order filter is applied to search the exact parameter vector using the a priori estimate and the covariance obtained in the previous process. The estimate

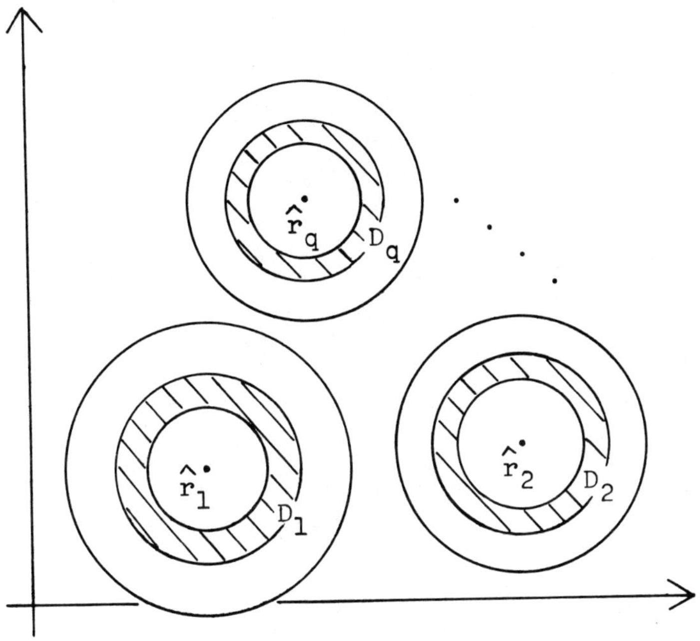

Fig. 1. Observation states to be selected for estimation.

($\hat{X}_{./.}$) and its error covariance ($C_{./.}$) are derived from eqs. (3) and (4) as follows,

$$\hat{\underline{X}}_{k+1/k} = \hat{\underline{X}}_{k/k} \quad (17)$$

$$C_{k+1/k} = C_{k/k} \quad (18)$$

$$\hat{\underline{X}}_{k+1/k+1} = \hat{\underline{X}}_{k+1/k} + K_{k+1}(y_{k+1} - f(\underline{x}_{k+1}, \hat{\underline{X}}_{k+1/k}) - \pi_{k+1}) \quad (19)$$

$$C_{k+1/k+1} = C_{k+1/k} - K_{k+1}\underline{f}_X(\underline{x}_{k+1}, \hat{\underline{X}}_{k+1/k})C_{k+1/k} \quad (20)$$

$$K_{k+1} = C_{k+1/k}\underline{f}_X'(\underline{x}_{k+1}, \hat{\underline{X}}_{k+1/k})(\underline{f}_X(\underline{x}_{k+1}, \hat{\underline{X}}_{k+1/k})C_{k+1/k} \cdot \underline{f}_X'(\underline{x}_{k+1}, \hat{\underline{X}}_{k+1/k}) + L_{k+1} + R_{k+1})^{-1} \quad (21)$$

$$L_{k+1} = \frac{1}{2}tr(f_{XX}(\underline{x}_{k+1}, \hat{\underline{X}}_{k+1/k})C_{k+1/k}f_{XX}(\underline{x}_{k+1}, \hat{\underline{X}}_{k+1/k}) \cdot C_{k+1/k}) \quad (22)$$

$$\pi_{k+1} = \frac{1}{2}tr(f_{XX}(\underline{x}_{k+1}, \hat{\underline{X}}_{k+1/k})C_{k+1/k}) \quad (23)$$

where $\underline{f}_X(\underline{x}_{k+1}, \hat{\underline{X}}_{k+1/k})^j = \left.\frac{\partial f(\underline{x}, \underline{X})}{\partial X^j}\right|_{\underline{x}=\underline{x}_{k+1}, \underline{X}=\hat{\underline{X}}_{k+1/k}}$

$$f_{XX}(\underline{x}_{k+1}, \hat{\underline{X}}_{k+1/k})^{ij} = \left.\frac{\partial^2 f(\underline{x}, \underline{X})}{\partial X^i \partial X^j}\right|_{\underline{x}=\underline{x}_{k+1}, \underline{X}=\hat{\underline{X}}_{k+1/k}}$$

This estimate of $\underline{X}$ by the second order filter may probably converge to its true value, if the estimate and covariance obtained through the previous processes are used as a priori values, which seems to have been sufficiently improved via preliminary and medium processes from the initial period with vague information.

Moreover, this process may be pursued more effectively if the following observation state selections are adopted in each estimation stage. At first, in the pre-determined multi-modal function, we consider the each modal point $\hat{\underline{r}}_i$ and proper region D_i which locates in the neighborhood of the state with the steepest gradient as shown in Fig. 1. Up to the first q-th estimation stage, the searching point is selected sequentially in each region D_i, so as to minimize the trace of the estimation error covariance. After that, in the next q stages, each modal points $\hat{r}_i (i=1,2,\ldots,q)$ are taken successively as the searching points. Then, these state selections are repeated alternately with the reconstruction of multi-modal function according to the improvement of the estimation.

Thus, the unknown multi-modal system has been identified by the consistent function learning processes. As a result, this means that the global extrema search of the multi-modal system has been performed.

IV. NUMERICAL EXAMPLE

As a numerical example, let us consider the two-dimensional, tri-modal system identification in the space $\Omega(\underline{x})$:

$$\Omega(\underline{x}) = (\underline{x} | \underline{x} \in R^2 \; ; \quad 0 \leqq x^1 \leqq 10, \quad 0 \leqq x^2 \leqq 10)$$

The true values of unknown parameters are given in Table 1, and the real aspect of this system is shown in Fig. 2.

4-1 First Process

By considering the order of unevenness of the real object, we take in formula (5) as follows,

$$m = 25, \qquad ((s_i^1, s_i^2);\ s_i^1, s_i^2 = 1,3,5,7,9)$$

$$A = \begin{bmatrix} 1.39 & 0 \\ 0 & 1.39 \end{bmatrix}$$

Such a value of A will make the shape of formula (5) almost flat in $\Omega(\underline{x})$ (see Fig. 3). By Kalman filter, the estimate results of h_i are obtained using 25 times' observations taken from each lattice point in the $\Omega(\underline{x})$, with arbitrary <u>a priori</u> values such as

$$\hat{\underline{h}}_{0/0} = (0.5\ 0.5\ \ldots\ 0.5)'$$

$$S_{0/0} = \text{dia. } (1,1,\ldots,1)$$

$$E_k = 0.032$$

Table 1. Parameter estimation result in each process.

mode		true	$\bar{X}$	$\hat{X}_{0/0}$	$diaC_{0/0}$	$\hat{X}_{15/15}$
1	r_1^1	7.00	7.07	6.88	0.050	6.92
	r_1^2	3.00	2.91	3.10	0.019	3.10
	a_1	1.00	0.74	0.96	0.001	1.00
	P_1^{11}	1.00	0.81	0.88	0.003	0.89
	P_1^{12}	0.0	-0.01	-0.09	0.004	-0.10
	P_1^{22}	0.50	0.47	0.48	0.005	0.48
2	r_2^1	3.00	2.95	2.86	0.073	2.89
	r_2^2	2.00	2.34	2.10	0.010	2.10
	a_2	1.00	0.60	1.01	0.064	1.00
	P_2^{11}	1.00	0.80	1.76	0.638	1.16
	P_2^{12}	0.0	0.06	0.15	0.301	0.15
	P_2^{22}	1.00	0.43	0.85	0.550	1.12
3	r_3^1	5.00	5.21	4.99	0.066	5.02
	r_3^2	6.00	5.92	6.03	0.027	6.04
	a_3	1.00	0.66	1.58	1.989	1.03
	P_3^{11}	0.50	0.45	0.58	0.004	0.58
	P_3^{12}	0.0	0.03	-0.08	0.016	-0.07
	P_3^{22}	1.00	0.52	1.87	1.706	1.04

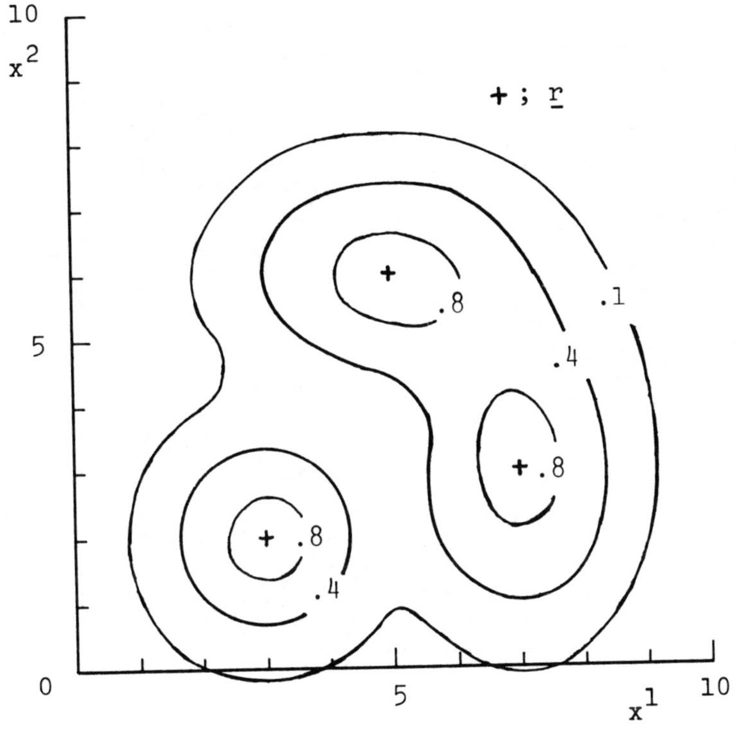

Fig. 2. Real aspect of an object.

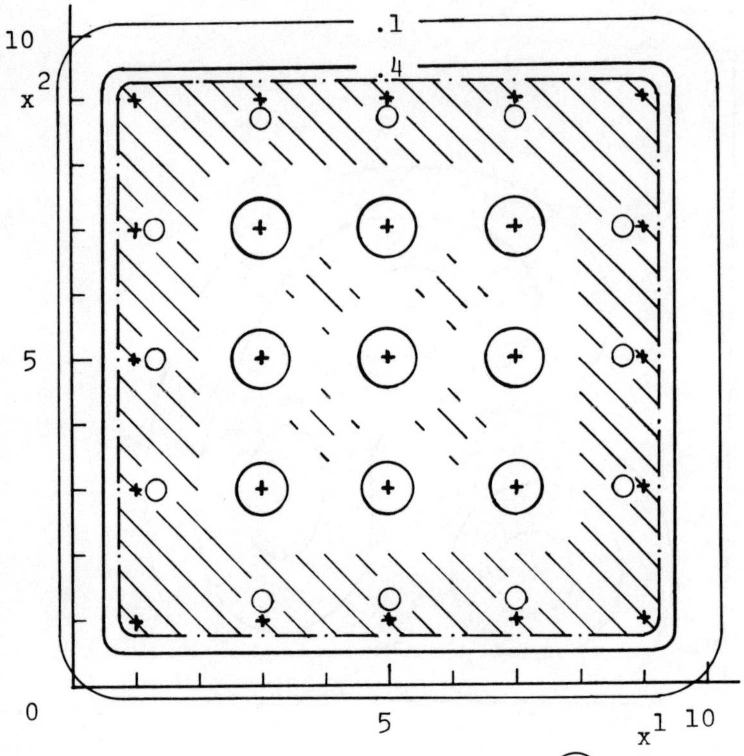

+ ; $\underline{s}_i$, ; .5 , ○ , ◯ ; .6

Fig. 3. Initial aspect in the first process.

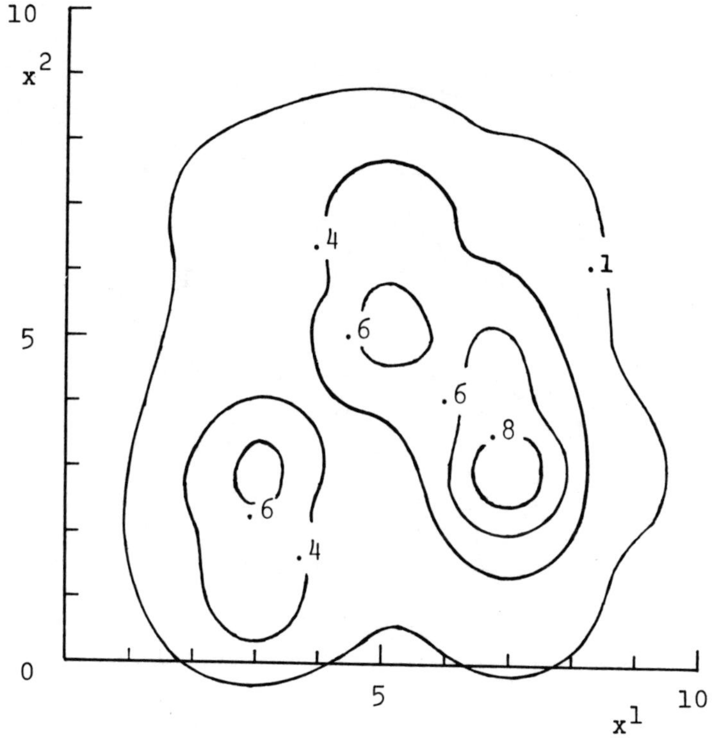

Fig. 4. Learning result by multi-gaussian functions.

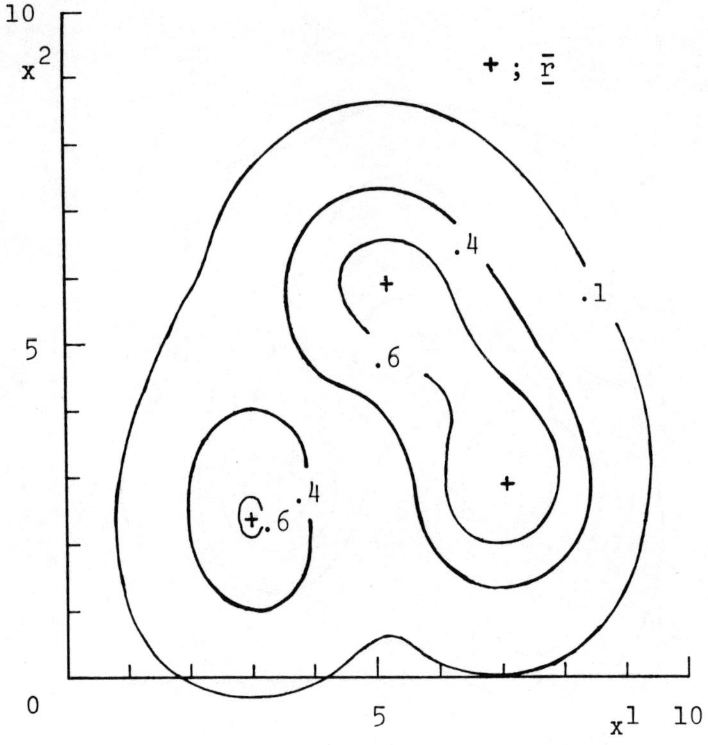

Fig. 5. Aggregated aspect in the first process.

This preliminary estimation gives a roughly learned result as shown in Fig. 4. And these 25 estimated Gaussian functions are aggregated into the tri-modal function (see Fig.5), of which parameter value $\underline{\bar{X}}$ is obtained as given in Table 1.

4-2 Second Process

The a priori values $\underline{\hat{X}}_{0/0}$, $C_{0/0}$ for the final process are calculated by NLP as shown in Table 1 using 25 renewed observation data which are taken from the same lattice points as in the previous process. In the NLP calculation with SUMT, main items are as follows,

$\underline{X}_0 = \underline{\bar{X}}$, $H_0 = I$ (NxN unit matrix);(initial values for minimization in SUMT)

$c_1 = 10^{-5}$, $c_2 = 10^{-10}$, $\varepsilon = 10^{-4}$

In the above calculation, the search on c_2 is divided into two portions. The first is the usual one with $\varepsilon=10^{-3}$ and the second is to seek $\underline{X}^*$ and C^* more finely with $\varepsilon=10^{-4}$. And a second order interpolation method is used for the search of α_i which requires about 80% of NLP computing time.

4-3 Third Process

The satisfactory estimation of unknown parameters is performed by the second order filter in 15 steps with a priori values obtained in the second process. One of the estimation processes is shown in Fig. 7 and other results are given in Table 1. Thus the tri-modal unknown object has been globally learned well enough as shown in Fig. 6.

The total computing time required through these learning processes is 193 seconds (FACOM 230-60), i.e. 27 seconds in the first process, 55 seconds in the second process and 111 seconds in the final process.

V. CONCLUSIONS

Regarding the multi-modal searching as the estimation problem of parameters by a nonlinear filter, it has been shown that, initiating the preliminary search, the proper a priori values for the filter can be obtained through a maximum likelihood estimation. This means that the actual calculation for global searching may

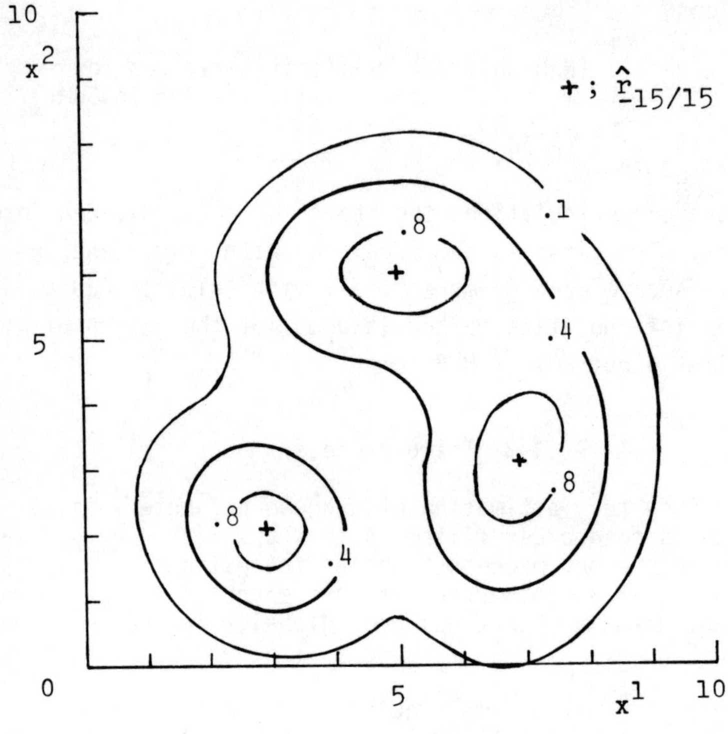

Fig. 6. Final learning result

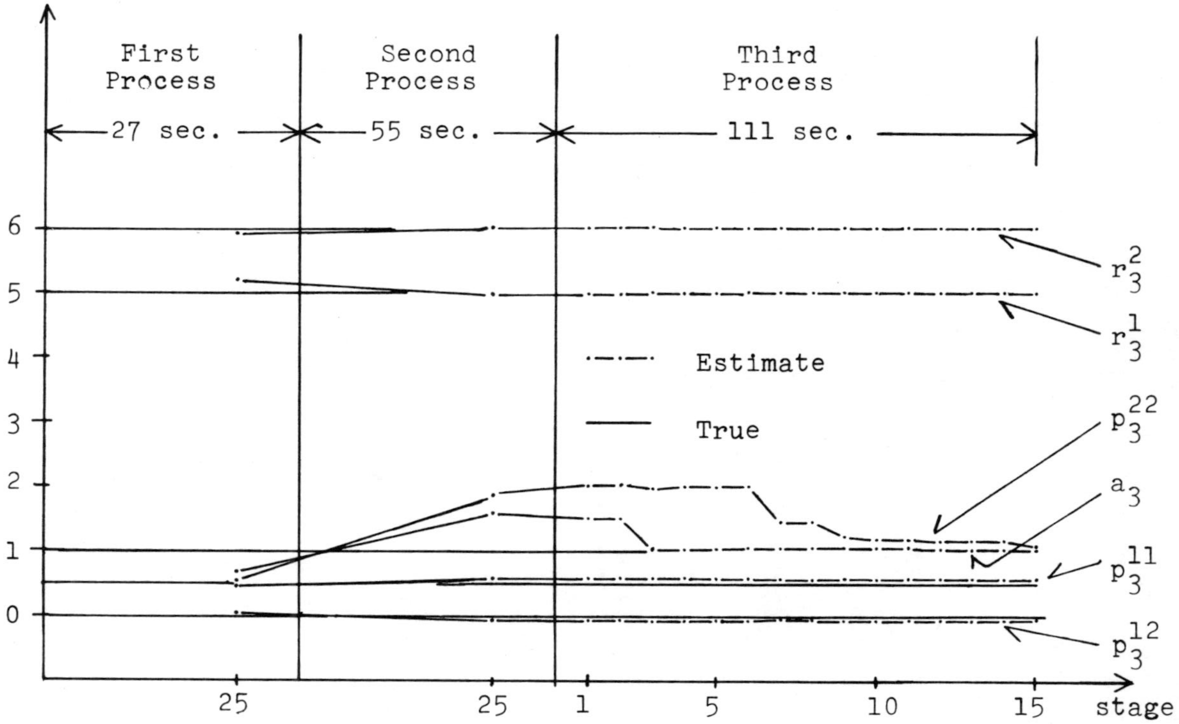

Fig. 7. Estimation process of parameters in mode 3. ($R_k = 10^{-3}$)

be pursued systematically by using our consistent procedure. The stochastic information is easily available to each process since it is based upon the stochastic estimation.

Depending upon the level of confidence about the accuracy of given a priori values, we may take any process in our learning procedure as the initial process from which the estimation will begin. However, when the outline of object is so vague that it is difficult to select a proper initial value, these three processes are effectively applied in succession because these processes have their own special features for our parameter estimation.

When the gradient values of performance are extra observed besides the only value of performance, it is supposed that the learning identification will be more improved by the increase of available information.

APPENDIX

I. Aggregation of Gaussian Functions

We consider the aggregation of a linear combination of M Gaussian functions (A1) into one Gaussian function (A2),

$$\sum_{i=1}^{M} h_i \exp(-\tfrac{1}{2} \|\underline{x}-\underline{s}_i\|_A^2) \tag{A1}$$

$$h^* \exp(-\tfrac{1}{2} \|\underline{x}-\underline{s}^*\|_{A^*}^2) \tag{A2}$$

where $\underline{x}$ is a nx1 vector, A is a nxn positive definite symmetric matrix, and h_i are nonnegative coefficients (i=1,2,...,M).

Normalizing formulae (A1) and (A2) by multiplying the factor $\frac{(\det A)^{1/2}}{(2\pi)^{n/2} h_s}$ and $\frac{(\det A^*)^{1/2}}{(2\pi)^{n/2} h^*}$ respectively yields

$$p(\underline{x}) = \sum_{i=1}^{M} \frac{h_i}{h_s} (2\pi)^{-n/2} (\det A)^{1/2} \exp(-\tfrac{1}{2}\|\underline{x}-\underline{s}_i\|_A^2) \tag{A3}$$

$$p^*(\underline{x}) = (2\pi)^{-n/2} (\det A^*)^{1/2} \exp(-\tfrac{1}{2}\|\underline{x}-\underline{s}^*\|_{A^*}^2) \tag{A4}$$

where $h_s = \sum_{i=1}^{M} h_i$

By regarding $p(\underline{x})$ as a probability density function of $\underline{x}$ equivalent to $p^*(\underline{x})$ which is in Gaussian form, we obtain $\underline{s}^*$, A^* and h^* as follows,

$$\underline{s}^* = \frac{1}{h_s} \sum_{i=1}^{M} h_i \underline{s}_i \tag{A5}$$

$$A^* = (A^{-1} - \underline{s}^* \underline{s}^{*'} + \frac{1}{h_s} \sum_{i=1}^{M} h_i \underline{s}_i \underline{s}_i')^{-1} \tag{A6}$$

$$h^* = h_s (\det A)^{-1/2} (\det A^*)^{1/2} \tag{A7}$$

II. Fletcher-Powell's Minimization Method

If the cost function $J(\underline{X})$ is twice continuously differentiable and strictly convex, $\underline{X}^*$ which minimizes $J(\underline{X})$ can be obtained by the following algorithm,

1 $\underline{s}_i = -H_i \underline{g}_i$

2 find $\alpha_i (>0)$ which minimizes $J(\underline{X}_i + \alpha_i \underline{s}_i)$

3 $\underline{\sigma}_i = \alpha_i \underline{s}_i$, $\underline{X}_{i+1} = \underline{X}_i + \underline{\sigma}_i$, $\underline{y}_i = \underline{g}_{i+1} - \underline{g}_i$

4 $H_{i+1} = H_i + \underline{\sigma}_i \underline{\sigma}_i' / \underline{\sigma}_i' \underline{y}_i - H_i \underline{y}_i \underline{y}_i' H_i / \underline{y}_i' H_i \underline{y}_i$

where $i=0,1,2,\ldots$, $\underline{X}_0$ (Nx1 vector), the initial estimate of $\underline{X}^*$, must be given and H_0 (NxN matrix) is set to unit matrix in general and $\underline{g}_i = \underline{J}_X'(\underline{X}_i)$ i.e. Jacobian of $J(\underline{X}_i)$.

In the above algorithm, as i tends to infinity, the sequences $\{\underline{X}_i\}$ and $\{H_i\}$ converge to $\underline{X}^*$ and an approximated value of $(J_{XX}(\underline{X}^*))^{-1}$ respectively. In the practical application, we stop the calculation at the proper step such that the condition $\|\underline{s}_i\| \leq \varepsilon$ is satisfied for a given small positive real number ε.

REFERENCES

(1) K. Nakamura, M. Oda: "Heuristic and Learning Control", Proceedings of the Japan-U.S. Seminar on Learning Process in Control Systems, Pattern Recognition and Machine Learning, Plenum Pub. Co. (1971).

(2) K. Asai, S. Kitajima: "Learning Control of Multimodal Systems by Fuzzy Automata", ditto.

(3) N. Okino, Y. Sakai: "A Method for Separating Every Peak from Multimodal Function Using Monte Carlo Integration", Trans. of the Society of Instrument and Control Engineers, Vol. 8, No. 6, (1972).

(4) K. Kubota: "A New Maximum Searching Method for Unknown One-dimensional Multimodal Function", Trans. of the Society of Instrument and Control Engineers, Vol. 9, No. 4, (1973).

(5) S. Tsuji, K. Kumamaru: "System Identifications by a Nonlinear Filter", Proceedings of the Japan-U.S. Seminar on Learning Process in Control Systems, Pattern Recognition and Machine Learning, Plenum Pub. Co. (1971).

(6) S. Tsuji, K. Kumamaru: "System Identifications and Function Learnings by a Nonlinear Filter", Memoires of the Faculty of Engineering, Kyushu University, Vol. 30, No. 4, (1971).

(7) H. Cox: "On the Estimation of State Variables and Parameters for Noisy Dynamic Systems", IEEE Trans. on AC, Vol. AC-9, No. 1, (1964).

(8) R. Fletcher, M.J.D. Powell: "A Rapidly Convergent Method for Minimization", Computer J., Vol. 6, No. 2, (1964).

(9) A.V. Fiacco, G.P. McCormic: "The Sequential Unconstrained Minimization Technique for Nonlinear Programming", Management Science, Vol. 10, No. 2, (1964).

(10) J. Kowalik, M.R. Osbone: "Methods for Unconstrained Optimization Problems", American Elserver Pub. Co. (1968).

(11) H.J. Kushner: "On the Differential Equations Satisfied by Conditional Probability Densities of Markov Processes", SIAM J. Control, Series A, Vol. 2, No. 1, (1967).

(12) H.J. Kushner: "Dynamical Equations for Optimal Nonlinear Filtering", J. of Differential Equations, Vol. 3, (1967).

(13) H.J. Kushner: "Approximations to Optimal Nonlinear Filters", IEEE Trans. on AC, Vol. AC-12, No. 5, (1967).

(14) D.L. Alspac, H.W. Sorenson: "Nonlinear Bayesian Estimation Using Gaussian Sum Approximation", IEEE Trans. on AC, Vol. AC-17, No. 4, (1972).

(15) S. Tsuji, H. Takata: "The Discrete Nonlinear Estimation Theory by Using the Orthogonal Projection Method", Memoires of the Faculty of Engineering, Kyushu University, Vol. 22, No. 1, (1972).

(16) S. Tsuji, M. Nakamura: "Nonlinear Filter Using a Statistical Processing Technique", Trans. of the Institute of Electrical Engineers of Japan, Vol. 93-C, No. 5, (1973).

(17) S. Tsuji, N. Maeda: "On the Extrema Search of Unknown Multimodal Function by Nonlinear Filter", Technology Reports of the Kyushu University, Vol. 46, No. 4, (1973).

LEARNING DUAL CONTROL UNDER COMPLETE STATE INFORMATION

Kahei Nakamura and Yoshimasa Yoshida

Nagoya University, Nagoya, Japan

Nagoya Institute of Technology, Nagoya, Japan

1. INTRODUCTION

This paper concerns a unified study on learning dual control of constant but unknown parameter systems subject to random noise.

The learning dual control strategy is a real-time convergent control process which executes simultaneously the two different functions of identification and control in one step of a learning process, and is principally constructed by two parts. One part is an identifier-estimator which generates plant parameter estimates and their associated covariances, as well as plant state estimates and their associated covariances from observed data. The other part is an adaptive controller which involves a feedback controller generating control from state estimates and a self-organizing higher level controller renewing the feedback control law at each step where the plant parameter estimates and their covariances are obtained.

The functions of identification, estimation and control in the dual control is essentially interrelated. Therefore, the general solution of the dual control problem is a very difficult problem required to solve nonlinear equations having a product of two unknown random variables. The realistic clue to solving the dual control lies in how to realize approximately the separation of those functions to simplify the problem. The stochastic approximation approach for dual control developed by Y.Z. Tsypkin[6)7)8)] yields a scheme which is available even for the case of incomplete state information. However the learning rate of this approach is very slow, and it is unavoidable for the intolerably poor operation to appear in the early state[9)]. This paper intends to realize the separation under the following assumptions:

a) Assumption of uncorrelated residual error: - The linearity condition on identification is assured at each step of the learning process by the assumption that the prediction error which is the difference between the actual plant output and the estimate of plant output predicted by using past measurements is represented as an uncorrelated random process.

b) Assumption of complete state information: - The linearity condition on control is assured under the assumption that the state of the system is completely computed from past measurements.

Under the above two assumptions, the existence of a linearity condition for dual control is secured, and hence the separation becomes possible to be realized. The first assumption is satisfied by setting an uncorrelation property between the control (u) and the plant disturbance (e) in the input-output difference equation. Then, the least mean squares method can be adopted as the identification procedure. The second assumption makes it possible to eliminate the state estimator and the observation noise. Therefore, the transferability between the state equation model and the difference equation model is really feasible.

The objection of this paper is, in sum, to make the optimal algorithm of stochastic dual control scheme for the plant of special structure with unknown parameters under the above two conditions. This problem is principally solved by constructing a convergent successive process which simultaneously executes the identification by least mean squares method and the control determination by dynamic programming method with the criterion of the ensemble average of a quadratic function of a state variable and a control variable. By virtue of the two assumptions, the controller finally becomes a linear feedback structure as the conventional stochastic control problem. The control law of which is renewed adaptively by the higher-level self-organizing controller.

The features of the author's scheme for dual control comparing with the existing approaches of References [1] $\sim$ [5] is further explained here. The introduction of both the parameter estimate and its covariance is studied in [1], [4] and others. Horowits's work[4] is not consistent of the fact that the covariance of parameter estimates decreases gradually with the progress of identification. This seems to originate from the use of a fixed probability distribution function specified a priori. As for the optimization scheme, References [1] and [4] consider only an immediate one-step optimization but not the optimization over the full step process. Although M. Mendes considered the determination of optimum control by a dynamic programming method, he introduces only the parameter estimate but not its covariance. The work of J. Wieslander[1] is distinguishable by the consideration of both the parameter estimate and its covariance, although he does not adopt the full-step criterion. K. J. Åstrom[3] points out that the dynamic optimization with full-step criterion becomes a very difficult nonlinear problem because it needs to consider both the state estimate and the parameter estimate with its covariance for all the future steps.

A remarkable feature of this paper is that in spite of the introduction of the parameter estimate with its covariance for the optimization by dynamic programming method, the difficulty pointed out by Åstrom has been avoided by the careful consideration of complete state information for the future steps.

2. STATEMENT OF THE PROBLEM

2.1 Plant Equation

As the state equation model of the plant is considered here a stochastic, discrete, linear, time-invariant and single-input/single-output system with system noise but no observation noise, as the following:

$$x(t+1) = \Phi x(t) + b\,u(t) + \xi(t)\,, \tag{1}$$

$$y(t) = c^T x(t)\,,$$

$$t \in \{\ldots, -1, 0, 1, 2, \ldots\}\,, \tag{2}$$

where $x(t) \in R^n$, $u(t)$ is a plant scalar input, $\xi(t)$ is a equally distributed normal random vector variable independent of $x(t)$, $\xi(t) \in R^n$, $E[\xi(t)] = 0$, $\mathrm{cov}[\xi(t), \xi(s)] = \Xi\,\delta(t-s)$, $\Xi = \mathrm{diag}[\sigma_1^2, \sigma_2^2, \ldots \sigma_n^2]$, $\mathrm{tr}\,\Xi$ is known, $\delta(t-s)$ is the Kronecker delta function, and $y(t)$ is defined as a scalar observation. The initial value $x(0)$ is assumed a Gaussian random vector with mean m and covariance matrix $P(0)$, $P(0)$ is non-negative definite.

$$\Phi = \begin{bmatrix} -a_1 & 1 & 0 & \cdots & 0 \\ -a_2 & 0 & 1 & \cdots & 0 \\ \vdots & & & & \vdots \\ -a_{n-1} & 0 & 0 & \cdots & 1 \\ -a_n & 0 & 0 & \cdots & 0 \end{bmatrix},$$

$$a^T = (a_1, a_2, \ldots\ldots, a_n),$$
$$b^T = (b_1, b_2, \ldots\ldots, b_n),$$
$$c^T = (1, 0, \ldots\ldots, 0),$$
$$\xi^T = (\xi_1(t), \xi_2(t), \ldots\ldots, \xi_n(t)). \qquad (3)$$

$a_i, b_i;\ i=1, \ldots, n$ are unknown constants. Note that under the above assumption $x(t)$ and $y(t)$ will be Gaussian for all t.

It is not difficult to verify that (1) has a corresponding simple difference equation model. For simplicity of representation the following notation $e(t)$ can be adopted to represent the driving noise of the difference equation.

$$e(t) \stackrel{\Delta}{=} \xi_1(t-1) + \xi_2(t-2) + \ldots + \xi_n(t-n),$$

where $e(t) \in R^1$, $E[e(t)] = 0$, $var[e(t), e(s)]$

$$= tr\ \Xi\delta\ (t-s). \qquad (4)$$

Then we yield the equivalent difference equation.

$$y(t) + a_1 y(t-1) + \ldots + a_n y(t-n)$$
$$= b_1 u(t-1) + \ldots + b_n u(t-n) + e(t)\ . \qquad (5)$$

Let us adopt the following notation.

$$g^T(t-1) \stackrel{\Delta}{=} [-y(t-1), \ldots, -y(t-n), u(t-1), \ldots, u(t-n)]$$

$$\theta^T = (a_1, \ldots, a_n, b_1, \ldots, b_n) \qquad (6)$$

Using this notation, (5) becomes

$$y(t) = g^T(t-1)\theta + e(t)\ . \qquad (7)$$

Conversely, another equivalent state equation may be derived from (5). The derivation is trivial and gives

$$x(t+1) = \Phi x(t) + b\,u(t) + c\,e(t+1)\ , \tag{8}$$

$$y(t) = c^T x(t) \tag{9}$$

This is an equivalent representation of ((1), (2)) with respect to the input-output relation.

Now, we can rewrite the system ((8), (9)) with the time shifted t to t-1 as follows :

$$x_1(t) = y(t)$$

$$x_2(t) = -a_2 y(t-1) \dots -a_n y(t-n+1) + b_2 u(t-1) + \dots + b_n u(t-n+1)$$

$$x_3(t) = -a_3 y(t-1) \dots -a_n y(t-n+2) + b_3 u(t-1) + \dots + b_n u(t-n+2)$$

$$\cdot\ \cdot\ \cdot\ \cdot\ \cdot$$

$$x_n(t) = -a_n y(t-1) + b_n u(t-1) \tag{10}$$

Thus we have the closed form

$$x(t) = G\,h(t)\ , \tag{11}$$

where

$$G = \begin{bmatrix} 1 & 0 & 0 & \dots & 0 & 0 & 0 & \dots & 0 \\ 0 & -a_2 & -a_3 & \dots & -a_n & b_2 & b_3 & \dots & b_n \\ \vdots & \vdots & \vdots & & \vdots & \vdots & \vdots & & \vdots \\ 0 & -a_{n-1} & -a_n & \dots & 0 & b_{n-1} & b_n & \dots & 0 \\ 0 & -a_n & 0 & \dots & 0 & b_n & 0 & \dots & 0 \end{bmatrix}_{n\times(2n-1)}$$

$$h^T(t) = [y(t),\dots,y(t-n+1),\ u(t-1),\dots,u(t-n+1)].$$

Introducing the transform relation of (11), the new state equation on $h(t)$ is described as follows, instead of (8):

$$h(t+1) = Ah(t) + d\,u(t) + c_1\, e(t+1) \tag{12}$$

where

$$A = \left[\begin{array}{cccccc|cccc}
-a_1 & -a_2 & \cdots & -a_{n-1} & -a_n & & b_2 & \cdots & b_{n-1} & b_n \\
1 & 0 & \cdots & 0 & 0 & & 0 & \cdots & 0 & 0 \\
0 & 1 & \cdots & 0 & 0 & & 0 & \cdots & 0 & 0 \\
\vdots & \vdots & & \vdots & \vdots & & \vdots & & \vdots & \vdots \\
0 & 0 & \cdots & 1 & 0 & & 0 & \cdots & 0 & 0 \\
\hline
0 & 0 & \cdots & 0 & 0 & & 0 & \cdots & 0 & 0 \\
0 & 0 & \cdots & 0 & 0 & & 1 & \cdots & 0 & 0 \\
\vdots & \vdots & & \vdots & \vdots & & \vdots & & \vdots & \vdots \\
0 & 0 & \cdots & 0 & 0 & & 0 & \cdots & 1 & 0
\end{array}\right]_{(2n-1)\times(2n-1)}$$

$$d^T = [b_1 \quad 0 \quad \cdots \quad 0 \quad 0 \quad 0 \quad \cdots \quad 0 \quad 0] \quad (2n-1)$$

$$c_1 = [1 \quad 0 \quad \cdots \quad 0 \quad 0 \quad 0 \quad \cdots \quad 0 \quad 0] \quad (2n-1)$$

Note that the rank of A holds $2n-2$ even if the dimension is $(2n-1)\times(2n-1)$.

2.2 Criterion Function

Consider the situation at time t, where the past measurements $y(0), \ldots, y(t-1), u(0), \ldots, u(t-1)$ are in hand.

We attempt to decide the control $u(t)$ based on the equivalent state model ((8), (9)) minimizing the following conditional expected quadratic index of performance:

$$E[J_t|y,\mathcal{Y}_{t-1},\mathcal{U}_{t-2}]$$

$$= E[x^T(t_f)\,Q_{t_f}\,x(t_f) + \sum_{j=t}^{t_f-1} x^T(j+1)\,Q_t\,x(j+1)$$

$$+ u^2(j|t-1) \mid y(j),\mathcal{Y}_{t-1},\mathcal{U}_{t-2}]\,,\quad t=0,\ldots,t_f-1, \qquad (13)$$

where $\mathcal{Y}_t$ and $\mathcal{U}_t$ is a sequence vector of observation defined as follows:

$$\mathcal{Y}_t^T \triangleq [y(0),\ldots,y(t)]$$

$$\mathcal{U}_t^T \triangleq [u(0),\ldots,u(t)]\ . \qquad (14)$$

The formulation of (13) obliges to estimate the state x and this leads to a nonlinear problem by the reason of incomplete state and parameter informations. We will thus reformulate the problem. Introducing (11), (13) becomes

$$E[J_t|y,\mathcal{Y}_{t-1},\mathcal{U}_{t-2}]$$

$$= E[h^T(t_f)\,G^T\,Q_{t_f}\,G\,h(t_f) + \sum_{j=t}^{t_f-1} h^T(j+1)\,G^T\,Q_t G\,h(j+1)$$

$$+ u^2(j|t-1) \mid y(j),\mathcal{Y}_{t-1},\mathcal{U}_{t-2}]\,,\ t=0,\ldots,t_f-1. \qquad (15)$$

Let us define

$$Q_0 \triangleq G^T\,Q_{t_f}\,G$$

$$Q \triangleq G^T\,Q_t\,G. \qquad (16)$$

Rewriting (15) we have

$$E[J_t|y,\mathcal{Y}_{t-1},\mathcal{U}_{t-2}]$$

$$= E[h^T(T_f)\,Q_0\,h(t_f) + \sum_{j=t}^{t_f-1} h^T(j+1)\,Q\,h(j+1)$$

$$+ u^2(j|t-1) \mid h(j),\mathcal{Y}_{t-1},\mathcal{U}_{t-2}]\,,\quad t=0,\ldots,t_f-1. \qquad (17)$$

The relation between (Q_0, Q) and (Q_{t_f}, Q_t) is described as (16) by G with constant parameters. If we choose (Q_{t_f}, Q_t) a definite value, we could readily determine (Q_0, Q) for known parameters. In the case of unknown parameters, we could know the relation between (Q_0, Q) and (Q_{t_f}, Q_t) after the identification has been completed. It should be noted that ranks of Q_0 and Q are less than or equal to n.

This feature is the penalty that must be paid for avoiding the nonlinear problem that arises in the case of ordinary state models. The matrices Q_0 and Q are nonnegative definite. Q may depend on time. The double argument is used to emphasize that $u(j|t-1)$ is the value of u(j) based on measurements up to time t-1.

2.3 Procedure for Dual Control

The problem for dual control considered here is to obtain online estimates of the parameters a_i, b_i ; i=1, ... , n and their covariances, and to decide an online optimal control based on past measurements up to time t-1. Thus the algorithm can be summarized as follows.

1. Estimate the plant parameter $\theta(\cdot|t-1)$ and its covariance $\Sigma(\cdot|t-1)$ with least mean square method by using measured vector $\mathcal{Y}_{t-1}$, $\mathcal{U}_{t-2}$.
2. Calculate the state estimate h(j) and its covariance P(j) as functions of $u(t|t-1)$, $j \in [t+1, t_f-1]$.
3. Decide the optimal control so that the criterion (17) is minimized, in other words, the state estimates h(j) predicted from past data is in best coincidence with the desired state (zero state) as possible as by applying dynamic programming to (t_f-t+1) steps process after t.

3. ALGORITHM FOR ONLINE IDENTIFICATION

Consider estimating the parameter θ and its covariance Σ in (7), which is the observation equation with the observation noise of e(t). The state of a dynamical system may be regarded as θ. Considering θ is time invariant, we have the following state equation.

$$\theta(t+1) = \theta(t) \qquad (18)$$

$$y(t) = g^T(t-1)\ \theta + e(t) \tag{19}$$

The state transition matrix is the unity matrix I. The observation matrix is a 1×2n time-variable row vector consisted of measured values of input and output signals. e(t) is a stationary independent random Gaussian variable. A Kalman filter may be used to estimate the state θ.

$$\hat{\theta}(\cdot|t) = E\,[\theta|\mathcal{Y}_t,\ \mathcal{U}_{t-1}]$$

$$= \hat{\theta}(\cdot|t-1) + \Sigma(\cdot|t-2)\ g(t-1)[\mathrm{tr}\ \Xi\ + g^T(t-1)$$

$$\Sigma(\cdot|t-2)\ g(t-1)]^{-1}[y(t) - g^T(t-1)\theta(\cdot|t-1)] \tag{20}$$

$$\Sigma(\cdot|t-1) = \Sigma(\cdot|t-2) - \Sigma(\cdot|t-2)\ g(t-1)[\mathrm{tr}\ \Xi$$

$$+ g^T(t-1)\ \Sigma(\cdot|t-2)\ g(t-1)]^{-1}\ g^T(t-1)\ \Sigma(\cdot|t-2) \tag{21}$$

Such a method is already available in the literature [1]. The least squares method gives the same result too.

4. DECISION OF OPTIMAL CONTROL

4.1 Estimation of State

To decide the optimal control sequence under the criterion (17), the estimate of one step ahead state $\hat{h}(j+1|j)$ must be calculated by using the result of identification. We can find $\hat{h}(j+1|j)$ basing on measurements up to time t-1 as follows.

$$\hat{h}(j+1|j) \triangleq E[h(j+1)|h(j),\ \mathcal{Y}_{t-1},\ \mathcal{U}_{t-2}]$$

$$= \hat{A}(\cdot|t-1)\ h(j) + \hat{d}(\cdot|t-1)\ u(j|t-1), \tag{22}$$

where $\hat{A}(\cdot|t-1) \triangleq E\ [A|\ \mathcal{Y}_{t-1},\ \mathcal{U}_{t-2}]$ and $\hat{d}(\cdot|t-1) \triangleq E\ [d|\ \mathcal{Y}_{t-1},\ \mathcal{U}_{t-2}]$.

Next, we will derive the covariance function of

$$\tilde{h}(j+1|j) \triangleq h(j+1) - \hat{h}(j+1|j)\ .$$

Defining $\tilde{A}(\cdot|t-1)$ and $\tilde{d}(\cdot|t-1)$ as follows;

$$\tilde{A}(\cdot|t-1) \triangleq A - \hat{A}(\cdot|t-1)$$

$$\tilde{d}(\cdot|t-1) \triangleq d - \hat{d}(\cdot|t-1) \quad ,$$

and subtracting (22) from (12) we find

$$\tilde{h}(j+1|j) = \tilde{A}(\cdot|t-1)+h(j)+\tilde{d}(\cdot|t-1)u(j|t-1)+c\ e(j+1) \ . \quad (23)$$

The covariance function of $h(j+1|j)$

$$P(j+1|j) \triangleq \text{cov}\ [\tilde{h}(j+1|j),\ \tilde{h}(j+1|j)] \quad (24)$$

can be proved to satisfy the following relation after some calculations[14].

$$\text{tr}\ K\ P(j+1|j) = E\ [\tilde{h}^T(j+1|j)\ K\ \tilde{h}(j+1|j)]$$

$$= h^T(j)\ k_{11}\Sigma_a\ h(j) + u(j|t-1)\ k_{11}\sigma_{ad}{}^T\ h(j)$$

$$+ h^T(j)\ k_{11}\sigma_{ad}\ u(j|t-1) + u(j|t-1)\ k_{11}\sigma_d^2$$

$$u(j|t-1) + k_{11}\text{tr}\ \Xi \quad (25)$$

where k_{11} is a (1,1) element of K which is an arbitrary $(2n-1)\times(2n-1)$ matrix, and

$$\sigma_{ad} \triangleq \begin{bmatrix} E(\tilde{a}\ \tilde{b}_1) \\ \hline E(\tilde{b}'\ \tilde{b}_1) \end{bmatrix} = \begin{bmatrix} \sigma_{ad,1} \\ \hline \sigma_{ad,2} \end{bmatrix} , \qquad \Sigma_a = \begin{bmatrix} \Sigma_{11} & \Sigma_{12} \\ \Sigma_{21} & \Sigma_{22} \end{bmatrix}$$

$$\sigma^2{}_d \triangleq E(\tilde{b}_1{}^2)$$

$$(b')^T \triangleq [b_2\ b_3\ \dots\ b_n]$$

$$\Sigma = \begin{bmatrix} \Sigma_{11} & \sigma_{ad,1} & \Sigma_{12} \\ \sigma^T_{ad,1} & \sigma^2_d & \sigma^T_{ad,2} \\ \Sigma_{21} & \Sigma_{ad,2} & \Sigma_{22} \end{bmatrix} \quad (26)$$

4.2 Functional Equation for Incomplete Parameter Information

Here consider the optimization of the (t_f-t+1) step-process after t. Now situating at future time $k>t$, we will decide the predictive optimum control of $u(k|t-1)$. The expected criterion (17) can be written as a sum of two terms.

$$E[h^T(t_f)\,Q_0\,h(t_f) + \sum_{j=k}^{t_f-1} h^T(j+1)\,Q\,h(j+1) + u^2(j|t-1)\,|\,h(j),$$

$$\mathcal{Y}_{t-1}, \mathcal{U}_{t-2}] + E[\sum_{j=t}^{k-1} h^T(j+1)Q\,h(j+1) + u(j|t-1)$$

$$|\,h(j), \mathcal{Y}_{t-1}, \mathcal{U}_{t-2}]\,;$$

$$t = 0 \sim t_f-1 \qquad (27)$$

The second term does not depend on $u(k|t-1)$, $u(k+1|t-1)$, ... , $u(t_f-1|t-1)$. Assuming that a unique minimum is taken with respect to all strategies which are expressed as a function of $(\mathcal{Y}_{t-1}, \mathcal{U}_{t-2})$. Minimizing the first term of (27) with respect to $u(t_f-1|t-1)$, $u(t_f-2|t-1)$, ... , $u(k|t-1)$, we define a functional f as follows.

$$\min_{u(k|t-1),\ldots,u(t_f-1|t-1)} E[h^T(t_f)Q_0\,h(t_f) + \sum_{j=k}^{t_f-1} h^T(j+1)$$

$$\times\,Q\;h\,(j+1) + u^2(j|t-1)\,|\;h(j), \mathcal{Y}_{t-1}, \mathcal{U}_{t-2}]$$

$$\triangleq f[h(k), \mathcal{Y}_{t-1}, \mathcal{U}_{t-2}, k] \quad . \qquad (28)$$

Here the function f satisfies the following equation.

$$f[h(k), \mathcal{Y}_{t-1}, \mathcal{U}_{t-2}, k]$$

$$= \min_{u(k|t-1)} E[h^T(k+1)\,Q\,h(k+1) + u^2(k|t-1)$$

$$+ f[h(k+1), \mathcal{Y}_{t-1}, \mathcal{U}_{t-2}, k+1]\,|\,h(k), \mathcal{Y}_{t-1}, \mathcal{U}_{t-2}] \qquad (29)$$

(29) is the Bellman equation for the case of incomplete information.

Considering that the past information $\mathcal{Y}_{t-1}$ and $\mathcal{U}_{t-2}$ is equivalently expressed by $\hat{\theta}(\cdot|t-1)$ and $\Sigma(\cdot|t-2)$ which is the result of identification, the functional V is then defined as follows:

$$V[h(k),\ \hat{\theta}(\cdot|t-1),\ \Sigma(\cdot|t-2),\ k]$$

$$\triangleq f[h(k),\ \mathcal{Y}_{t-1},\ \mathcal{U}_{t-2},\ k] \tag{30}$$

Using the function V, we now find that the functional (29) can be written as

$$V[h(k),\ \hat{\theta}(\cdot|t-1),\ \Sigma(\cdot|t-2),\ k]$$

$$= \min_{u(k|t-1)} E[h^T(k+1)\ Q\ h(k+1) + u^2(k|t-1) + V[h(k+1),$$

$$\hat{\theta}(\cdot|t-1),\ \Sigma(\cdot|t-2),\ k+1]\ |\ h(k),\ \hat{\theta}(\cdot|t-1),\ \Sigma(\cdot|t-2)] \tag{31}$$

The initial condition for the functional (31) is

$$V[h(t_f-1),\ \hat{\theta}(\cdot|t-1),\ \Sigma(\cdot|t-2),\ t_f-1]$$

$$= \min_{u(t_f-1|t-1)} E[h^T(t_f)\ (Q_0+Q)\ h(t_f) + u^2(t_f-1|t-1)|h(t_f-1),$$

$$\hat{\theta}(\cdot|t-1)\ ,\ \Sigma(\cdot|t-2)]\ ;$$

$$t \in \{0,1,\ldots,t_f-1\}\ . \tag{32}$$

4.3 Solution of the Functional Equation

We will now show that the solution of the functional equation (31) with the initial condition (32) is a quadratic function

$$V\ [h(k),\ \hat{\theta}(\cdot|t-1),\ \Sigma(\cdot|t-2),\ k]$$

$$= h^T(k)\ S(k|t-1)\ h(k) + s(k|t-1)\ . \tag{33}$$

Now, by applying the mathematical induction method, it will be proved that if (33) holds for k+1, then it also holds for k and the quadratic form holds for $k=t_f-1$. We assume that

$$V[h(k+1),\ \hat{\theta}(\cdot|t-1),\ \Sigma(\cdot|t-2),\ k+1]$$

$$= h^T(k+1)\ S(k+1|t-1)\ h(k+1) + s(k+1|t-1)\ , \qquad (34)$$
$$S(k+1|t-1) \geqq 0\ .$$

To evaluate the functional (31), the conditional distribution of $h(k+1)$ given $h(k)$ must be calculated. Substituting (34) into (31) and utilizing the relations

$$E[h^T(k+1)\ Q\ h(k+1)\,|h(k),\ \hat{\theta}\ ,\Sigma]$$

$$= \hat{h}^T(k+1|k)\ Q\ \hat{h}(k+1|k) + \mathrm{tr}\ Q\ P(k+1|k) \qquad (35)$$

$$E[h^T(k+1)\ S(k+1|t-1)\ h(k+1)\,|h(k),\ \hat{\theta}\ ,\ \Sigma]$$

$$= \hat{h}^T(k+1|k)\ S(k+1|t-1)\ \hat{h}(k+1|k) + \mathrm{tr}\ S(k+1|t-1)\ P(k+1|k) \qquad (36)$$

which are easily derived by the well-known Lemma, we have

$$V[h(k),\ \hat{\theta}(\cdot|t-1),\ \Sigma(\cdot|t-2),\ k]$$

$$= \min_{u(k|t-1)} \{\ \hat{h}^T(k+1|k)\ K(k+1|t-1)\ \hat{h}(k+1|k) + \mathrm{tr}\ K(k+1|t-1)$$
$$+ \mathrm{tr}\ K(k+1|t-1) \times P(k+1|k) + u^2(k|t-1) + s(k+1|t-1) \qquad (37)$$

where

$$K(k|t-1) \triangleq Q + S(k|t-1)\ , \qquad k \in [t,t_f]\ ;$$

$$K(t_f|t-1) \triangleq Q + Q_0 \geqq 0$$

and

$$S(t_f|t-1) = Q_0 \geqq 0\ . \qquad (38)$$

Substituting (22) and (23) for $\hat{h}(k+1|k)$ and $P(k+1|k)$ in (37), we have finally

$$V[h(k), \hat{\theta}(\cdot|t-1), \Sigma(\cdot|t-2), k]$$

$$= \min_{u(k|t-1)} \{h^T(k) [\hat{A}^T(\cdot|t-1) K(k+1|t-1) \hat{A}(\cdot|t-1)$$

$$+ k_{11}(k+1|t-1) \Sigma_a(\cdot|t-2)$$

$$- L^T(k|t-1)(1 + \hat{d}^T(\cdot|t-1) K(k+1|t-1) \hat{d}(\cdot|t-1)$$

$$+ k_{11}(k+1|t-1) \sigma_d^2(\cdot|t-2) L(k|t-1)] h(k)$$

$$+ [u(k|t-1) + L(k|t-1) h(k)]^T [1 + \hat{d}^T(\cdot|t-1) K(k+1|t-1)$$

$$\times \hat{d}(\cdot|t-1) + k_{11}(k+1|t-1) \sigma_d^2 (\cdot|t-2)] [u(k|t-1)$$

$$+ L(k|t-1) h(k)] + k_{11}(k+1|t-1) \operatorname{tr} \Xi + s(k+1|t-1)\} \tag{39}$$

where

$$L(k|t-1) \triangleq [1 + \hat{d}^T(\cdot|t-1) K(k+1|t-1) \hat{d}(\cdot|t-1)$$

$$+ k_{11}(k+1|t-1) \sigma_d^2 (\cdot|t-2)]^{-1} [\hat{d}^T(\cdot|t-1) K(k+1|t-1)$$

$$\hat{A}(\cdot|t-1) - k_{11}(k+1|t-1) \sigma_{ad}^T (\cdot|t-2)] \quad . \tag{40}$$

(39) is obtained by completing the squares. The inverse in (40) would exist if $K(k+1|t-1) \geqq 0$. The minimum is thus obtained for

$$u(k|t-1) = -L(k|t-1) h(k) \tag{41}$$

and the optimal strategy is a linear function represented by the row vector $L(k|t-1)$ which is obtained from past data before time t-1.

Hence

$$V[h(k), \hat{\theta}(\cdot|t-1), \Sigma(\cdot|t-2), k]$$

$$= h^T(k)[\hat{A}^T(\cdot|t-1)\ K(k+1|t-1)\ \hat{A}(\cdot|t-1) + k_{11}(k+1|t-1)$$

$$\times \Sigma_a(\cdot|t-2) - L^T(k|t-1)(1 + \hat{d}^T(\cdot|t-1)\ K(k+1|t-1)\hat{d}(\cdot|t-1)$$

$$+ k_{11}(k+1|t-1)\ \sigma_d^2(\cdot|t-2))\ L(k|t-1)]h(k)$$

$$+ k_{11}(k+1|t-1)\ \mathrm{tr}\ \Xi + s(k+1|t-1)\ . \qquad (42)$$

We thus have found that the functional (31) has a solution of the form (33) with

$$S(k|t-1) = \hat{A}^T(\cdot|t-1)\ K(k+1|t-1)\ \hat{A}(\cdot|t-1) + k_{11}(k+1|t-1)$$

$$\times \Sigma_a(\cdot|t-2) - L^T(k|t-1)\ [1 + \hat{d}^T(\cdot|t-1)\ K(k+1|t-1)$$

$$\times \hat{d}(\cdot|t-1) + k_{11}(k+1|t-1)\ \sigma_d^2(\cdot|t-2)]\ L(k|t-1)$$

$$s(k|t-1) = k_{11}(k+1|t-1)\ \mathrm{tr}\ \Xi + s(k+1|t-1)\ . \qquad (43)$$

4.4 Discrete-time Riccati Equation for Random Parameter Systems

Furthermore, we will derive a matrix Riccati equation to calculate $K(k|t-1)$ for a class of random parameter systems. Using $K(k|t-1)=Q+S(k|t-1)$ and making some modifications (43) yields

$$K(k|t-1) = [\hat{A}(\cdot|t-1) - \hat{d}(\cdot|t-1)\ L(k|t-1)]^T\ K(k+1|t-1)$$

$$\times [\hat{A}(\cdot|t-1) - \hat{d}(\cdot|t-1)\ L(k|t-1)]$$

$$+ E[\tilde{A}(\cdot|t-1) - \tilde{d}(\cdot|t-1)\ L(k|t-1)]^T\ K(k+1|t-1)$$

$$\times [\tilde{A}(\cdot|t-1) - \tilde{d}(\cdot|t-1)\ L(k|t-1)]$$

$$+ L^T(k|t-1)\ L(k|t-1) + Q. \qquad (44)$$

Considering A and d as random variables, we can rewrite (40) and (44) as follows:

$$L(k|t-1) = [1 + E(d^T K(k+1|t-1) d \mid \mathcal{Y}_{t-1}, \mathcal{U}_{t-2})]^{-1}$$

$$\times E(d^T K(k+1|t-1) A \mid \mathcal{Y}_{t-1}, \mathcal{U}_{t-2}). \qquad (45)$$

$$K(k|t-1) = E([A - d L(k|t-1)]^T K(k+1|t-1)$$

$$\times [A - d L(k|t-1)] \mid \mathcal{Y}_{t-1}, \mathcal{U}_{t-2})$$

$$+ L^T(k|t-1) L(k|t-1) + Q \qquad (46)$$

Applying (45) to (46)

$$K(k|t-1) = E(A^T K(k+1|t-1) A \mid \mathcal{Y}_{t-1}, \mathcal{U}_{t-2})$$

$$- E(d^T K(k+1|t-1) A \mid \mathcal{Y}_{t-1}, \mathcal{U}_{t-2})^T$$

$$\times [1 + E(d^T K(k+1|t-1) d \mid \mathcal{Y}_{t-1}, \mathcal{U}_{t-2})]^{-1}$$

$$\times E(d^T K(k+1|t-1) A \mid \mathcal{Y}_{t-1}, \mathcal{U}_{t-2}) + Q, \qquad (47)$$

with the initial condition $K(t_f|t-1) = Q_0 + Q \geqq 0$. (47) is a discrete version of the matrix Riccati equation for a class of random parameter systems.

These formulas (45), (47) indicate that when the identification process would be completed the control strategy tends to the ordinary one for known parameter systems.

5. ALGORITHM OF DUAL CONTROL

The algorithm of the identification is given by the following recursive form:

$$\hat{\theta}(\cdot|t-1) = \hat{\theta}(\cdot|t-2) + \Sigma(\cdot|t-3)\, g(t-2)$$

$$\times\, [\mathrm{tr}\ \Xi + g(t-2)\, \Sigma(\cdot|t-3)\, g(t-2)]^{-1}$$

$$\times\, [y(t-1) - g^T(t-2)\, \theta(\cdot|t-2)] \qquad (48)$$

$$\Sigma(\cdot|t-2) = \Sigma(\cdot|t-3) - \Sigma(\cdot|t-3)\, g(t-2)\, [\mathrm{tr}\ \Xi$$

$$+ g^T(t-2)\, \Sigma(\cdot|t-3)\, g(t-2)]^{-1}\, g^T(t-2)\, \Sigma(\cdot|t-3) \qquad (49)$$

Let the admissible control strategies be such that u(t) is a function of $h(t)$, $\hat{\theta}(\cdot|t-1)$, $\Sigma(\cdot|t-2)$ and t. Assume that Q_0 and Q are nonnegative difinite. Then the control sequence that minimizes the criterion (17) for the linear stochastic system with unknown parameters is then given by

$$u(t|t-1) = -L(t|t-1)\, h(t) \qquad (50)$$

$$t \in [0,\ t_f-1]\ ,$$

$$L(k|t-1) = [1 + E(d^T\ K(k+1|t-1)|d\ \hat{\theta}(\cdot|t-1),\ \Sigma(\cdot|t-2)]^{-1}$$

$$\times\, [E(d^T\ K(k+1|t-1)\ A\ |\hat{\theta}(\cdot|t-1),\ \Sigma(\cdot|t-2))]$$

$$K(k|t-1) = E(A^T\ K(k+1|t-1)\ A\ |\hat{\theta}(\cdot|t-1),\ \Sigma(\cdot|t-2))$$

$$- E(d^T\ K(k+1|t-1)\ A|\ \hat{\theta}(\cdot|t-1),\ \Sigma(\cdot|t-2))$$

$$\times\, [1 + E(d^T\ K(k+1|t-1)\ d\ |\hat{\theta}(\cdot|t-1),\ \Sigma(\cdot|t-2))]^{-1}$$

$$\times\, E(d^T\ K(k+1|t-1)\ A|\ \hat{\theta}(\cdot|t-1),\ \Sigma(\cdot|t-2),$$

$$K(t_f|t-1) = Q_0 + Q,\quad k \in [t, t_f-1]\quad . \qquad (51)$$

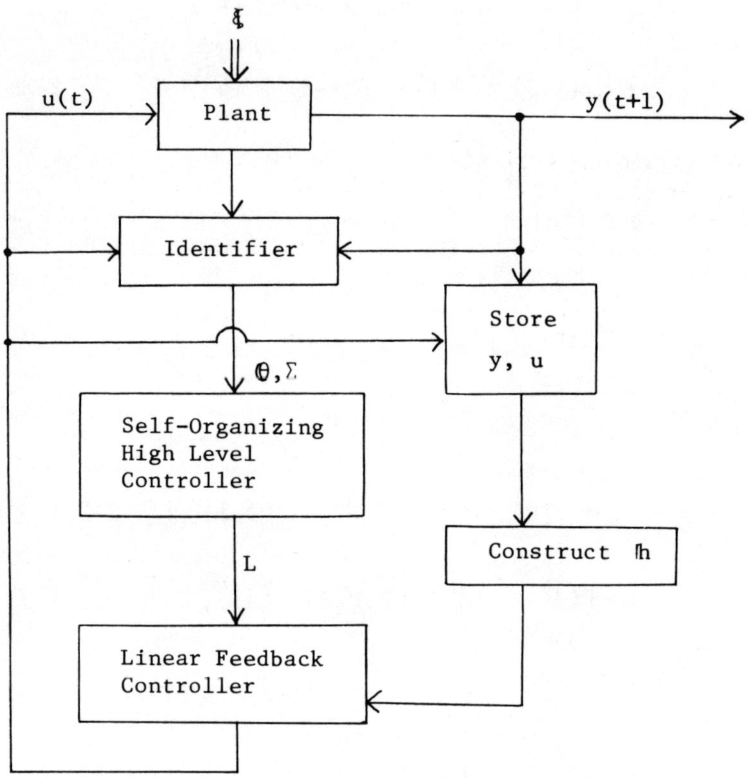

Fig. 1. Block diagram for the proposed dual control scheme.

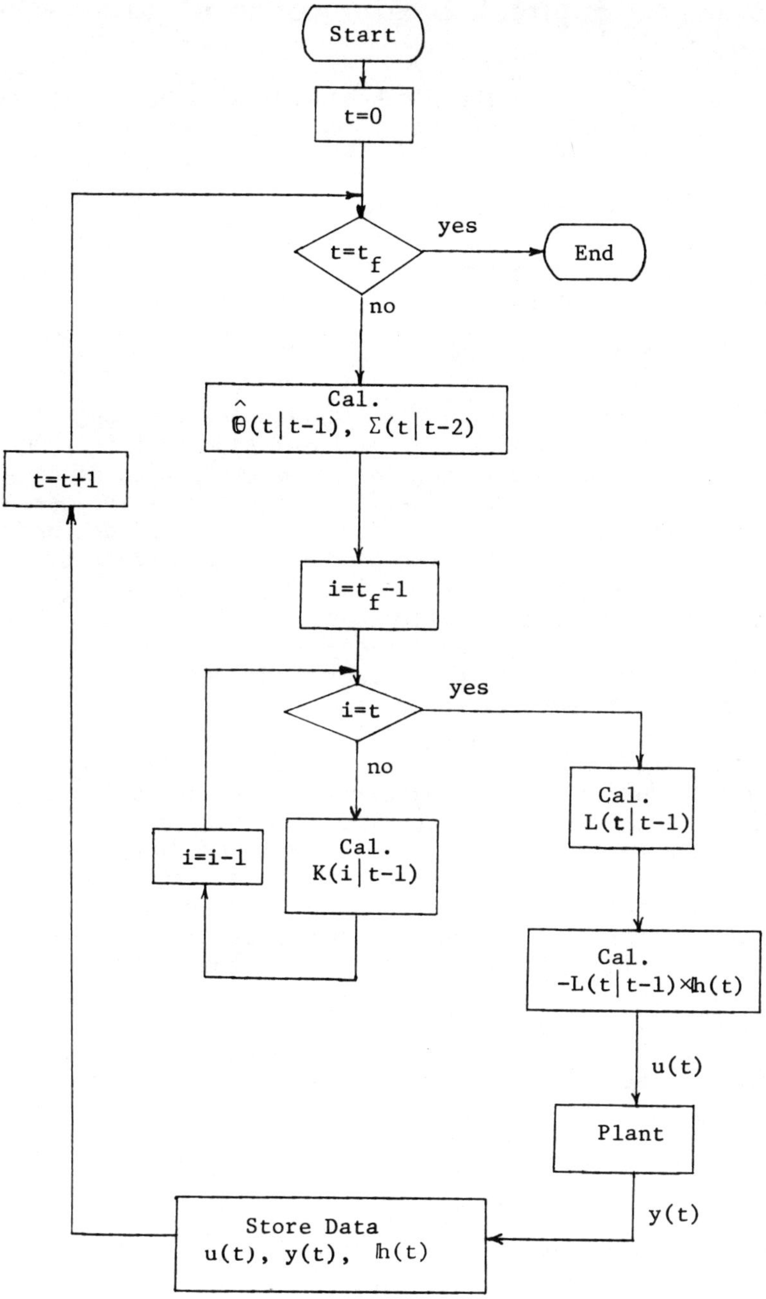

Fig. 2. Flow chart for the dual control system of Fig. 1.

Note that, as is clear by (50) and (11), an instantaneous value of the present observation $y(t)$, i.e. $x_1(t)$ must be at once transferred to the current control.

Fig. 1. and Fig. 2. are the block diagram and the simplified flow chart of the proposed dual control system respectively.

6. CONCLUSIONS

A learning scheme for optimal dual control of unknown parameter systems for an expected quadratic index of performance is presented, under assumptions of the uncorrelation property of identification error and the complete state information.

The separation between the identification and the control in the presented procedure is given. From the reason that the complete state information is enforced in the determination of control, the expected performance criterion requires only the parameter estimate with its covariance as the past information, and the control signal is always uncorrelated with the current disturbance of the difference equation. Through the realization of the separation between identification and control, the dual control procedure can take a noticeable form where the identification is first performed and then the control is decided by using the parameter estimate and its covariance given as the identification result. The identification of the unknown parameter system is implemented by use of Kalman filter technique, while the dynamic programming procedure is recursively employed for the optimization of control. Through the careful consideration of the assumption of complete state information for the full step of dynamic programming process, the perfect linear feedback law has become realizable. Moreover, the optimal control strategy is affected by the additional terms involving the covariance of parameter estimation error. These are different points from the existing scheme for dual control. The precise examination of the result exposes that the linear feedback law and the matrix Riccati equation for unknown systems are extensions of the ordinary linear feedback law and the discrete matrix Riccati equation to the random parameter class, respectively.

REFERENCES

1) J. Wieslander and B. Witternmark: "An Approach to Adaptive Control Using Real Time Identification", Automatica, 7-2, pp. 211-217 (1971).
2) M. Mendes: "An On Line Adaptive Control Method", Automatica, 7-3, pp. 323-332 (1971).
3) K.J. Astrom and B. Wittenmark: "Problems of Identification

and Control", Jour. of Mathematical Analysis and Application 34, pp. 91-113 (1971).
4) B.M. Horowitz: "A New Approach to Certain Problems of Identification and Control", IEEE Trans., Ac-15-4, pp. 475-477 (1970).
5) G.N. Saridis and R.N. Lobbia: "Parameter Identification and Control of Linear Descrete-time Systems", IEEE Trans., AC 17-1, pp. 52-60 (1972).
6) Ya.Z. Tsypkin: "Optimization, Adaptation, and Learning in Automatic Systems", in Computer and Information Sciences-II, J.T. Tou, Ed. pp. 21-29; Academic Press (1967).
7) Ya.Z. Tsypkin: "Adaptation, Training and Self-Organization in Automatic Systems", Automatic and Remote Control, Vol. 27-1, pp. 16-50 (1966).
8) Ya.Z. Tsypkin: "Learning Systems in Advances" in Information Systems Science, Vol. 2, J.T. Tou, Ed. pp. 1-35, Plenum Press (1969).
9) B.M. Horowitz: "On Some Aspects of Dual Control Theory", A Thesis, New York University, School of Eng. and Science, Dept. of Elect. Eng. (1969).
10) Y. Yoshida: "On the Construction of Learning Dual Control", Preprint of the 10th SICE Conference in Japan (Aug., 1971).
11) Y. Yoshida: "A Construction Method of Learning Dual Control", Preprint of the 14th Joint Auto. Control Conference in Japan (Nov., 1971).
12) Y. Yoshida: "On-Line Identification of Phase Variable Linear Systems of Unknown Parameters for Dual Control", Bulletin of Nagoya Institute of Technology, 23 (1971).
13) Y. Yoshida and K. Nakamura: "A Synthesis of Recursive Stochastic Dual Control", Res. Rept. of Auto. Cont. Lab., Nagoya Univ., Vol. 19 (1972).
14) Y. Yoshida and K. Nakamura: "A Synthesis of Learning Dual Control for Single-Input / Single-Output Systems", Intl. Jour. of Control, to appear.
15) Y. Yoshida and K. Nakamura: "Learning Dual Control with Complete State Information", Res. Rept. of Auto. Cont. Lab., Nagoya Univ., Vol. 20 (1973).

ON A CLASS OF VARIABLE-STRUCTURE SYSTEMS*

Surender K. Gupta, Kuduvally N. Swamy,
Tzyh-Jong Tarn, and John Zaborszky

Washington University
St. Louis, Missouri 63130

ABSTRACT

The purpose of this paper is to investigate the optimal control of a class of discrete, time-invariant, variable-structure systems. Both deterministic and stochastic problems are considered for unbounded control and the cost functional quadratic in state. Solutions are obtained in closed-form.

In the deterministic case, it is seen that the regular path and singular paths satisfying the functional equation of dynamic programming may exist simultaneously. It is shown then that the regular path is optimal. Both additive and multiplicative plant noise are considered in the stochastic problem. It is shown that the presence of noise considerably simplifies the analysis since the cases with singularities are contained in sets of measure zero.

I. INTRODUCTION

Recently there has been a spurt of interest in variable-structure systems since several natural phenomena can be modeled as systems of this class [5]. Several aspects like controllability [6-8], reachability [9], identification [10], and structural properties [11] have been explored.

The subject of this paper is optimal control of a class of variable structure systems where the scalar control enters

*This research was supported in part by the National Science Foundation Grant GK-36531 and GK-22905A#2.

linearly but with a nonlinearly state dependent gain. This work is a generalization of earlier work [1-4] on a more restricted class where the control gain is linear in the state. Interestingly, the result for this wider class differs only slightly from the results for the bilinear class but unfortunately some of the elegance is lost.

In this paper solutions are obtained in closed form for the deterministic and stochastic control problem for the cost functional quadratic in state, and the control unbounded. In the deterministic case, the optimality of the regular path, when regular and singular paths exist simultaneously is shown. The stochastic control problem considered here assumes complete state information, where the state equation is corrupted by noise which enters linearly through a potentially state dependent gain. The expression for optimal control and the structure of the recursive equations are similar to those in the deterministic case, but have different implications. The presence of noise makes the calculation of optimal control easier in the stochastic case because singular paths form sets of measure zero. The noise with state dependent gain influences the recursive equation directly by contributing an additive term while the noise results in added cost even when it enters through a constant gain.

II. DETERMINISTIC OPTIMAL CONTROL

The free-end-point problem is considered in this section for the deterministic system. A closed-form solution is obtained for the optimal control problem by the method of dynamic programming. It is shown that the solution of the dynamic programming functional equation may be nonunique, with a multiplicity of paths satisfying the functional equation.

First we state some definitions:

Definition 2.1: Path of a dynamical system generated by applying at each time the uniquely determined optimal control is called the regular path of the system.

Definition 2.2: A singularity is said to exist at k if the optimal control at k is not unique.

Definition 2.3: Optimal path of a dynamical system which has at least one singularity is called a singular path of the system.

For the variable-structure systems considered, a unique regular path, and singular paths may exist simultaneously. The main result of this section is that if the regular path and singular paths exist simultaneously, it is the regular path that is optimal.

2.1 Problem Statement

Consider a discrete, variable-structure system described by

$$\underline{x}_{k+1} = \Phi\, \underline{x}_k + \underline{d}\, u_k g_k(\underline{x}_k), \quad 0 \le k \le N-1, \tag{2.1a}$$

where $\underline{x}_k = (x_k^1, x_k^2, \dots, x_k^n) \in R^n$, is the state vector, $u_k \in R^1$ is the scalar control, $g_k : R^n \to R^1$ is any nonlinear scalar valued bounded function, Φ and $\underline{d}$ have the following form

$$\Phi = \begin{bmatrix} 0 & 1 & 0 & \dots & 0 \\ 0 & 0 & 1 & \dots & 0 \\ \cdot & & & & \\ \cdot & & & & \\ 0 & 0 & 0 & \dots & 1 \\ a_1 & a_2 & a_3 & & a_n \end{bmatrix} \qquad \underline{d} = \begin{bmatrix} 0 \\ 0 \\ \cdot \\ \cdot \\ 0 \\ 1 \end{bmatrix} \quad . \tag{2.1b}$$

The problem is to choose the sequence of controls $u_0, \dots, u_{N-1}$ so that the cost functional

$$J = \Big(\sum_{k=1}^{N-1} \underline{x}_k^T\, Q\, \underline{x}_k \Big) + \underline{x}_N^T\, Q'\, \underline{x}_N \tag{2.2}$$

is a minimum, subject to (2.1) and the condition $\underline{x}_0 = \underline{x}(0)$. Furthermore, $\underline{x}_N$ may be specified (fixed-end-point problem), or may not be specified (free-end-point problem). Q and Q', the $(n \times n)$ constant cost matrices, are symmetric, and nonnegative. The superscript ^{T} denotes transpose. N is a fixed, finite positive integer.

Remark 2.1: Consider the system

$$\underline{y}_{k+1} = A\, \underline{y}_k + D\, \hat{\underline{g}}_k\, (\underline{y}_k) u_k$$

when $D = \underline{c}\, \underline{h}^T$, is an $(n \times m)$ matrix of rank one. If $(A, \underline{c})$ is a controllable pair, the above equation can be transformed by the transformation $\underline{y}_k = P\underline{x}_k$ to the phase-variable form

$$\underline{x}_{k+1} = \Phi \underline{x}_k + \underline{d}[\underline{h}^T \hat{\underline{g}}_k\, (P\underline{x}_k)]\, u_k \tag{2.3}$$

(2.3) is of the form (2.1). When $\hat{\underline{g}}_k(\underline{y}_k) = \underline{y}_k$, we obtain the bilinear system as discussed in [1-4].

2.2 Preliminaries

Define for any symmetric ($n^x n$) matrix $Q_k = (q^k_{ij})$ with $q^k_{nn} \neq 0$ a symmetric matrix $C_k = (C^k_{ij})$

$$C^k_{ij} \triangleq \frac{(q^k_{nn}\, a_i + q^k_{i-1,n})\,(q^k_{nn} a_j + q^k_{j-1,n})}{q^k_{nn}} \tag{2.4}$$

where $q^k_{0,n} \triangleq 0$ and a_i, $1 \leq i \leq n$ are as given in (2.1). Also define for any symmetric, ($n^x n$) matrix $S = (s_{ij})$,

$$F(S) \triangleq \begin{bmatrix} 0 & \underline{0}^T \\ \underline{0} & \tilde{F}(S) \end{bmatrix} \tag{2.5}$$

where

$$\tilde{F}(S) = (s'_{ij}),\ i,j,\ = 1,2,\ldots,n-1$$

$$s'_{ij} = \begin{cases} s_{ij} - \dfrac{s_{in}\, s_{nj}}{s_{nn}} , & \text{if } s_{nn} \neq 0 \\ s_{ij} , & \text{if } s_{nn} = 0 \end{cases}$$

Lemma 2.1: Suppose $Q_k = (q^k_{ij})$ is symmetric and $q^k_{nn} \neq 0$. Then for C_k as in (2.4),

$$(\Phi^T Q_k \Phi - C_k) = F(Q_k).$$

Furthermore, if Q_k is nonnegative, so is $F(Q_k)$.

Proof: See [1] and [2].

Remark 2.2: When Q_k is symmetric and nonnegative with $q^k_{nn} = 0$, it is seen that $\Phi^T Q_k \Phi = F(Q_k)$.

2.3 Optimal Control

Theorem 2.2: The solution to the problem stated in Section 2.1 with $\underline{x}_N$ free is given by

$$\underset{0\le k\le N-1}{u_k} = \begin{cases} -\dfrac{\sum_{i=1}^{n} x_k^i\, q_{i-1,n}^k + q_{nn}^k\, \underline{a}^T\, \underline{x}_k}{q_{nn}^k\, g_k\,(\underline{x}_k)}, & \text{if } q_{nn}^k g_k(\underline{x}_k) \neq 0 \quad (2.6a) \\ \text{nonunique}, & \text{if } q_{nn}^k g_k(\underline{x}_k) = 0 \quad (2.6b) \end{cases}$$

where $\underline{a}^T = (a_1,a_2,\ldots,a_n)$is as given in (2.1), $Q_k = (q_{ij}^k)$ is given recursively by

$$\underset{0\le k\le N-2}{Q_k} = \begin{cases} Q + F(Q_{k+1}), & \text{if } q_{nn}^{(k+1)} g_{k+1}(\underline{x}_{k+1}) \neq 0 \quad (2.7a) \\ Q + \Phi^T Q_{k+1}\Phi, & \text{if } q_{nn}^{(k+1)} g_{k+1}(\underline{x}_{k+1}) = 0 \quad (2.7b) \end{cases}$$

with the initial condition $Q_{N-1} = Q'_{kk}$ is nonnegative for all k, $0 \le k \le N-1$. The optimal cost for the last (N-k) stages is given by

$$\underset{0\le k\le N-1}{V_{N-k}} = \begin{cases} \underline{x}_k^T\, F(Q_k)\, \underline{x}_k, & \text{if } q_{nn}^k\, g_k(\underline{x}_k) \neq 0 \quad (2.8a) \\ \underline{x}_k^T\, \Phi^T Q_k \Phi\, \underline{x}_k, & \text{if } q_{nn}^k\, g_k(\underline{x}_k) = 0 \quad (2.8b) \end{cases}$$

Remark 2.3: This result is quite similar to the result obtained in [1], [2] for the case of bilinear systems that is $g_k(\underline{x}_k) = \underline{b}^T\underline{x}_k$ with $\underline{b}^T = (b_1,b_2,\ldots,b_n)$ a constant n-vector. In fact, the only formal difference is in the denominator of the expanded form for u_k and in the condition for the singular case; $g_{nn}^k\, g_k\,(\underline{x}_k)$ replaces $g_{nn}^k\, \underline{b}^T\underline{x}_k$ in both instances. Unfortunately the neat matrix form for u_k does not apply to the variable structure case.

Proof: The proof is by induction. From the principle of optimality, we have the functional equations [12]

$$V_1 = \min_{u_{N-1}} (x_N^T\, Q'\, \underline{x}_N) \tag{2.9}$$

$$\underset{0\le k\le N-2}{V_{N-k}} = \min_{u_k} (\underline{x}_{k+1}^T\, Q\, \underline{x}_{k+1} + V_{N-k-1}) \tag{2.10}$$

(i) From (2.1) and (2.9), setting $Q_{N-1} = Q'$, we get u_{N-1} as in (2.6a) with $k = N-1$, if $q_{nn}^{(N-1)} g_{N-1}(\underline{x}_{N-1}) \neq 0$. If $q_{nn}^{(N-1)} g_{N-1}$

$(\underline{x}_{N-1}) = 0$, it is seen that u_{N-1} is nonunique.

Substituting for u_{N-1} in (2.9) and noting that

$$u_{N-1} g_{N-1}(\underline{x}_{N-1}) + \underline{a}^T \underline{x}_{N-1} = -\frac{1}{q_{nn}^{(N-1)}} \sum_{i=1}^{n} x_{N-1}^i \, q_{i-1,n}^{(N-1)}$$

we get (2.8) with $k = N-1$.

(ii) Assume that

$$V_{N-k-1} = \begin{cases} x_{k+1}^T F(Q_{k+1}) \underline{x}_{k+1}, & \text{if } q_{nn}^{(k+1)} g_{k+1}(x_{k+1}) \neq 0 \\ \underline{x}_{k+1}^T \Phi^T Q_{k+1} \Phi \, x_{k+1}, & \text{if } q_{nn}^{(k+1)} g_{k+1}(\underline{x}_{k+1}) = 0 \end{cases}$$

where Q_{k+1} is nonnegative. From (2.1) and (2.10), proceeding as in (i) above, we obtain (2.6) and (2.7).

From Lemma 2.1, it follows that Q_k is nonnegative. If $q_{nn}^k g_k(\underline{x}_k) = 0$, from (2.1), $(\underline{x}_{k+1}^T Q \, \underline{x}_{k+1} + V_{N-k-1})$ in (2.10) becomes independent of u_k; hence, then u_k is nonunique. (Further discussion on the nonuniqueness of u_k when $q_{nn}^k g_k(\underline{x}_k) = 0$ in Remark 2.4). Substituting for $(u_k g_k(x_k) + \underline{a}^T \underline{x}_k)$ in (2.10), we obtain (2.8) which is of the form assumed. Hence, we have u_k, Q_k and V_{N-k} as in (2.6), (2.7) and (2.8) respectively. Since Q_{N-1} is nonnegative, from (2.7) and Lemma 2.1, Q_k is nonnegative, $0 \leq k \leq N-1$. This completes the proof of the theorem.

Remark 2.4:

(i) The optimal control u_k is uniquely determined from the above theorem at each k, if $q_{nn}^k g_k(\underline{x}_k) \neq 0$, $0 \leq k \leq N-1$. Hence by definition 2.1 and (2.1), the path, if it exists, is regular if $q_{nn}^k g_k(\underline{x}_k) \neq 0$ for all k, $0 \leq k \leq N-1$. From (2.6b) u_k is nonunique when $q_{nn}^k g_k(\underline{x}_k) = 0$, hence the path is singular if $q_{nn}^k g_k(\underline{x}_k) = 0$ for at least one value of k, $0 \leq k \leq N-1$.

(ii) If at any k, $q_{nn}^k g_k(\underline{x}_k) = 0$, the corresponding V_{N-k} is (2.8b)

$$V_{N-k} = x_k^T (\Phi^T Q_k \Phi) \underline{x}_k, \tag{2.11}$$

where Q_k is given by (2.7). Associated with Q_k is a distribution of singularities in the future (i.e., at k+1, k+2,...,N-1).

From (2.11), any u_k that does not alter the distribution of singularities in the future can be chosen for optimal control at k. Such a u_k is in general nonunique.

(iii) From (2.7) and (2.8) it is clear that the regular and singular paths have different sequences of V_k's associated with them, i.e., they have different costs.

(iv) Consider the discrete, nonlinear system described by

$$x_{k+1}^{i} = x_{k}^{i+1} \quad , \quad 1 \leq i \leq n-1 \tag{2.12}$$

$$x_{k+1}^{n} = f_k(\underline{x}_k) + u_k g_k(\underline{x}_k),$$

where $f_k(\cdot)$ and $g_k(\cdot)$ are any nonlinear scalar valued functions. If $g_k(\underline{x}_k) \neq 0$ for all k, $0 \leq k \leq N-1$, then the optimal control problem of (2.12) with cost functional (2.2) can be reduced to a linear optimal control problem by substituting $v_k = f_k(\underline{x}_k) + u_k g_k(\underline{x}_k)$.

2.4 Computation of Singular and Regular Paths

In computing the optimal control, a priori assumption must be made about the distribution of singularities. Suppose it is assumed that the singularities occur at $k = k_1, k_2, \ldots, k_r$. From (2.7), Q_k, $0 \leq k \leq N-1$, can be obtained recursively beginning with $Q_{N-1} = Q'$. Then from (2.6) and (2.1) u_k and $\underline{x}_k$, $0 \leq k \leq N-1$ can be obtained. However, the path with the assumed singularities exists only if $q_{nn}^{k} g_k(\underline{x}_k) = 0$ for $k = k_1, k_2, \ldots, k_r$, and $q_{nn}^{k} g_k(\underline{x}_k) \neq 0$ for all other k, $0 \leq k \leq N-1$. These conditions then must be tested for $\underline{x}_k$ on the computed path, $0 \leq k \leq N-1$. By this procedure, we would have generated a path satisfying the functional equation (2.9) and (2.10) as desired. If solution to the optimal control problem exists, then there exists a path which necessarily satisfies the functional equation. This path can be obtained by the procedure described. If $q_{nn}^{k} g_k(\underline{x}_k) = 0$, and hence the control is nonunique, two cases may occur: (i) $g_k(\underline{x}_k) = 0$, then from (2.1) we see that the control u_k does not affect the future states, thus an arbitrary u_k may be chosen. The obvious choice $u_k = 0$; (ii) $q_{nn}^{k} = 0$ but $g_k(\underline{x}_k) \neq 0$, then from (2.1) we see that the control u_k does affect the future states which in turn will affect the distribution of the singularities in the future. Thus in this case we need to choose a u_k which does not alter the distribution of singularities in the future.

Example 2.1: This example illustrates the method of finding a path with a certain distribution of singularities, it it exists.

$$\text{Let } \Phi = \begin{bmatrix} 0 & 1 & 0 \\ 0 & 0 & 1 \\ 1 & 0 & 1 \end{bmatrix}, \qquad Q = Q' = \begin{bmatrix} 1 & -1 & 0 \\ -1 & 1 & 0 \\ 0 & 0 & 1 \end{bmatrix}$$

$\underline{x}_0^T = (1,1,1)$, $N = 4$, $\underline{x}_4$ be free. For simplicity we let $g_k(\underline{x}_k) = \underline{b}^T\underline{x}_k$, where $\underline{b}^T = (1,0,1)$.

(i) Assume that there are no singularities at k, $0 \leq k \leq 3$. We compute first the sequence Q_k from (2.7) then the sequence u_k and $\underline{x}_k$ from (2.6) and (2.1) and finally check the value of $q_{33}^k\underline{b}^T\underline{x}_k$:

k	0	1	2	3	4
x_k^1	1.0	1.0	1.0	0.38	0.152
x_k^2	1.0	1.0	0.38	0.152	0.069
x_k^3	1.0	0.38	0.152	0.069	0.0
$q^k_{33}\underline{b}^T\underline{x}_k$	$\neq 0$	$\neq 0$	$\neq 0$	$\neq 0$	
u_k	-0.81	-0.89	-0.94	-1.0	

Hence, the regular path exists and its cost is $V_4^R = 0.62$.

(ii) Assume that a singular path exists with a singularity at k = 1. Then it is seen that $q_{33}^0\underline{b}^T\underline{x}_0 \neq 0$, $u_0 = -15/14$, $\underline{x}_1^T = (1,1,-1/7)$. However $q_{33}^1\underline{b}^T\underline{x}_1 \neq 0$. Hence, there is no singularity at k = 1. The proposed singular path does not exist.

2.5 Optimality of Singular and Regular Paths

It was shown in Section 2.4 how to compute regular and singular paths if they exist. Since the sequence of controls to be used (and hence the states produced) is different for regular and singular paths, it is conceivable that for a given problem regular and any number of the various possible singular paths can exist simultaneously, each with a sequence of $\{V_k\}_{k=1}^{N}$. This implies that the solution to (2.9) and (2.10) can be nonunique. Now, satisfaction of (2.9) and (2.10) is both necessary and sufficient for optimality; i.e., the optimal path satisfies the functional equation and every path generated by it is a possible optimal path. Each optimal path generated by (2.9) and (2.10) is unique among all other possible paths which belong to the same class. Class here is defined as consisting of all paths having singularities at the same k.

The overall optimal path is pinpointed by the following theorem which is the central result on optimality.

<u>Theorem 2.3</u>: (Optimality Theorem) For the free-end-point problem stated in Section 2.1, if the regular path and singular paths exist simultaneously, the regular path is optimal.

<u>Proof</u>: The idea of the proof is as follows: It is possible to construct for any singular path S, the regular path R and their corresponding costs V_N^S, V_N^R a finite sequence of numbers $V_N^{S_1}$, $V_N^{S_2}, \ldots, V_N^{S_m}$ such that $V_N^R \le V_N^{S_m} \le \ldots \le V_N^{S_2} \le V_N^{S_1} \le V_N^S$, hence V $V_N^R \le V_N^S$, which establishes the optimality of the regular path. For details of the proof see [1,2].

<u>Remark 2.5:</u> The optimal control problem stated in Sectin 2.1 for the case when $\underline{x}_N$ is specified to be $\underline{0}$ can be converted to an (N-n0) -stage problem with certain terminal constraints, which can be solved by the method developed here provided the constraints are met. Details of the reduction to (N-n)-stage problem are similar to bilinear systems as given in [1,2].

<u>Example 2.2:</u> This is an example where the regular path and a singular path both exist.

$$\text{Let } \Phi = \begin{bmatrix} 0 & 1 \\ 1 & 1 \end{bmatrix}, \qquad Q = Q' = \begin{bmatrix} 1 & 1 \\ 1 & 1 \end{bmatrix}$$

$g(\underline{x}_k) = x_k^1 + (x_k^2)^2$, $\underline{x}_0^T = (\alpha^2$, $-25/9)$, α a constant and N = 2.

(i) Assume that the path is regular. Proceeding as in Example 2.1

k	0	1	2
x_k^1	α^2	$-\frac{25}{9}$	$\frac{25}{9}$
x_k^2	$-\frac{25}{9}$	$\frac{25}{9}$	$-\frac{25}{9}$
$q_{22}^k g(\underline{x}_k)$	$\alpha^2 + \frac{625}{81} \neq 0$	$\frac{400}{81} \neq 0$	
u_k	$(\frac{50}{9} - \alpha^2) / (\frac{625}{81} + \alpha^2)$	$-\frac{9}{16}$	

Hence the regular path exists and $V_2^R = 0$.

(ii) Assume that $k = 0$ is a regular point and $k = 1$ is a singular point

k	0	1	2
x_k^1	α^2	$-\frac{25}{9}$	$\frac{5}{3}$
x_k^2	$-\frac{25}{9}$	$\frac{5}{3}$	$-\frac{10}{9}$
$q_{22}^k g(\underline{x}_k)$	$5(\alpha^2 + \frac{625}{81}) \neq 0$	0	
u_k	$(\frac{40}{9} - \alpha^2) / (\frac{625}{81} + \alpha^2)$	arbitrary (say 0)	

The proposed singular path exists and $V_2^S = 125/81$. It is clear that $V_2^R < V_2^S$.

III. STOCHASTIC OPTIMAL CONTROL

This section extends the study of optimal control to the case when the variable structure system of the class considered in the preceding section is under the influence of additive noise, some of which enters through a gain which depends on the state linearly. The expression for the optimal control turns out to be similar in form to that in the deterministic case, but the recursive equation for the cost is affected by the noise statistics. It is observed that the presence of noise actually simplifies the analysis compared to that in the deterministic case. This work extends earlier bilinear results [1,3] to variable structure systems of a certain class.

3.1 Problem Statement

Consider a stochastic discrete variable-structure system

$$\underline{x}_{k+1} = \Phi\, \underline{x}_k + \underline{d}\, u_k g_k(\underline{x}_k) + H\, \underline{\xi}_k + B(\underline{x}_k)\underline{\alpha}_k\ , \quad 0 \le k \le N-1 \qquad (3.1)$$

where $\underline{x}_k \in R^n$, $u_k \in R^1$, Φ, $\underline{d}$ and $g_k(\cdot)$ are as in (2.1), H is an $(n \times d_1)$ constant matrix and $B(\underline{x}_k) = \sum_{i=1}^{n} x_k^i B_i$, where B_i, $1 \le i \le n$, are $(n \times d_2)$ constant matrices. The initial state $\underline{x}_0$ is Gaussian with $E(\underline{x}_0) = \underline{m}_0$, $\operatorname{cov}(\underline{x}_0) = \Sigma_0$. The noise sequences $\{\underline{\xi}_k : k = 0,1, \ldots, N-1\}$, $\{\underline{\alpha}_k : k = 0,1,\ldots,N-1\}$ are each Gaussian white, and independent of each other and of $\underline{x}_0$, with mean and covariance

$$E(\underline{\xi}_k) = 0, \qquad E(\underline{\xi}_k \underline{\xi}_j^T) = \Xi_k\, \delta_{kj} \quad ;$$

$$E(\underline{\alpha}_k) = 0, \qquad E(\underline{\alpha}_k \underline{\alpha}_j^T) = \Lambda_k\, \delta_{kj}$$

respectively, where δ_{kj} is the Kronecker delta function. Exact measurement of the state is assumed available at each k, $0 \le k \le N-1$. The problem is to choose the control policy $\{u_i(\underline{x}_i),\ 0 \le i \le N-1\}$ so that the (unconditional) cost functional

$$J = E\left\{ \sum_{k=1}^{N-1} \underline{x}_k^T\, Q\, \underline{x}_k + \underline{x}_N^T\, Q'\, \underline{x}_N \right\} \qquad (3.2)$$

is a minimum, where Q and Q' are symmetric and nonnegative. E is the expectation operator.

3.2 Optimal Control Policy

Theorem 3.1: The solution to the problem stated in Section 3.1 is given by

$$\underset{0 \le k \le N-1}{u_k} = \begin{cases} -\dfrac{\sum_{i=1}^{n} x_k p_{i-1,n}^k + p_{nn}^k \underline{a}^T \underline{x}_k}{p_{nn}^k g_k(\underline{x}_k)} & \text{if } p_{nn}^k g_k(\underline{x}_k) \ne 0 \\ \text{arbitrary}, & \text{if } p_{nn}^k g_k(\underline{x}_k) = 0 \end{cases} \qquad (3.3)$$

where, $p^k_{0,n} \triangleq 0$, $P_k = (p^k_{ij})$ is given recursively by

$$P_k = Q + F(P_{k+1}) + G_{k+1}, \quad 0 \leq k \leq N-2 \tag{3.4}$$

with initial condition $P_{N-1} = Q'$, where G_{k+1} is given by

$$(G_{k+1})_{ij} = tr(\Lambda_{k+1} B_i^T P_{k+1} B_j), \quad 0 \leq k \leq N-2$$

and $F(\cdot)$ is as in (2.5). G_k and P_k are nonnegative, $0 \leq k \leq N-1$. The (conditional) optimal cost for the last (N-k) stages given $\underline{x}_k$ is given by

$$V_{N-k} = \underline{x}_k^T \{(\Phi^T P_k \Phi - w_k) + G_k\}\underline{x}_k + tr \sum_{i=k}^{N-1} (H^T P_i H) \Xi_i, \quad k = 0,\ldots N-1 \tag{3.5}$$

where

$$\underset{0 \leq k \leq N-1}{w_k} = \begin{cases} \tilde{C}_k & , \text{ if } p^k_{nn} g_k(\underline{x}_k) \neq 0 \\ 0 & , \text{ if } p^k_{nn} g_k(\underline{x}_k) = 0 \end{cases} \tag{3.6}$$

$\tilde{C}_k = (\tilde{c}^k_{ij})$ for $P_k = (p^k_{ij})$ are related as in (2.4) for Q_k. Minimum value of J (unconditional) is given by

$$\underline{m}_0^T S_0 \underline{m}_0 + tr(S_0 \Sigma_0) + tr \sum_{i=0}^{N-1} (H^T P_i H) \Xi_i, \tag{3.7}$$

where $S_0 = (\Phi^T P_0 \Phi - w_0) + G_0$.

Note that the addition of noises not only increases the cost (3.5) which also occurs in linear systems but also alters the form of the expression for optimal control (3.3), (3.4) which is unlike linear systems. The noise $\underline{\alpha}$ even directly contributes a term to the control (3.4).

Before we prove the theorem, we need the following lemma.

<u>Lemma 3.2:</u> For $\underline{x}_k$ as in (3.1) the conditional expectation of $\underline{x}_k^T w_k \underline{x}_k$ is

$$E\{\underline{x}_k^T w_k \underline{x}_k | \underline{x}_{k-1}\} = E\{\underline{x}_k^T \tilde{C}_k \underline{x}_k | \underline{x}_{k-1}\} \tag{3.8}$$

with w_k as defined in (3.6).

Proof: The result follows by noting that

$$\int_{X_0^{(k)}} (\underline{x}_k^T \tilde{C}_k \underline{x}_k) \; f_{\underline{x}_k|\underline{x}_{k-1}} (\underline{x}_k | \underline{x}_{k-1}) \, d\underline{x}_k = 0 ,$$

where $f_{\underline{x}_k|\underline{x}_{k-1}} (\underline{x}_k | \underline{x}_{k-1})$ is the conditional density and $X_0^{(k)}$ is the set,

$$X_0^{(k)} = \{\underline{x}_k : g_k(\underline{x}_k) = 0\} .$$

which has measure zero.

Proof of the theorem: Proof of the theorem is inductive. We have the functional equation [13]

$$V_1 = \min_{u_{N-1}} E\{\underline{x}_N^T Q' \underline{x}_N | \underline{x}_{N-1}\} \tag{3.9}$$

$$V_{N-k} = \min_{u_k} E\{\underline{x}_{k+1}^T Q \underline{x}_{k+1} + V_{N-k-1} | \underline{x}_k\}, \; k = 0,\dots,N-2. \tag{3.10}$$

(i) From (3.9) and (3.1), setting $P_{N-1} = Q'$, proceeding as in the proof of Theorem 2.2, we get

$$u_{N-1} = - \frac{\sum_{i=1}^{n} x_{N-1}^i p_{i-1,n}^{(N-1)} + p_{nn}^{(N-1)} \underline{a}^T \underline{x}_{N-1}}{p_{nn}^{(N-1)} g_{N-1}(\underline{x}_{N-1})} \quad \text{if } p_{nn}^{(N-1)} g_{N-1}(\underline{x}_{N-1}) \neq 0 \tag{3.11}$$

when $p_{nn}^{(N-1)} g_{N-1}(\underline{x}_{N-1}) = 0$, u_{N-1} is arbitrary since then $E(\underline{x}_N^T Q' \underline{x}_N | \underline{x}_{N-1})$ becomes independent of u_{N-1}. From (3.11) and (3.9)

$$V_1 = \underline{x}_{N-1}^T \{(\Phi^T P_{N-1} \Phi - w_{N-1}) + G_{N-1}\} \underline{x}_{N-1} + tr(H^T P_{N-1} H) \Xi_{N-1}$$

where $(G_{N-1})_{ij} = tr(\Lambda_{N-1} B_i^T P_{N-1} B_j)$, and w_{N-1} as in (3.6).

From $\underline{x}_{N-1}^T G_{N-1} \underline{x}_{N-1} = \mathrm{tr}(\Lambda_{N-1} B^T(\underline{x}_{N-1}) P_{N-1} B(\underline{x}_{N-1}))$, G_{N-1} is nonnegative using the fact that tr (AB) $\geq$ 0 if A and B are nonnegative [14].

(ii) Now assume that for any k, $0 \leq k \leq N-2$,

$$V_{N-k-1} = \underline{x}_{k+1}^T \{(\Phi^T P_{k+1}\Phi - w_{k+1}) + G_{k+1}\}\underline{x}_{k+1} + \mathrm{tr} \sum_{i=k+1}^{N-1} (H^T P_i H)\Xi_i, \tag{3.12}$$

where w_{k+1} as defined in (3.6) and P_i, $k+1 \leq i \leq N-1$, are symmetric and nonnegative, and $(G_{k+1})_{ij} = \mathrm{tr}(\Lambda_{k+1} B_i^T P_{k+1} B_j)$. From [14], G_{k+1} is nonnegative. From (3.12), (3.10) and (3.8)

$$u_k = \begin{cases} -\dfrac{\sum_{i=1}^{n} x_k^i p_{i-1,n}^k + p_{nn}^k \underline{a}^T \underline{x}_k}{p_{nn}^k g_k(\underline{x}_k)}, & \text{if } p_{nn}^k g_k(\underline{x}_k) \neq 0 \\ \text{arbitrary}, & \text{if } p_{nn}^k g_k(\underline{x}_k) = 0 \end{cases} \tag{3.13}$$

where

$$P_k = Q + F(P_{k+1}) + G_{k+1} \,.$$

Clearly, P_k is nonnegative. Substituting (3.13) into (3.10), (as in the proof of Theorem 2.2), we get the expressions for V_{N-k} and w_k as in (3.5) and (3.6) respectively, and $(G_k)_{ij} = \mathrm{tr}(\Lambda_k B_i^T P_k B_j)$. Again, from [14], G_k is nonnegative.

Combining (i) and (ii) of the proof, we have u_k, V_{N-k}, P_k and G_k, $0 \leq k \leq N-1$, as stated in the theorem.

Noting that $J_{min} = E(V_N)$ and that $E(\underline{x}^T S \underline{x}) = \underline{m}^T S \underline{m} + \mathrm{tr}\, SR$, where $E(\underline{x}) = \underline{m}$, cov $(\underline{x}) = R$, we get (3.7). This completes the proof of the theorem.

Remarks 3.1: (i) If at any k, $p_{nn}^k = 0$, then $p_{nn}^k g_k(\underline{x}_k) = 0$ for all $\underline{x}_k$. Hence u_k is arbitrary for all $\underline{x}_k$ and in (3.5) $w_k = 0$ for all $\underline{x}_k$ (see (3.6)). Of course, in this special case the set of singular controls is not of measure zero. However, $\Phi^T P_k \Phi = F(P_k)$ (see remark 2.1). Hence, the recursive equation is still given by (3.4).

(ii) It is assumed that each component of $\underline{x}_k$, $1 \leq k \leq N-1$ is random.

IV. CONCLUSIONS

The subject of this paper is the optimal control of a class of single-input, discrete, variable-structure systems. Closed-form solutions have been obtained for both the deterministic and stochastic problems which, to the authors' knowledge, is the first such solution for any nonlinear system other than the bilinear.

As opposed to the optimal control of linear deterministic systems, regular and singular optimal paths from a given state can exist simultaneously in deterministic variable structure systems but then the regular path is optimal. This effect is similar to what happens in bilinear systems but the form of the equations changes. When additive noise entering directly through a constant gain is present this will affect the control sequence only indirectly though altering the form of the recursive equations. If the noise enters through a linearly state dependent gain then this noise component also directly contributes a term to the control. This again is a situation similar to that in bilinear systems, while in linear systems the form of the control equation is unaffected by additive noise. The noise, however, does increase the cost in all cases. One can say that in a sense the presence of noise facilitates the solution because the singular paths form a set of measure zero, hence the complications arising in the deterministic case from the existence of multiple paths are wiped out by taking expectations.

REFERENCES

(1) Swamy, K.N., "Optimal Control of Single-Input Discrete Bilinear Systems," D.Sc. Dissertation, Washington University, St. Louis, Mo., December 1973.

(2) Swamy, K.N., and Tarn, T.J., "Deterministic Optimal Control of Single-Input Discrete Bilinear Systems," submitted for publication.

(3) Swamy, K.N., and Tarn, T.J., "Stochastic Optimal Control of Single-Input Discrete Bilinear Systems," submitted for publication.

(4) Swamy, K.N., and Tarn, T.J., "Optimal Control of Discrete Bilinear Systems," in "Proceedings of NATO Advanced Study Institute on Geometric and Algebraic Methods for Nonlinear Systems," Aug.-Sept. 1973, Imperial College, London.

(5) Mohler, R.R., and Ruberti, A., ed., "Theory and Applications of Variable Structure Systems," Academic Press, New York, 1972.
(6) Goka, T., Tarn, T.J., and Zaborszky, J., "On the Controllability of a Class of Discrete Bilinear Systems," Automatica, Vol. 9, No. 5, September 1973, pp. 615-622.
(7) Tarn, T.J., Elliott, D.L., and Goka, T., "Controllability of Discrete Bilinear Systems with Bounded Control," IEEE Transactions on Automatic Control, Vol. AC-18, June 1973, pp. 298-301.
(8) Rink, R.E., and Mohler, R.R., "Completely Controllable Bilinear Systems," SIAM J. Contr., Vol. 6, No. 3, pp. 477-486, 1968.
(9) d'Alessandro, P., "Structural Properties of Bilinear Discrete-Time Systems," Ric. Automatica, Vol. 3, Aug. 1972.
(10) Bruni, C., Di Pillo, G., and Koch, G., "Mathematical Models and Identification of Bilinear Systems," in "Theory and Applications of Variable Structure Systems" (Mohler, R.R. and Ruberti, A., ed.), Academic Press, New York, 1972, pp. 137-152.
(11) Brockett, R.W., "On the Algebraic Structure of Bilinear Systems," in "Theory and Applications of Variable Structure Systems" (Mohler, R.R., and Ruberti, A., ed.), Academic Press, New York, 1972, pp. 153-168.
(12) Bellman, R., "Adaptive Control Processes: A Guided Tour," Princeton University Press, Princeton, 1961, p. 56.
(13) Jacobson, D.H., and Mayne, D.Q., "Differential Dynamic Programming," American Elsevier Publishing Company, Inc., New York, 1970, p. 136.
(14) Graybill, F.A., "Introduction to Matrices with Applications in Statistics," Wadsworth Publishing Company, Belmont, Calif., 1969, p. 231.

A METHOD OF LEARNING CONTROL VARYING SEARCH DOMAIN BY FUZZY AUTOMATA

Seizo Kitajima and Kiyoji Asai
Osaka City University — University of Osaka Perfecture
Osaka, Japan — Sakai, Osaka, Japan

1. INTRODUCTION

The learning control method is used for the purpose of finding and holding the optimum of controlled systems whose characteristics are unknown. There are the sequential decision method and the method by the stochastic automata [1] or the fuzzy automata[2] as this learning control method.

The authors have developed the learning control system using the Moore type fuzzy automata in which a state has multi-outputs and in which the higher-order transition and the operation of self-organization are performed, and discussed on the behaviors of the system [3][4]. In this system, the domain of the objective function for optimization is divided into the number of subdomains corresponding to the number of states in the fuzzy automaton, and every subdomain is also divided into some unit domains corresponding to the output from a state. A global and a local search are executed respectively.

In this paper, the partition method of the objective function domain and decision method of the subdomain size are proposed in order to improve the control characteristics. The learning behaviors of the system are discussed on the basis of the results of simulation using the digital computer.

The principle of this learning search method is shown as follows:

Every point of the objective function is assigned to every state of the fuzzy automaton, and a subdomain is formed as a set of some points. The membership function is given in every state so as to indicate the degree of optimum. The state with the maximum value of the membership function is selected from among the states, each of which corresponds to every point of a subdomain. The point selected thus is named the candidate point, and a new subdomain is formed from points around the candidate point. The size of the subdomain grows larger when the value of objective function at the candidate point is smaller than the values of objective function at the other points in the subdomain and grows smaller when the value of objective function at the candidate point is not so that the optimum may be found rapidly. When a new subdomain and another overlap each other and two candidate points exist in the same subdomain, two subdomains will be separated if the degree of separation is large, or be unified if that is small.

Thus, the candidate point is decided after the local search is executed in the subdomain, and then the global search in which the search is executed among the candidate points is executed in order to find the optimum point over the whole domain of objective function. The global search and the local search are executed alternately, so that the optimum point is found rapidly. Consequently, the control system can hold the optimum at a small hunting loss for the multimodal systems.

2. VARIABLE STRUCTURE FUZZY AUTOMATA

The variable structure fuzzy automaton is the Moore type automaton in which the transition from a state to another state is executed through the states selected on the basis of the membership function of each state.

This automaton is expressed as follows.

$$M = < S, X, U, \{F(x)\}, g_s, H_l > \quad (1)$$

where

$S = \{s_1, s_2, \ldots, s_\xi\}$: set of ξ states,

$X = \{x_1, x_2, \ldots, x_\mu\}$: set of μ inputs,

$U = u_1, u_2, \ldots, u_\xi\}$: set of ξ outputs,

$F(x)$: $\xi \times \xi$ fundamental fuzzy transition matrix,

g_s : membership function for indicating the degree of the optimum at the state ($0 \le g_s \le 1$),

H_l : $\xi \times \xi$ selection matrix ($l = 1,2$).

The matrix H_1 (i.e. matrices H_1 and H_2) is characterized by the membership function, whose values are only 1 or 0, for the transition from a specified state to the same state or another specified state. The behavior of the transition may be changed by using the matrix H_1. This automaton is named the variable structure automaton.

In order to execute the transition through some specified states, for example s_i, s_j and s_k, the entries of matrix H_1 may be given as follows.

$$H_1 = \begin{array}{c} \\ s_i \\ \\ s_j \\ \\ s_k \\ \\ \end{array} \left(\begin{array}{ccccccc} & s_i & & s_j & & s_k & \\ \cdots & 1 & \cdots & 1 & \cdots & 1 & \cdots \\ & \vdots & & \vdots & & \vdots & \\ \cdots & 1 & \cdots & 1 & \cdots & 1 & \cdots \\ & \vdots & & \vdots & & \vdots & \\ \cdots & 1 & \cdots & 1 & \cdots & 1 & \cdots \\ & \vdots & & \vdots & & \vdots & \end{array} \right) \qquad (2)$$

where the entries except 1 are 0.

The transition matrix $F'(x)$ by which the transition is executed actually is given by

$$F'(x) = F(x) \otimes H_1 \qquad (3)$$

where $\otimes$ is fuzzy product, and the entries of $F'(x)$ are given by $f'_{ij} = \min(f_{ij}, h_{1,ij})$.

In order to execute the transition from a state to a specified state in cyclic, for example $s_i \to s_j \to s_k \to s_i$, the entries of selection matrix H_2 may be given as follows.

$$H_2 = \begin{array}{c} \\ s_i \\ \\ s_j \\ \\ s_k \\ \\ \end{array} \left(\begin{array}{ccccccc} & s_i & & s_j & & s_k & \\ \cdots & \cdot & \cdots & 1 & \cdots & \cdot & \cdots \\ & \vdots & & \vdots & & \vdots & \\ \cdots & \cdot & \cdots & \cdot & \cdots & 1 & \cdots \\ & \vdots & & \vdots & & \vdots & \\ \cdots & 1 & \cdots & \cdot & \cdots & \cdot & \cdots \\ & \vdots & & \vdots & & \vdots & \end{array} \right) \qquad (4)$$

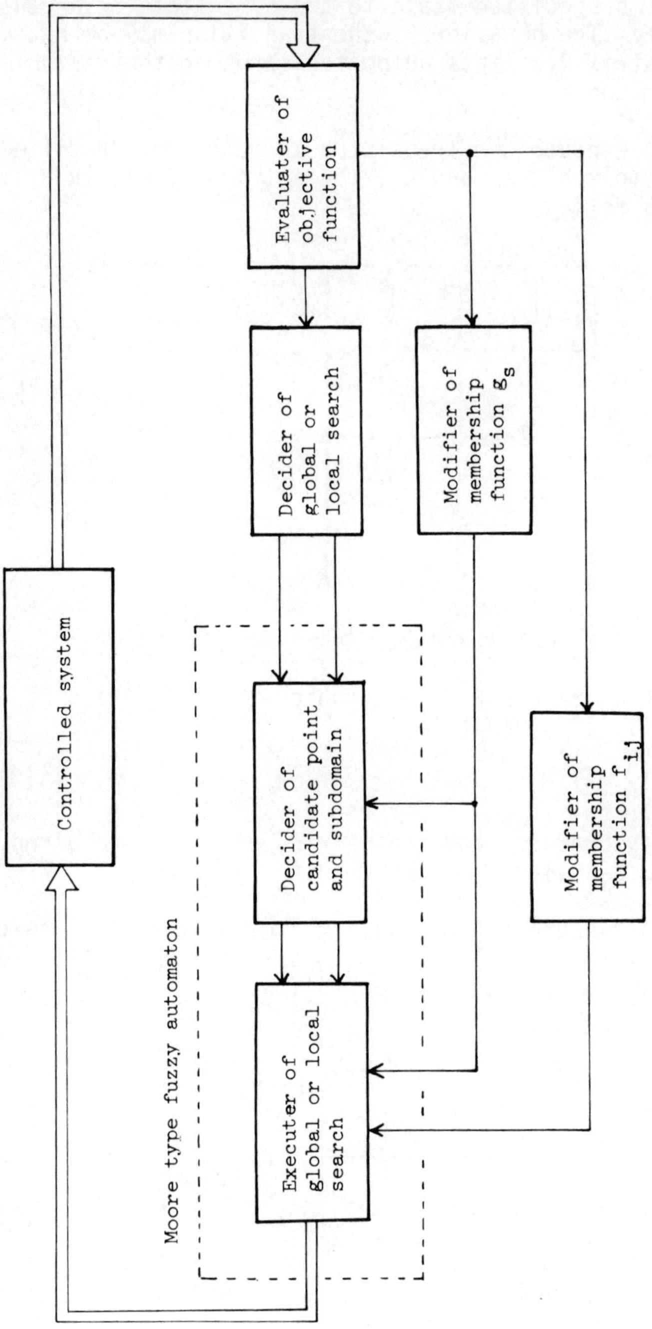

Fig. 1. Block diagram of learning control system.

where the entries except 1 are 0.

The transition matrix $F'(x)$ by which the transition is executed actually is given by

$$F'(x) = \{F(x) \otimes H_2\} \oplus H_2 \tag{5}$$

where $\oplus$ is fuzzy sum, and the entries of $F'(x)$ are given by f'_{ij} = max {min $(f_{ij}, h_{2,ij}, h_{2,ij}\}$.

In the following, the performance of a learning control system using this automaton as learning controller will be described.

3. LEARNING CONTROL SYSTEMS

The block diagram of a learning control system using the variable structure fuzzy automaton is shown in Fig. 1.

This system consists of the controlled system with unknown characteristics, the evaluater of objective function with multi-modal response surface, and the Moore type fuzzy automaton with ξ states and ξ outputs.

The domain of the objective function is partitioned by ξ lattice points corresponding to the number of states of the automaton. A subdomain is formed as a set of some lattice points, and these subdomains correspond to subsets of the automaton. The lattice point corresponding to the state S_m^i with the maximum value g_m^i of membership function is named the candidate point of the subdomain.

3.1 Performance of System

The flow chart of the performance of the control system is shown in Fig. 2, and the performance is explained as follows.

(1) In the fuzzy automaton, the values of entries f_{ij} (membership function of the transition) of the fundamental transition matrix and the value of the membership function g_s are set in the middle value 0.5 between 0 and 1, respectively. On the other hand, the candidate points are selected, and the size of subdomain and the values of entries of the selection matrices H_1 and H_2 are decided. The transition is executed on the basis of the membership function $f_{ex,ij}$ given as follows.

In the case of the global search:

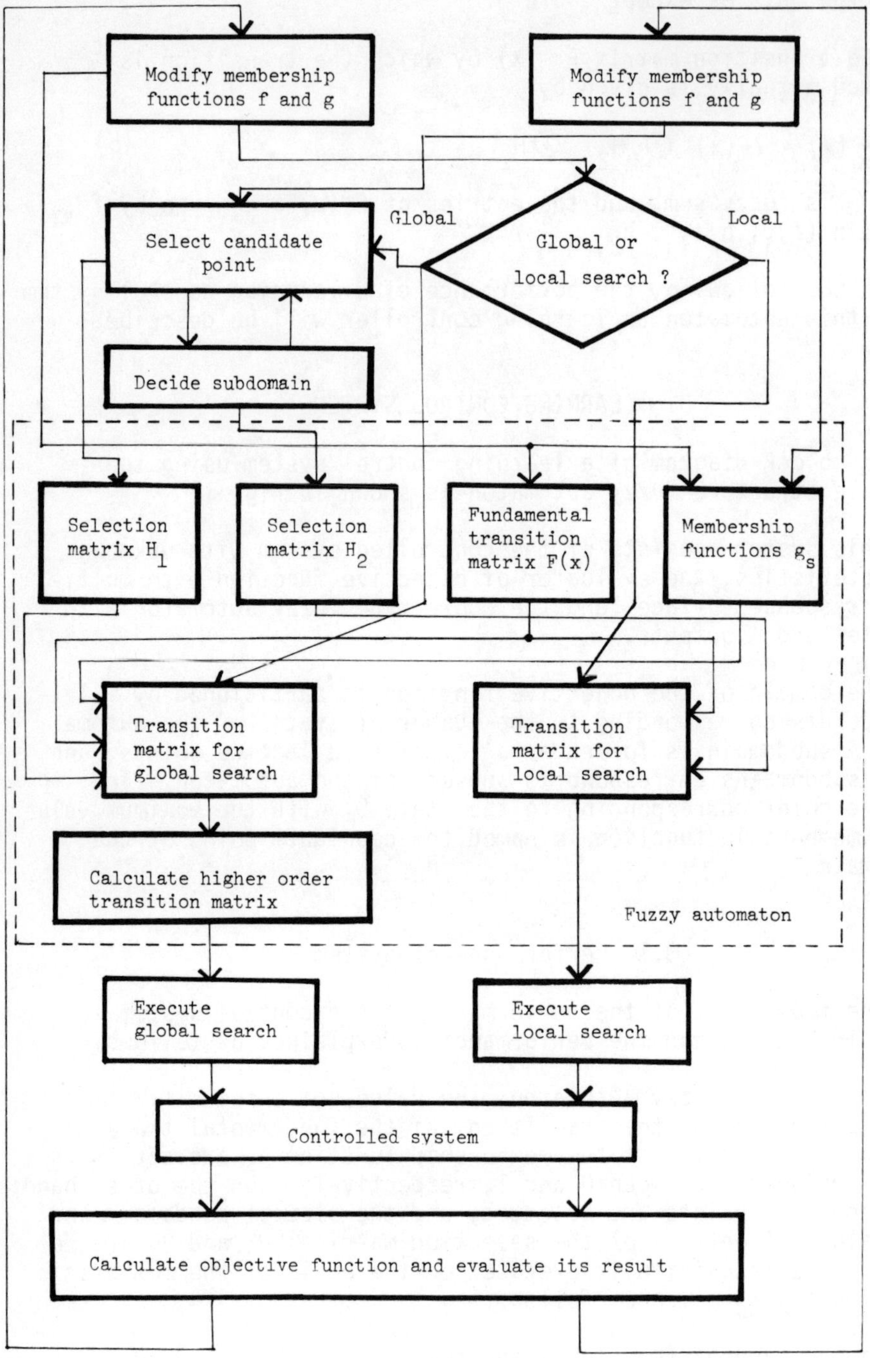

Fig. 2. Flow chart of performance of learning control system.

$$F'(x) = \{F(x) \otimes H_1\} , \quad (6)$$

$$f_{ex,ij} = \min(f'_{ij}, g_j). \quad (7)$$

In the case of the local search:

$$F'(x) = \{F(x) \otimes H_2\} \oplus H_2 , \quad (8)$$

$$f_{ex,ij} = \min(f'_{ij}, g_j). \quad (9)$$

In the case of the global search, a new transition matrix whose entries are given as $f_{ex,ij}$ is formed, and its k-th order transition matrix is obtained. When the k-th order transition is executed by using this matrix, the outputs of the automaton will be sent out.

(2) The outputs of the automaton are applied to the controlled system from which the outputs will be sent out. These outputs are applied to the evaluater for the objective function.

(3) In the evaluater, the present value I(n) of the objective function is compared with the mean value $\overline{I}(n)$ of the values of the objective function thus far obtained, and the result of the trial is evaluated.

(4) On the basis of the evaluation, the membership functions f_{ij} and g_s are modified as follows.

(i) the case of the global search

If $I(n) > \overline{I}(n)$, i.e. the trial is success,

$$f_{ij}(n + 1) = \alpha\, f_{ij}(n) + (1 - \alpha) \cdot 1 , \quad (10)$$

If $I(n) \leq \overline{I}(n)$, i.e. the trial is failure,

$$g_m^j(n + 1) = \alpha\, g_m^j(n) \quad (11)$$

where $\alpha = 1 - |(I(n) - \overline{I}(n))/\overline{I}(n)|$ and n is the trial number.

(ii) the case of the local search

If $I(n) > \overline{I}(n)$ (success),

$$g_k^j(n+1) = \alpha \cdot g_k^j(n) + (1-\alpha) \cdot 1 , \qquad (12)$$

If $I(n) \leq \bar{I}(n)$ (failure),

$$g_k^j(n+1) = \alpha\, g_k^j(n). \qquad (13)$$

If $f_{ij}(n+1) \geq \theta_1$ (θ_1 is the threshold value) after the global or the local search has been executed, the membership functions of transition are given as follows.

$$f_{lj}(n+1) = \{\max_k g_k^j\}\ \phi\,[f_{ij}(n+1) - \theta_1] , \qquad (14)$$

where $l \neq i,j$,

ϕ : function whose value is 1 if [ ] is positive and 0 if [ ] is negative.

$$f_{jj}(n+1) = \{\alpha\, f_{jj}(n) + (1-\alpha) \cdot 1\}\ \phi\,[f_{ij}(n+1) - \theta_2] , \qquad (15)$$

where $\theta_2 > \theta_1$.

Thus the transition to the subset in which the optimum may probably be included can be executed easily. Also the value of the membership function for the transition from the state s_m^j to the state s_m^j may grow larger when $f_{ij}(n+1) \geq \theta_2 > \theta_1$.

(5) After the membership functions f_{ij} and g_k^j have been modified, the local search will be executed if the following equation is satisfied,

$$f_{ij}(n+1) \geq \theta_1 . \qquad (16)$$

In the subset, all the states which the following equation satisfies are searched.

$$\phi[f_{ij}(n+1) - \theta_1] + \phi[g_k^j(n) - \bar{g}_k^j] = 2 , \qquad (17)$$

where $\bar{g}_k^j$ is the mean value of g_k^j in subset s_j.

The global search will be executed after the local search.

(6) Before the global search, the candidate points will be selected and the selection matrix H_1 will be obtained. And also, after the candidate points have been selected, the position and the size of subdomain may be decided by the method discussed later, and the selection matrix H_2 may be obtained.

(7) After the fundamental transition matrix, the selection matrices and etc. have been obtained, the global or the local search will be executed.

By repeating the procedure mentioned above, the optimum may be searched.

3.2 Method of Decision of Subdomain

If a lattice point P^i_m is given, a subdomain may be defined as the space including all points in the circle whose center is the point P^i_m and radius is A, that is,

$$\{P^i \mid d(P^i, P^i_m) \leq A\} \equiv S_i, \tag{18}$$

where $d(P^i, P^i_m)$: distance between P^i_m and P^i,
S_i : ith subdomain.

Therefore, if the candidate point P^i_m is selected before the global search, the subdomain may be decided by the above equation.

In this case the size of subdomain may be changed according to the various values of A. In this paper, three sizes of the subdomain are chosen, and the size will be changed as follows.

(i) The size of the subdomain will grow larger by one step, when the value of objective function at the candidate point is smaller than the values of objective function at the other points in the subdomain.

(ii) The size of the subdomain will grow smaller by one step, when the value of objective function at the candidate point is larger than the values at the other points in the subdomain.

When two subdomains S_i and s_j overlap each other and the distance between the points P^i_m and P^j_m is smaller than a threshold value, these domains will be separated or unified according to the degree of separation given by the following equation,

$$D = \frac{|g^i_m - g^j_m|}{g^i_m + g^j_m} \tag{19}$$

where g^i_m and g^j_m : membership functions of the states s^i_m and s^j_m of the automaton corresponding to P^i_m and P^j_m, respectively.

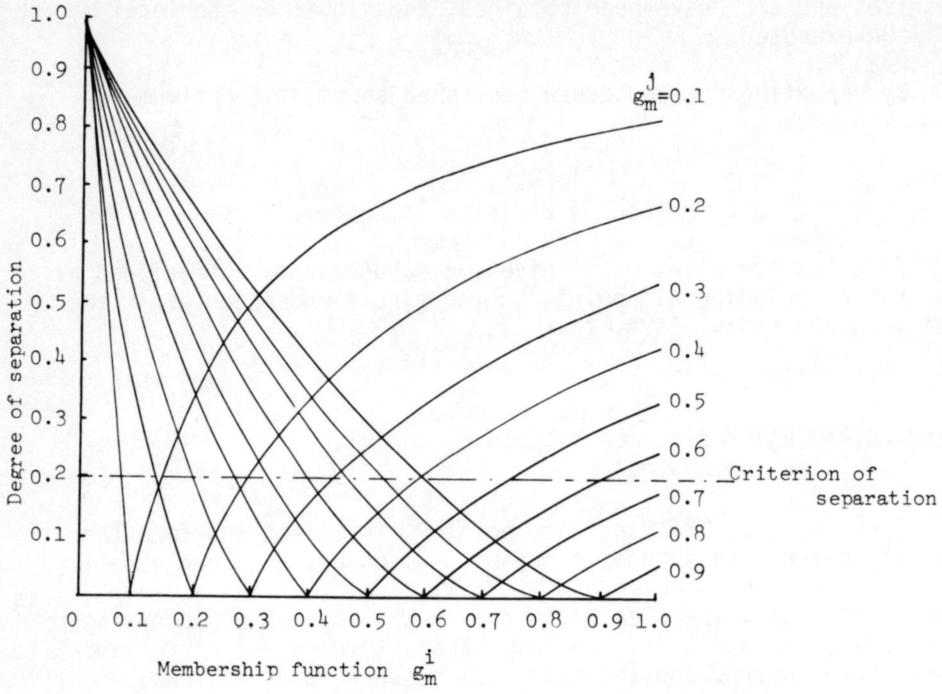

Fig. 3. Characteristics of degree of separation.

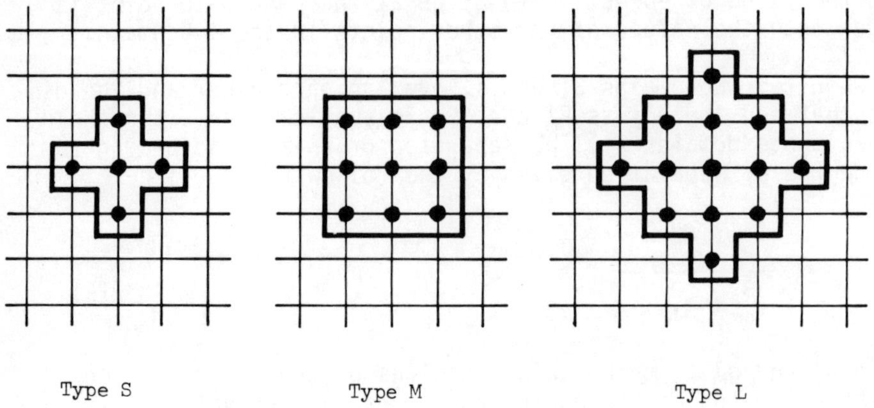

Fig. 4. Types of subdomain.

In the case of $D \geq \theta_3$, two subdomains may be unified. The candidate point corresponding to the state whose membership function g_m is the largest may be selected, and then the subdomain whose center puts on this point may be obtained. The size of the subdomain is larger by one step than the previous size.

In the case of $D > \theta_3$, two subdomains may be separated.

The graph of the equation given above is shown in Fig. 3. In Fig. 3, the dotted line shows a criterion of separation. Two subdomains may be unified if the value D is smaller than the value of the criterion of separation. From Fig. 3, it is seen that it is difficult to unify two subdomains even if the difference between the membership functions g_m^i and g_m^j is small when the values of g_m^i and g_m^j are small.

4. SIMULATION RESULTS

A simulation study has been carried out for the purpose of investigating the behaviors of the learning control using the method described above. The objective function for the optimization is a function of two variables, u_1 and u_2; and there are the optimum, one local optimum and one saddle point. The domain of the objective function was divided by 9×9 points of lattice corresponding to the number of states in the fuzzy automaton. Three types of subdomain S, M and L shown in Fig. 4 were chosen for the convenience of the simulation study, and the type M was picked out at the beginning of the simulation.

The procedure for the simulation study will be explained below by Fig. 5.

Search No. 1: The global search is executed over 9 candidate points.

Search No. 2: The global search is again executed over the other 9 points.

Search No. 3: The local search will be executed since the value of membership function of the transition path from the subdomain ③ to the subdomain ④ is larger than the threshold value θ_1 ($\theta_1 = 0.8$). After the local search, the size of the subdomain ④ will be made larger by one step since the value of the objective function at the candidate point is smaller than the values at the other points in the subdomain. The candidate point of every subdomain will be selected from among the points with the maximum value of the membership function in each subdomain.

Search No. 4: The global search is executed over the candidate points of the subdomains ①, ②, ③, ④, and ⑤ on the basis of the algorithm of the higher-order transition.

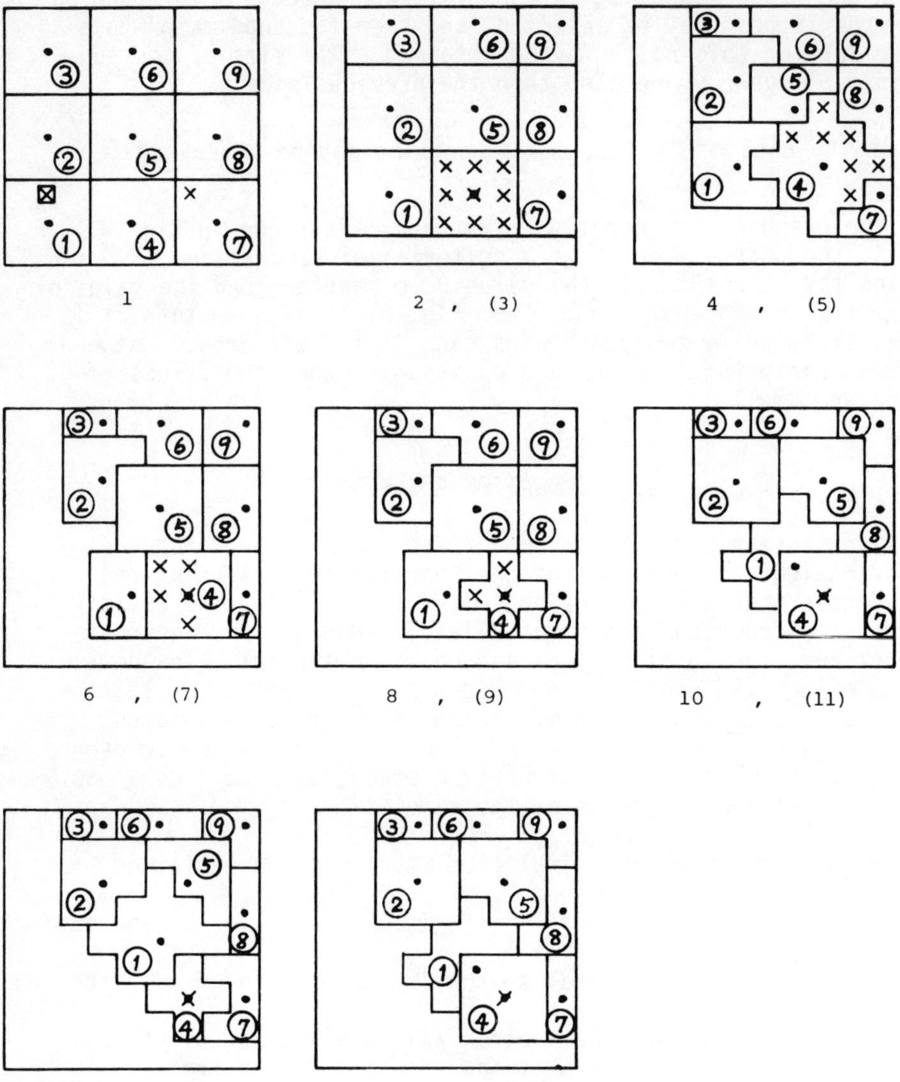

⊗ : Optimum , ⊠ : Local optimum , ● : Candidate points ,
× : Trial points in local search , ①,②,···: Number of subdomain
1,2,--- : Number of search , ( ) : Local search .

Fig. 5. Behavior of movement of subdomain.

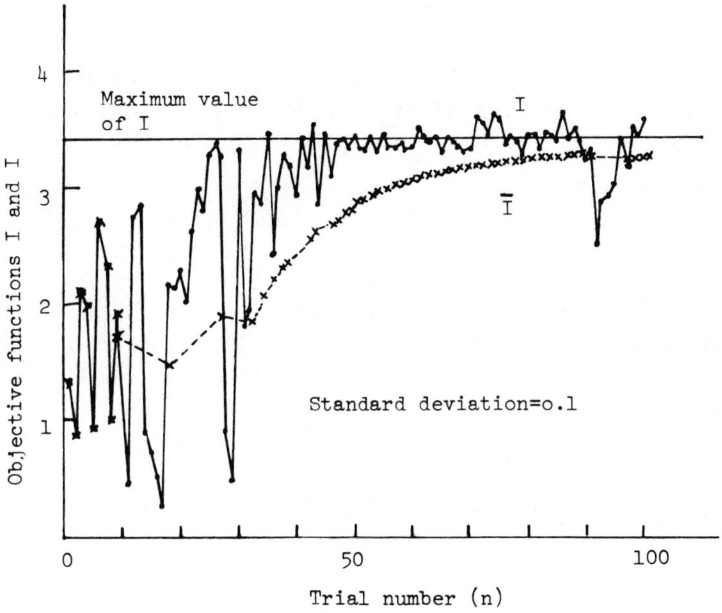

Fig. 6. An example of results of simulation study.

Search No. 5: The local search is executed. After that, the size of the subdomain ④ will be made smaller by one step since the value of the objective function at the candidate point is larger than the values at the other points in the subdomain.

Search No. 6: The global search is executed over the candidate points of the subdomains ① and ④.

Search No. 10. The subdomains ① and ④ are unified by the method mentioned before.

Thus, the search is executed over the whole domain of objective function at the early stage of the learning, but all subdomains approach to the optimum with the progress of the learning, and the search is executed nearer the optimum at the final stage of the learning.

Fig. 6 shows the convergent characteristics of the objective function with the increasing of trial number, and shows that the objective function may be in its maximum value at about the 50th trial.

5. CONCLUSION

In the preceding paragraphs, the constitution of the learning control system using the fuzzy automata in which the position and the size of subdomains are varied and its learning characteristics have been described.

From the results of simulation study, it has been concluded that the learning control system by this method is able to realize the characteristics of rapid response and small hunting loss for the multimodal systems.

6. REFERENCES

(1) K.S. Fu, "Stochastic Automata as a Models of Learning Systems", The Second Symposium on Computer and Information Sciences, Aug. 22-24, 1966.

(2) W.G. Wee & K.S. Fu, "A Formulation of Fuzzy Automata and its Application as a Model of Learning Systems", IEEE Trans., Vol. SSC-5, No. 3, July, 1969, pp. 215-223.

(3) K. Asai and S. Kitajima, "Learning Control of Multimodal Systems by Fuzzy Automata", Pattern Recognition and Machine Learning, Plenum Press, 1971, pp. 195-203.

(4) K. Asai and S. Kitajima, "A Method for Optimizing Control of Multimodal Systems using Fuzzy Automata", Information Sciences, Vol. 3, No. 4, Oct., 1971, pp. 343-353.

ADAPTIVE COMPUTER AIDING IN DYNAMIC DECISION PROCESSES

Amos Freedy and Gershon Weltman

Perceptronics, Inc.

Woodland Hills, California

SUMMARY

This report describes in brief a research program directed toward the application of adaptive computer techniques for aiding the human decision maker in dynamic decision processes. Aiding information of several types comes from the on-line acquisition of the decision maker's value structure by a trainable computer system. A maximum-likelihood model of real-world behavior is used to predict environment-state transitions, and an expected utility model of decision-maker behavior is used to predict, evaluate, and modify or automate operator decisions. The overall system models information-acquisition strategy, as well as action choices. It is presently being implemented on an interactive minicomputer, and applied to a simulated intelligence operation involving surveillance of a mobile fishing fleet using sensors of varying costs and reliabilities. Research goals include experimental investigation of the factors which influence optimal decision aiding in complex, realistic and open intelligence-gathering and decision-making tasks. A major concern is to identify aiding techniques which best match the judgmental abilities of man with the discriminative capacity of an adaptive machine.

CONCEPT

In a large class of real-world problems the decision maker (DM) must respond to a dynamic input environment of multivariate data. These data come from sources of differing reliabilities and costs, and have different values in the achievement of decision objectives. Decisions are made sequentially, and their consequences are likely to affect future choices. The ability of the operator to build a satisfactory strategy for relating the poorly defined inputs to his successive decisions is a major determinate of success. Learning may be a significant part of this process, particularly in non-stationary decision environments.

Examples of such situations in the military range from the global to the highly specific. They include DM responses to international and regional intelligence reports, to local command and control needs (such as deposition of air, sea, and ground forces), to photo image interpretation, and to noisy signals characteristic of sonar and radar returns. Numerous examples occur outside of the military as well. Besides national intelligence, these include crime prevention, air and highway traffic control, population and environmental planning, and that quintessential decision problem--the stock market.

Previous studies have shown that human decision makers do not perform optimally in such tasks with respect to either objective or subjective criteria (Rouse; 1972; Rapoport, 1972). It has been suggested that the discrepancies between the optimal and actual performance are due to two major sources: (1) cognitive constraints of the operator, and (2) the subjective optimality problem. The first is unique to dynamic decision process and involves mainly the limitations on information processing rate, memory and judgment of probabilities (Rapoport, 1967). The second is due to subjective biases, random errors in aggregating information and the operator's difficulties in adopting a well-defined criterion of performance to guide behavior (Tversky, Slovic, Lichtenstein, 1972). It is also evident in static and multistage decision processes.

The high-speed digital computer has opened several important avenues to operator decision aiding in these complex tasks (Vaughan and Mavor, 1972). Computer aiding can improve performance by (1) supplying auxiliary-information processing functions which reduce the "inferential gap"; (2) automating some decision aspects on a mathematically optimal basis; and (3) analyzing operator behavior and offsetting counterproductive tendencies. Successful computer aided decision making has been reported in a number of cases (e.g., Miller, et al., 1967; Howell, 1967; and Kelly and Peterson, 1971). Most aiding systems, however, use the human as a sporatic transducer of probabilities and (in some cases) utilities, and are limited to performing computational functions. A few systems (e.g., Goldberg, 1970; and Dawes, 1970) have attempted to "adapt" the parameters of a decision maker model by observing DM behavior. However, these systems have been limited to off-line, static decision making tasks.

The approach described here extends the potential of adaptive decision aiding by making available the means for on-line observation of the decision maker, continuous estimation of his multiattribute utilities for information and outcomes, and simultaneous assessment of the decision environment. This enables immediate aiding in those decision situations where relevance and value must be assessed by a human while he is confronted with the decision task. The approach combines the discipline of learning-network theory with previous research on decision aiding and modeling, introducing computational algorithms and computer techniques which are uniquely suited to on-line estimation of decision-model parameters in multivariate processes.

The general concept is realized by means of a trainable, parallel, decision-maker model, based on an extension of the Autonomous Control Subsystem (ACS), a computer learning program previously developed by Perceptronics. The computer model continuously tracks the decision environment and the DM's responses, learning his value structure and decision strategy, as well as the conditional probabilities of state transistions. In effect, the experienced decision maker "shows" the computer how to evaluate the situation in his own terms. The machine in turn aids the decision maker or substitutes for him, as policy dictates.

Information generated by the model is used to aid the decision maker in several ways; these include:

(1) Elucidating his immediate value structure. This permits outside and self assessment of the DM's response to various components of the decision environment, as well as comparison with other value standards, such as organizational rules or expert opinion.

(2) Highlighting "important" incoming data. Data summaries based on previous in-task responses can keep the operator from missing events deemed critical, and help him minimize over-estimations and under-estimations of input information.

(3) Indicating changes or inconsistencies in value structure. While the adaptive model can follow changes in utility, information that a change has occurred may signal a significant happening in the decision environment, or a major reassessment by the DM of important decision criteria.

(4) Quantifying expected utility. In intelligence gathering, for example, the decision to continue data acquisition or to report the estimated state may rest on whether the instantaneous expected utility (EU) is under or over a preestablished level.

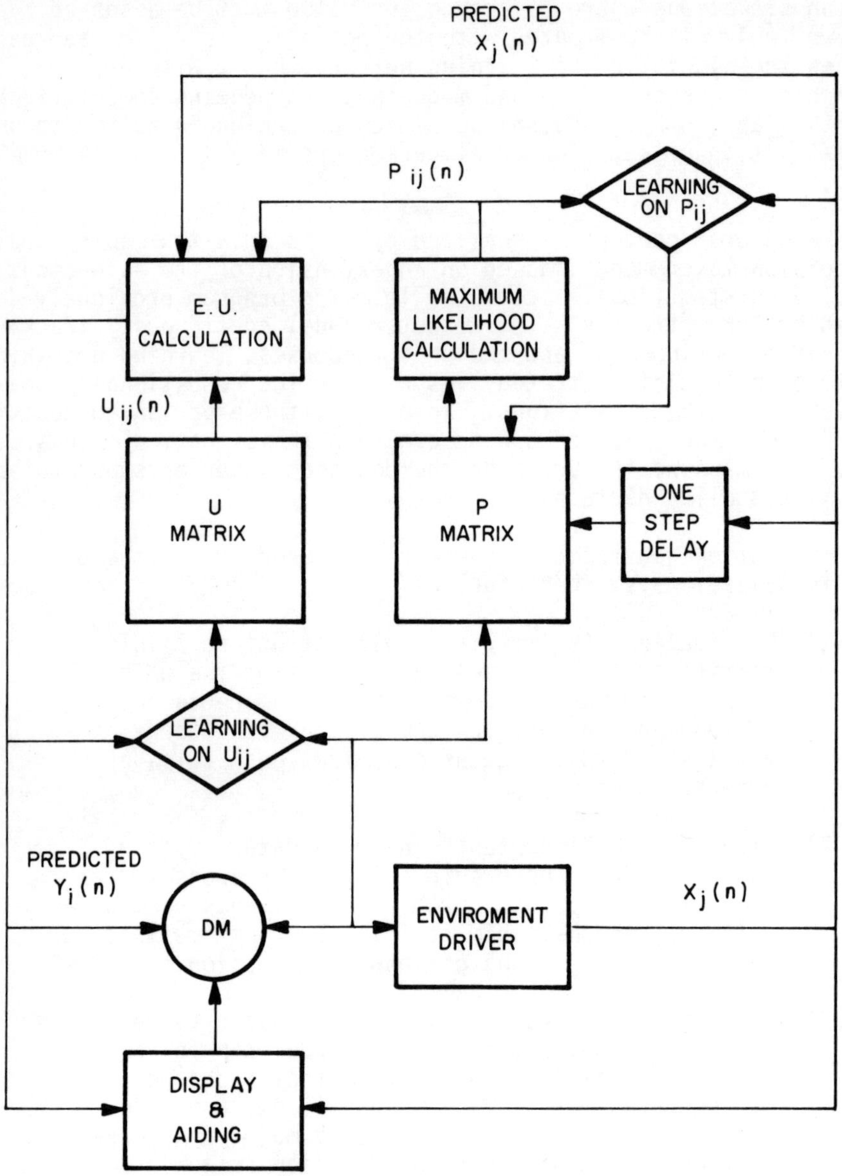

Fig. 1. ADDAM - Adaptive Dynamic Decision Aiding Machine

(5) Suggesting optimum decisions. The DM can be presented with decisions based on machine optimization which incorporates his recognized value structure and/or that of other identifiable sources (e.g., published policy, expert consensus, etc.). Continued acceptance of machine decisions may lead to progressive transfer of the task to the computer, with the human DM retaining the means of review and override.

The on-line model makes available a new type of measure--which we might term "dynamic utility". It is immediately evident that dynamic utility has major implications for decision research and decision training, as well as for operational aiding. It is of interest, for example, to determine the factors which influence how rapidly the DM's value structures converge after introduction to a decision environment, or reconverge after a major permutation. Convergence, itself, or the lack of it, gives us important information about the task and about the decision maker. Similarly, the movement of utilities toward stable values provides valuable feedback in decision training, while continuous comparison with multivariate "expert" standards allows the timely and precise insertion of remedial or corrective training material.

SYSTEM ORGANIZATION

The Adaptive Dynamic Decision Aiding Machine (ADDAM) models the behavior of a human decision maker in a complex, non-stationary decision environment. The model learns the behavior of the environment by adjusting the conditional probabilities of state transitions. It learns the behavior of the human by monitoring his actions and adjusting the parameters of an expected utility model. Probabilities and utilities are adaptively adjusted in real time by comparing predicted behavior with actual behavior and positively or negatively reinforcing the appropriate parameters. Decision aiding is accomplished by making available to the decision maker information contained in the model and derived from heuristic procedures.

The organization of ADDAM, shown in Figure 1, is centered around a decision-outcome likelihood matrix and a decision utility matrix. The decision-outcome likelihood matrix (P Matrix) is composed of conditional probability elements which relate the future state of the system ($\overline{X}(\underline{n}+1)$), at time (n+1), given the decision ($Y(\underline{n})$) and the present state ($X(n)_i^{b-b}$) both at time (n). The state vector (X) defines the status of the simulated real world. In the present task this includes a description of the elements of the fishing fleet and their location in a two-dimensional plane. The components of the vector Y define the action taken by the operator. In the present task, this includes deployment of sensors, their location, and status reports concerning changes in the state of the fishing fleet.

Elements of the decision utility matrix (U Matrix) are estimates of subjective utilities for each portion of the state and for each possible decision alternative. Elements of the P and U matrices are multiplied to form an expected utility aggregation model. Objective conditional probabilities were selected for use in the P Matrix because they are simple to calculate and because several studies indicate that, regardless of the decomposition rule, the same decision is usually predicated (Fischer, 1972). Future versions of the model may use heuristic modification of the objective probabilities to reflect conservation estimates in multistage decision tasks (Rapoport, 1972). Non-additive models will also be investigated.

In his interaction with ADDAM, the decision maker provides an action (Y(n)) which: (1) yields a new system state (X(n+1)) which is generated by the environment driver, (2) is used to adjust the elements of the U Matrix, and (3) is used by the P Matrix to predict the outcome. The new system state is: (1) displayed to the operator for his next decision through the sensors he selects, (2) used to adjust the elements of the decision-outcome matrix by comparing it to the predicated outcome, and (3) used in the E.U. calculation to predict the next decision. This predicated decision and the system state are compared to provide aiding information.

Learning estimation of utilities and probabilities takes place in two stages. The conditional-probability elements are rewarded and punished according to the agreement between the predicted outcome and the observed outcome. The method uses an estimation technique which allow the probabilities to adapt to changing environmental statistics (Minsky, 1969). The technique for estimating utilities involves a "non-parametric" training procedure for an N-category discriminant function (Nilsson, 1965).

Under this approach the E.U. model is considered as a linear discriminant function, with the utilitites being the weighing parameters and the probabilities the input-data features. Estimation techniques for the weighing parameters of such linear discriminant functions have been widely discussed in learning-system literature (Nilsson, 1965; Meisel, 1972). Application of these techniques in the context of behavioral decision making is new here; however, it was considered earlier by other investigators (Slagle, 1971). A first version of the technique has been realized, and is currently being used in an experiment-1 study to determine on-line, operator value structure in a simpler, man/computer control system (Freedy et al., 1973).

The environment driver generates the simulated behavior of the fishing fleet (described further in the following section).

The driver is a dynamic, statistical model, based on a probability matrix which describes state transitions of the elements of the fleet. This includes movement of fleet elements and their interaction. The matrix probabilities are provided a priori by experts. This approach is advantageous in scenario preparation, since complex behavioral patterns can be automatically generated by the computer with no additional programming.

Aiding information is presented to the operator via a display and aiding terminal. The aiding function is supported by a set of programs which are used to: (1) extract relevant data from the utility and probability matrices, (2) perform analyses to generate aiding information, and (3) format and display the information to the operator.

The aiding programs themselves depend on the specific aiding scheme under investigation, and may include: (1) convergence tests of utilities, (2) interrogation of the U matrix and selection of utilities of highest value, (3) comparison of estimated utilities with a given standard, (4) selection of events of high conditional probabilities, and (5) calculation of E.U. Results of the program operations are used as the basis for generating the aiding functions outlined in the previous section.

TASK SIMULATION

In general terms the task of the operator is to deploy sensors of varying costs to gain information about a dynamically-varying, hierarchical organization. Acquired information is combined with gratis information and used by the operator to determine and indicate the status of the organization. Payoff is based on the success and cost of the intelligence operation.

In our present simulation, the organization is a mobile fishing fleet. The elements of the organization are the factory ships, scouting boats, trawlers, and nets. The characteristics of the elements consist of heading, location, and function. Characteristics of the organization also include spatial and temporal patterns in the inter-element relationships, and the effect of environmental conditions on these relationships. The sensors provide more or less specific information about fleet elements. More specific sensors generally have lower reliability.

The task environment consists of an area of open sea and an island port with territorial waters. The environmental conditions are: day or night, phase of the moon, stormy or clear, and exposed or submerged reef. The task environment and the fishing fleet organization changes with each discrete jump of the simulated real-world clock. Between transitions the operator receives the sensor outputs, reads status classification messages from ADDAM, makes status classification decisions, receives aiding information from ADDAM, and makes sensor deployment decisions. The operator may decide to deploy any number of sensors during this period. Also, at this time, gratis sensors automatically give information about time, storm conditions, phase of the moon, etc. Physical interaction of the operator with the task is through a CRT graphics display and keyboard terminal. Environmental space is represented by a 5x5 matrix of squares. Marginal areas are reserved for "bookkeeping" notes and aiding messages. The operator (or ADDAM) assigns sensors to squares and receives for each sensor an indication of detected fleet elements; multiple sensors may be assigned for corroboration or consensus. Using the sensor information, the operator reports a status for each square to reflect his best estimate of the whereabouts and makeup of the fleet. Fleet elements and movement directions, as well as sensors, are represented symbolically to conserve display space and permit integration of data.

The fishing-fleet task can be mapped into many other intelligence-gathering and decision-making contexts. For example, the organizational elements and the procedural rules would fit the situation of guerrilla activity, ASW, and other military operations, as well as non-military activities such as freeway traffic control, medical diagnosis, and environmental monitoring. Thus, while the present task has sufficient realism to maintain operator interest, its generality is such that results can be extrapolated, and other, real-world applications can be served by much the same computer programs.

REFERENCES

1. Dawes, R.M., Graduate Admission; A Case Study, Oregon Research Institute, Tech. Report 10 (1), 1970.
2. Freedy, A., Hull, F.C., Lucaccini, L.F., and Lyman, J., A Computer-Based Learning System for Remote Manipulator Control, IEEE Trans. on S.M.C., SMC-1, 356-363, 1971.
3. Freedy, A., Weisbrod, R., and Weltman, G., "A Technique for Self-Optimization of Shared Man-Computer Decision and Control", Proceedings of the IEEE Conference on Decision and Control, (to be held December 1973).
4. Goldberg, L.R., Man vs. Model of Man: A Rationale Plus Some Evidence for a Method of Improving Upon Clinical Inferences", Psychological Bulletin, 73: 422-432, 1970.
5. Howell, W., Some Principles for the Design of Decision Systems: A Review of Six Years of Research on a Command-Control System Simulation, AMRL-TR-67-136, Aerospace Medical Research Laboratories, Wright-Patterson Air Force Base, Ohio, September, 1967.
6. Kelly, C.W., and C.R. Peterson, Probability Estimates and

Probabilistic Procedures in Current Intelligence Analysis, Report FSG 5047, Federal Systems Division, IBM Corp. Jan. 1971.
7. Miller, L.W., Kaplan, R.J., and Edwards, W., "JUDGE: A Value-Judgment-Based Tactical Command System", Organizational Behavior and Human Performance, 2, 329-374, 1967.
8. Minsky, M., and Papert, Perceptrons, MIT Press, Cambridge, MA., 1969.
9. Rapoport, A., "A Study of Human Control in a Stochastic Multistage Decision Task", J. Math. Psych., 4: 18-32, 1967.
10. Rouse, W.B., Cognitive Sources of Suboptimal Human Prediction, Ph.D. Dissertation MIT, Report DSR 70283-19, September, 1972.
11. Slagle, R.J., Artificial Intelligence, The Heuristic Programming Approach, McGraw-Hill Book Company, New York, 1971.
12. Fischer, G.W., Multi-Dimensional Value Assessment For Decision Making, Engineering Psychology Lab., University of Michigan, Tech. Report 03-7230-2-T, 1972.
13. Meisel, W.S., Computer-Oriented Approaches to Pattern Recognition, Academic Press, N.Y., 1972.
14. Nilsson, N.J., Learning Machines, New York: McGraw-Hill, 1965.
15. Rapoport, A., and T.S. Wallsten, Individual Decision Behavior, Annual Review of Psychology, 23: 131-176, 1972.
16. Tversky, A., P. Slovic, and S. Lichtenstein, Subjective Optimality: Conceptual Issues, Empirical Results, and Decision Aids, Office of Naval Research Conference on Applications of Decision and Information Processing Research, Monterey, California, October 4, 1972.
17. Vaughan, W.S., Jr. and Anne Schumacher Mavor, "Behavioral Characteristics of Men in the Performance of Some Decision-Making Task Components", Ergonomics, 1972.

OPTIMAL LEARNING RECOGNIZER FOR UNKNOWN SIGNAL SETS IN A CHANNEL WITH FEEDBACK LINK

Kokichi Tanaka, Shinichi Tamura

Osaka University

Toyonaka, Osaka 560, Japan

ABSTRACT

A general Kalman filer type supervised optimum quadric dichotomizer for an on-off sequence of unknown signal is presented. Then, the optimum analogue feedback is obtained which makes the average of forward signal transmission energy minimum. The extension of the proposed machine to nonsupervised problems is also discussed. The effectiveness of a feedback method over a non-feedback method is clarified. It is also tested by computer simulation.

I. INTRODUCTION

In some communication channels such as in space communications, it is a good assumption that the forward channel in the satellite-to-ground direction may be considered to be limited in the available transmission energy and be liable to be contaminated by additive noises. On the contrary, the ground power in the reverse direction can be taken much larger than in the forward direction. For such a case, feedback communication methods have been proposed so as to enhance the efficiency of the forward channel. In this paper, an optimum communication method in an on-off communication system with feedback link for an unknown signal waveform is described. Since a secondary radar for aircraft, etc. may also be regarded as a feedback communication system, the proposed method can also be applied to such problems.

In the dichotomization of an unknown signal is estimated from noisy measurements and then the measurement is dichotomized based

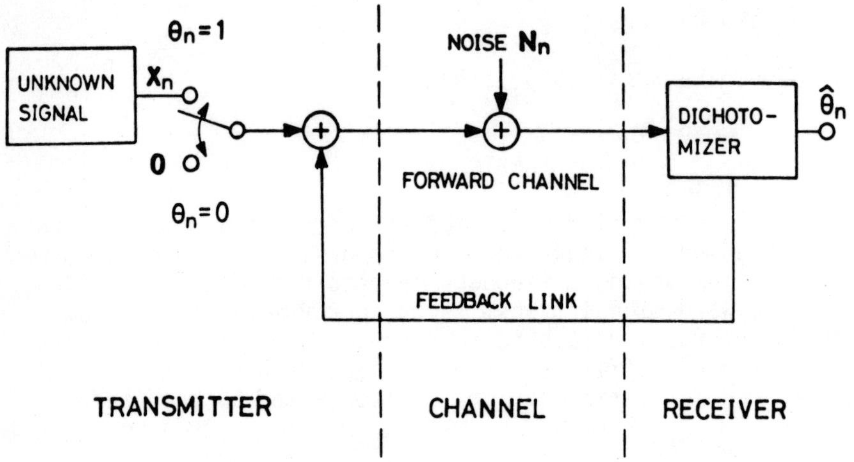

Fig. 1. Feedback communication system.

on the estimated signal. In this case if the unknown signal is estimated sequentially, such a process is called learning. The learning is classified into two types, i.e., supervised and non-supervised according to whether the learning system has a supervisor or not [1]. Further, the method of learning is also classified into parametric approach and non-parametric approach according to whether the measurement has a known probability distribution form or not. The representative one of the former is an adaptive linear filter by Jakowatz et al [2], Hinch 3 , and the author et al [4], and the representative one of the latter is a Kalman filter type quadric dichotomizer by Glaser [5], Abramson et al [6], Scudder [7], [8], and the author et al [9].

On the other hand, the concept of channel capacity has been made clear by Shannon [10]. As it is known well, however, an ideal system tha can attain the upper bound of the channel capacity will have a very complicated structure; and therefore for practical reasons, such a system is not used. In practice, highly effective communication methods, notwithstanding their relatively simple structures, have recently become of major interest to researchers. One such communication system is a feedback communication system that is made to have high performance by utilizing the feedback link. Especially, the analogue feedback communication system is now attracting the attention, and has been studied by Schalkwijk et al [11], Kashyap [12], Turin [13], Omura [14], and others. Further, Schalkwijk [15] has proposed the feedback method that makes transmission energy minimum by sending back the center-of-gravity that is a common component of each known signal.

In this paper, a general Kalman filter type supervised optimum quadric dichotomizer is presented in the first place. Next, the optimum feedback is obtained which makes the expected value of signal transmission energy minimum. It may be said that this feedback for minimum transmission energy is based on the same principle as the center-of-gravity feedback. Then, this proposed feedback method is compared with an optimum method without feedback, and the effectiveness of the feedback is shown. Further, these methods are extended to the nonsupervised problem and some results of computer simulation are presented.

II. DICHOTOMIZATION SYSTEM WITH FEEDBACK LINK

The system to be considered in this paper is illustrated in Fig. 1. Since a noisey feedback makes the analysis very cumbersome, we assume that the noise on the feedback link is negligible small. This condition is true when the feedback transmission power is sufficiently large. Further, we assume that the clocks of the transmitter and of the receiver have been already synchronized [16].

The measurement Y_n of the receiver in the n-th interval is given by

$$\underset{\sim}{Y}_n = \theta_n \underset{\sim}{X}_n - \underset{\sim}{u}_n + \underset{\sim}{N}_n , \qquad (1)$$

where θ_n is a binary message that takes value 0 or 1 in random fashion, and it conveys information to be transmitted. In the secondary radar case, its sequence corresponds to an identification code that is proper to each aircraft, etc. $\underset{\sim}{X}_n$ is an unknown signal that varies with time, obeying an ℓ-th order autoregressive process. $\underset{\sim}{u}_n$ is a feedback signal to be optimized, and $\underset{\sim}{N}_n$ is a forward channel noise. Approximating a time-varying stochastic process by an autoregressive process corresponds to the approximation of an autocorrelation function of the former process by a finite summation of exponential functions [9]. In this paper, the Bayes' optimum supervised dichotomizing method of θ_n by Kalman filter approach is presented in Section II and III, and further, the optimum feedback signal $\underset{\sim}{u}_n$ that makes the forward transmission energy minimum is obtained in Section IV.

Let us assume that the following conditions are given to the dichotomizer a priori. The binary random variable θ_n is independent of θ_k for $k \neq n$, and it takes the value 1 with the probability p and takes the value 0 with the probability 1-p. The L-dimensional additive noise vector $\underset{\sim}{N}_n$ has a Gaussian distribution, mutually independent of $\underset{\sim}{N}_k$ for $k \neq n$, with zero mean and covariance $\underset{\sim}{\Sigma}_n$, i.e.;

$$\underset{\sim}{N}_n \sim N(\underset{\sim}{0}, \underset{\sim}{\Sigma}_n), \qquad (2)$$

where $\underset{\sim}{0}$ is the null matrix (or null vector). The L-dimensional signal vector $\underset{\sim}{X}_n$ is a member of ℓ-th order autoregressive process having a Gaussian distribution with mean $\underset{\sim}{M}_n$ and covariance $\underset{\sim}{\Phi}_n$, i.e.;

$$\underset{\sim}{X}_n \sim N(\underset{\sim}{M}_n, \underset{\sim}{\Phi}_n)$$

and

$$\sum_{j=0}^{\ell} \underset{\sim}{a}_{nj} (\underset{\sim}{X}_{n-j} - \underset{\sim}{M}_{n-j}) = \underset{\sim}{Z}_{n-1}, \quad n, \ell = 1,2,... \qquad (3)$$

where $\underset{\sim}{Z}_n$ is a Gaussian random variable, and

$$\underset{\sim}{a}_{nj} = \underset{\sim}{0}, \qquad n \leq j \leq \ell$$

$$\underset{\sim}{a}_{n0} = \underset{\sim}{I}_L \qquad \text{(L x L unit matrix)}$$

$$E(\underset{\sim}{Z}_n) = \underset{\sim}{0}$$

$$E(\underset{\sim}{X}_{n-j}\, \underset{\sim}{Z}_n^T) = \underset{\sim}{0}, \qquad j = 0,1,2,\ldots$$

$$E(\underset{\sim}{Z}_n\, \underset{\sim}{Z}_k^T) = \underset{\sim}{0}, \qquad n \neq k$$

$$= \underset{\sim}{\psi}_n, \qquad n = k$$

$$\underset{\sim}{Z}_0 \sim N(\underset{\sim}{0}, \underset{\sim}{\Phi}_1).$$

The superscript T means transpose. The covariance matrices of $\underset{\sim}{X}_n$ are

$$\underset{\sim}{\Phi}_{nj} = E[(\underset{\sim}{X}_n - \underset{\sim}{M}_n)(\underset{\sim}{X}_{n-j} - \underset{\sim}{M}_{n-j})^T], \quad j = 0,1,2,\ldots \tag{4}$$

$$\underset{\sim}{\Phi}_n = \underset{\sim}{\Phi}_{n0}.$$

Similarly to the stationary process case [9], we have a certain realtion between these covariance matrices and the regression coefficients $\underset{\sim}{a}_{nj}$; and if one is given, the other can be determined.

Here, we define

$$\underset{\sim}{V}_n = \begin{matrix} \underset{\sim}{X}_n - \underset{\sim}{M}_n \\ \underset{\sim}{X}_{n-1} - \underset{\sim}{M}_{n-1} \\ \vdots \\ \underset{\sim}{X}_1 - \underset{\sim}{M}_1 \end{matrix} \qquad \text{(nL x 1 vector)}$$

$$\eta_n = [Z_n^T \; 0 \cdots 0]^T \qquad \text{((n+1)L x 1 vector)}$$

$$\Gamma_n = \begin{bmatrix} -a_{(n+1)1} & -a_{(n+1)2} & \cdots & -a_{(n+1)n} \\ I_L & 0 & \cdots & 0 \\ 0 & I_L & \ddots & \vdots \\ \vdots & \ddots & \ddots & 0 \\ 0 & \cdots & 0 & I_L \end{bmatrix}$$

((n+1)L x nL matrix)

for n = 1,2,..., ℓ-1, and

$$V_n = \begin{bmatrix} X_n - M_n \\ X_{n-1} - M_{n-1} \\ \vdots \\ X_{n-\ell+1} - M_{n-\ell+1} \end{bmatrix} \qquad \text{(ℓL x 1 vector)}$$

$$\eta_n = [Z_n^T \; 0 \cdots 0]^T \qquad \text{(L x 1 vector)}$$

$$\Gamma_n = \begin{bmatrix} -a_{(n+1)1} & -a_{(n+1)2} & \cdots & & -a_{(n+1)\ell} \\ I_L & 0 & \cdots & & 0 \\ 0 & I_L & \ddots & & \vdots \\ \vdots & \ddots & \ddots & \ddots & \vdots \\ 0 & \cdots & 0 & I_L & 0 \end{bmatrix}$$

(ℓL x ℓL matrix)

for $n = \ell, \ell+1, \cdots$. Then (3) yields

$$\underset{\sim}{V}_n = \underset{\sim}{\Gamma}_{n-1} \underset{\sim}{V}_{n-1} + \underset{\sim}{\eta}_{n-1} \cdot \tag{5}$$

Thus, the simple Markov representation of the ℓ-th order autoregressive process is obtained. Such a technique is called augmented state vector technique [17].

III. OPTIMUM SUPERVISED DICHOTOMIZATION

Let us obtain the optimum supervised learning and dichotomization rule. In this case, a teaching sequence $\theta^{n-1} = \{\theta_1, \theta_2, \theta_3, \ldots, \theta_{n-1}\}$ is given by a supervisor in the $\tilde{n}$-th interval. In the n-th interval, the optimum dichotomizer computes a posteriori probability as follows;

$$P(\theta_n | \underset{\sim}{Y}^n, \underset{\sim}{\theta}^{n-1}) = P(\theta_n | \underset{\sim}{Y}^{n-1}, \underset{\sim}{Y}_n, \underset{\sim}{\theta}^{n-1})$$

$$= \frac{P(\theta_n)\, p(\underset{\sim}{Y}_n | \underset{\sim}{Y}^{n-1}, \underset{\sim}{\theta}^n)}{p(\underset{\sim}{Y}_n | \underset{\sim}{Y}^{n-1}, \underset{\sim}{\theta}^{n-1})}, \tag{6}$$

where

$$\underset{\sim}{Y}^n = \{\underset{\sim}{Y}_1, \underset{\sim}{Y}_2, \ldots, \underset{\sim}{Y}_n\} .$$

The denominator of (6) is independent of θ_n. Though the numerator of (6) may be obtained directly by integration [9], in this paper we obtain by the Kalman theory.

The measurement of the system considered here is

$$\underset{\sim}{Y}_n = \theta_n \underset{\sim}{X}_n + \underset{\sim}{N}_n - \underset{\sim}{u}_n. \tag{7}$$

This may be rewritten as follows;

$$\underset{\sim}{Y}_n - \theta_n \underset{\sim}{M}_n + \underset{\sim}{u}_n = \theta_n \underset{\sim}{W}_n \underset{\sim}{V}_n + \underset{\sim}{N}_n$$

$$\triangleq \underset{\sim}{y}_n(\theta_n), \tag{8}$$

where

$$\underset{\sim}{W}_n = [\underset{\sim}{I}_L \ \underset{\sim}{0} \ \underset{\sim}{0} \cdots \underset{\sim}{0}] \qquad\qquad (L \times tL \text{ matrix})$$

$$t = \min (\ell, n).$$

If we observe by (8) the system that is driven by (5), the a posteriori probabilities of $\underset{\sim}{V}_n$, $\underset{\sim}{X}_n$, and $\underset{\sim}{Y}_n$ are given as follows [17];

$$p(\underset{\sim}{V}_n | \underset{\sim}{Y}^{n-1}, \underset{\sim}{\varrho}^n) = Q(\underset{\sim}{V}_n - \underset{\sim}{h}_n, \ \underset{\sim}{\Xi}_n)$$

$$p(\underset{\sim}{X}_n | \underset{\sim}{Y}^{n-1}, \underset{\sim}{\varrho}^n) = Q(\underset{\sim}{X}_n - \underset{\sim}{M}_n - \underset{\sim}{W}_n \, \underset{\sim}{h}_n, \ \underset{\sim}{W}_n \, \underset{\sim}{\Xi}_n \, \underset{\sim}{W}_n^T)$$

$$p(\underset{\sim}{Y}_n | \underset{\sim}{Y}^{n-1}, \underset{\sim}{\varrho}^n) = Q(\underset{\sim}{y}_n(\underset{\sim}{\theta}_n) - \theta_n \, \underset{\sim}{W}_n \, \underset{\sim}{h}_n, \ \underset{\sim}{\Sigma}_n + \theta_n \, \underset{\sim}{W}_n \, \underset{\sim}{\Xi}_n \, \underset{\sim}{W}_n^T), \quad (9)$$

where

$$Q(\underset{\sim}{A}, \underset{\sim}{B}) = |2\pi\underset{\sim}{B}|^{-1/2} \ \exp -\frac{1}{2} \, \underset{\sim}{A}^T \, \underset{\sim}{B}^{-1} \, \underset{\sim}{A}.$$

Here, $\underset{\sim}{h}_n$ is the a posteriori mean of $\underset{\sim}{V}_n$ given $\underset{\sim}{Y}^{n-1}$ and $\underset{\sim}{\theta}^{n-1}$, and $\underset{\sim}{\Xi}_n$ is its variance. Note that although the ordinary Kalman filter makes an estimate $\underset{\sim}{h}_{n|n}$ of the n-th stage given the measurements up to the n-th stage, our dichotomizer makes the n-th stage estimate $\underset{\sim}{h}_n$ given the measurements up to the (n-1)-th stage and that the optimal single-stage predicted estimate is

$$\underset{\sim}{h}_{n+1} = \underset{\sim}{\Gamma}_n \, \underset{\sim}{h}_{n|n},$$

We have an extension of the Kalman filter to the single-stage prediction as follows [9], [17].

$$\underset{\sim}{h}_1 = \underset{\sim}{0}, \qquad \underset{\sim}{\Xi}_1 = \underset{\sim}{\Phi}_1 \tag{10}$$

$$\underset{\sim}{h}_{n+1} = \underset{\sim}{\Gamma}_n \, [\underset{\sim}{h}_n + \underset{\sim}{K}_n \, (\underset{\sim}{y}_n(1) - \underset{\sim}{W}_n \, \underset{\sim}{h}_n)] \tag{11}$$

$$\underset{\sim}{K}_n = \theta_n \, \underset{\sim}{\Xi}_n \, \underset{\sim}{W}_n^T \, (\underset{\sim}{W}_n \, \underset{\sim}{\Xi}_n \, \underset{\sim}{W}_n^T + \underset{\sim}{\Sigma}_n)^{-1} \tag{12}$$

$$\underset{\sim}{\Xi}_{n+1} = \underset{\sim}{\Gamma}_n(\underset{\sim}{I}_{tL} - \underset{\sim}{K}_n \, \underset{\sim}{W}_n) \, \underset{\sim}{\Xi}_n \, \underset{\sim}{\Gamma}_n^T + \underset{\sim}{R}_n, \tag{13}$$

where

$$\underset{\sim}{R}_n = \begin{matrix} \underset{\sim}{\psi}_n & \underset{\sim}{0} \cdots & \underset{\sim}{0} \\ \underset{\sim}{0} & \underset{\sim}{0} \cdots & \vdots \\ \vdots & \ddots & \vdots \\ \underset{\sim}{0} & \cdots & \underset{\sim}{0} \end{matrix} \qquad \text{(t'L x t'L matrix)}$$

$t = \min(\ell, n+1).$

From (6), we have a likelihood ratio

$$\Lambda \triangleq \frac{P(\theta_n = 1 \mid \underset{\sim}{Y}^n, \underset{\sim}{\theta}^{n-1})}{P(\theta_n = 0 \mid \underset{\sim}{Y}^n, \underset{\sim}{\theta}^{n-1})}$$

$$= \frac{P(\theta_n = 1)\, p(\underset{\sim}{Y}_n \mid \underset{\sim}{Y}^{n-1}, \underset{\sim}{\theta}^{n-1}, \theta_n = 1)}{P(\theta_n = 0)\, p(Y_n \mid \underset{\sim}{Y}^{n-1}, \underset{\sim}{\theta}^{n-1}, \theta_n = 0)} . \tag{14}$$

Then if we rewrite the dichotomization rule of comparing Λ with 1/2, it becomes

$$\begin{aligned} \text{If} \quad D_n &\geq \delta_n \quad \text{then} \quad \hat{\theta}_n = 1 \\ &< \delta_n \quad \text{then} \quad \hat{\theta}_n = 0 \end{aligned} \tag{15}$$

for

$$D_n = \underset{\sim}{y}_n(0)^T \underset{\sim}{\Sigma}_n^{-1} \underset{\sim}{y}_n(0) - (\underset{\sim}{y}_n(1) - \underset{\sim}{W}_n \underset{\sim}{h}_n)^T (\underset{\sim}{\Sigma}_n + \underset{\sim}{W}_n \underset{\sim}{\Xi}_n \underset{\sim}{W}_n^T)^{-1}$$

$$\cdot\, (\underset{\sim}{y}_n(1) - \underset{\sim}{W}_n \underset{\sim}{h}_n) \tag{16}$$

and

$$\delta_n = 2 \log \frac{1-p}{p} + \log \frac{|\underset{\sim}{\Sigma}_n + \underset{\sim}{W}_n \underset{\sim}{\Xi}_n \underset{\sim}{W}_n^T|}{|\underset{\sim}{\Sigma}_n|} . \tag{17}$$

Here, $\underset{\sim}{W}_n \underset{\sim}{h}_n$ in (16) is the a posteriori mean of $\underset{\sim}{X}_n - \underset{\sim}{M}_n$, and $\underset{\sim}{W}_n \underset{\sim}{\Xi}_n \underset{\sim}{W}_n^T$ is its variance. These values are sequentially learned by (10) $\sim$ (13) with teaching sequence $\{\theta_n\}$ given. Therefore, we can see that the dichotomization is made based on the estimation of signal.

When $\ell=1$, we have

$$\underset{\sim}{V}_n = \underset{\sim}{X}_n - \underset{\sim}{M}_n, \qquad \underset{\sim}{\eta}_n = \underset{\sim}{Z}_n$$

$$\underset{\sim}{\Gamma}_n = - \underset{\sim}{a}_{(n+1)1}, \qquad \underset{\sim}{W}_n = \underset{\sim}{I}_L = \underset{\sim}{I}_{tL}.$$

and $(11) \sim (13)$ become

$$\underset{\sim}{h}_{n+1} = -\underset{\sim}{a}_{(n+1)1} [\underset{\sim}{h}_n + \theta_n \underset{\sim}{\Xi}_n (\underset{\sim}{\Xi}_n + \underset{\sim}{\Sigma}_n)^{-1} (\underset{\sim}{y}_n(1) - \underset{\sim}{h}_n)] \qquad (18)$$

$$\underset{\sim}{\Xi}_{n+1} = \underset{\sim}{a}_{(n+1)1} \underset{\sim}{\Sigma}_n (\theta_n \underset{\sim}{\Xi}_n + \underset{\sim}{\Sigma})^{-1} \underset{\sim}{\Xi}_n \underset{\sim}{a}_{(n+1)1}^T + \underset{\sim}{\psi}_n. \qquad (19)$$

If the stochastic process of the signal is stationary, i.e., when $\underset{\sim}{u}_n=\underset{\sim}{0}$, $\underset{\sim}{M}_n=\underset{\sim}{0}$, $\underset{\sim}{a}_{n1}=-a\underset{\sim}{I}_L$, $\underset{\sim}{\Sigma}_n = \underset{\sim}{\Sigma}$ and $\theta_n=1$ for all n, (18) and (19) coincide with the result of Abramson and Braverman [6]. Further, the signal is fixed, i.e., when $\underset{\sim}{u}_n=\underset{\sim}{0}$, $\underset{\sim}{M}_n=\underset{\sim}{M}$, $\underset{\sim}{a}_{n1}=-\underset{\sim}{I}_L$, $\underset{\sim}{\Sigma}_n=\underset{\sim}{\Sigma}$, $\underset{\sim}{\psi}_n=\underset{\sim}{0}$, and $\underset{\sim}{\Phi}_n=\underset{\sim}{\Phi}$ for all n, (10), (18), and (19) yield

$$\underset{\sim}{H}_1 = \underset{\sim}{M}, \qquad \underset{\sim}{\Xi}_1 = \underset{\sim}{\Phi}$$

$$\underset{\sim}{H}_{n+1} = \underset{\sim}{H}_n + \theta_n \underset{\sim}{\Xi}_{n+1} \underset{\sim}{\Sigma}^{-1}(\underset{\sim}{Y}_n - \underset{\sim}{H}_n) \qquad (20)$$

$$\underset{\sim}{\Xi}_{n+1} = (\underset{\sim}{\Xi}_n^{-1} + \theta_n \underset{\sim}{\Sigma}^{-1})^{-1}, \qquad (21)$$

where $\underset{\sim}{H}_n=\underset{\sim}{h}_n+\underset{\sim}{M}$ is the a posteriori mean of $\underset{\sim}{X}_n$ and $\underset{\sim}{\Xi}_n$ is also its variance. The recursive formula (20) and (21) coincide with the result of Scudder [7], [8].

IV. OPTIMUM FEEDBACK

As seen from the previous section when the additive noise in the feedback link is negligible, the value of the feedback does not affect the estimation error (a posteriori variance) and the error probability of dichotomization at all. Therefore, in this section, by adopting another performance measure, (that is, minimization of transmission energy), we try to make the average of transmission energy minimum. In other words, this minimization yields an improvement of the error probability for the same signal to noise ratio (SNR).

Now, let us define the transmission energy by

$$\underset{\sim}{\alpha}_n \overset{\Delta}{=} (\theta_n \underset{\sim}{X}_n - \underset{\sim}{u}_n)^T \; \underset{\sim}{A}_n \; (\theta_n \underset{\sim}{X}_n - \underset{\sim}{u}_n), \tag{22}$$

where $\underset{\sim}{A}_n$ is a positive definite symmetric matrix. If we adopt the a posteriori expectation of α_n as the performance measure, we have

$$J_n \overset{\Delta}{=} E_{\theta_n, \underset{\sim}{X}_n} (\alpha_n | \underset{\sim}{Y}^{n-1}, \underset{\sim}{\theta}^{n-1})$$

$$= \underset{\sim}{E}_{\theta_n, \underset{\sim}{X}_n} [(\theta_n | \underset{\sim}{X}_n - \underset{\sim}{u}_n)^T \underset{\sim}{A}_n (\theta_n \underset{\sim}{X}_n - \underset{\sim}{u}_n) | \underset{\sim}{Y}^{n-1}, \underset{\sim}{\theta}^{n-1}] . \tag{23}$$

Let us decide $\underset{\sim}{u}_n$ that makes J_n minimum. Taking the gradient of (23) with respect to $\underset{\sim}{u}_n$ and setting the result equal to zero, we obtain the following expression;

$$-2 \underset{\sim}{E}_{\theta_n, \underset{\sim}{X}_n} (\theta_n \underset{\sim}{X}_n^T | \underset{\sim}{Y}^{n-1}, \underset{\sim}{\theta}^{n-1}) \underset{\sim}{A}_n + 2 \underset{\sim}{u}_n^T \underset{\sim}{A}_n = 0.$$

Since $\underset{\sim}{A}_n$ is positive definite, we have the optimum value

$$\underset{\sim}{u}_n^* = \underset{\sim}{E}_{\theta_n, \underset{\sim}{X}_n} (\theta_n \underset{\sim}{X}_n | \underset{\sim}{Y}^{n-1}, \underset{\sim}{\theta}^{n-1})$$

$$= p(\underset{\sim}{W}_n \underset{\sim}{h}_n + \underset{\sim}{M}_n)$$

$$= p \underset{\sim}{H}_n, \tag{24}$$

where

$$\underset{\sim}{H}_n = \underset{\sim}{W}_n \underset{\sim}{h}_n + \underset{\sim}{M}_n$$

is the a posteriori mean of $\underset{\sim}{X}_n$. Then J_n for the optimum $\underset{\sim}{u}_n^*$ is given as follows;

$$J_n^* = E_{\theta_n, X_n} [(\theta_n X_n - p H_n)^T A_n (\theta_n X_n - p H_n) | Y^{n-1}, \theta^{n-1}]$$

$$= p E_{X_n} [(X_n - p H_n)^T A_n (X_n - p H_n) | Y^{n-1}, \theta^{n-1}]$$

$$+ (1-p) E_{X_n} (p^2 H_n^T A_n H_n) \tag{25.a}$$

$$= p E_{X_n} [\{(X_n - H_n) + (1-p) H_n\}^T A_n \{(X_n - H_n)$$

$$+ (1-p) H_n\} | Y^{n-1}, \theta^{n-1}] + (1-p) p^2 H_n^T A_n H_n$$

$$= p E_{X_n} [(X_n - H_n)^T A_n (X_n - H_n) | Y^{n-1}, \theta^{n-1}]$$

$$+ p (1-p)^2 H_n^T A_n H_n + (1-p) p^2 H_n^T A_n H_n$$

$$= p \, tr (\Xi_n A_n) + p (1-p) H_n^T A_n H_n , \tag{25.b}$$

where the identity

$$E(x^T A x) = E(tr A x x^T)$$

$$= tr[A E(x x^T)] \tag{26}$$

is used. The first term of (25.b) is caused by the estimation error of the signal, and the second term is caused by the on-off of the signal component.

Schalkwijk [15] has minimized the transmission energy by sending back the location of the center-of-gravity of known M-ary signals which is weighted by the a posteriori message probabilities. Since the method obtained in this section is also sending back the a posteriori mean of the unknown signal which is weighted by the a priori message probabilities, it can be said to be a sort of center-of-gravity feedback method.

Here, let us obtain the ensemble average $\overline{J}_n^{*F}$ of J_n^*, i.e.,

$$\overline{J}_n^{*F} = E_{X, H_n, \theta_n} (\alpha_n^*), \tag{27}$$

where α_n^* is the value of α_n for the optimum $\underset{\sim}{u}_n^*$ given by (24).
Assume for simplicity that the unknown signal is fixed, i.e.,
$\underset{\sim}{X}_n = \underset{\sim}{x}$, and its initial distribution be

$$\underset{\sim}{X} \sim N(M, \underset{\sim}{\Phi}).$$

Let

$$b = p(\underset{\sim}{X} - p \ \underset{\sim}{H}_n)^T \ \underset{\sim}{A}_n \ (\underset{\sim}{X} - p \ \underset{\sim}{H}_n) + (1-p) \ p^2 \ \underset{\sim}{H}_n^T \ \underset{\sim}{A}_n \ \underset{\sim}{H}_n. \qquad (28)$$

Now, similar to (25), we obtain

$$\bar{J}_n^{*F} = \underset{\underset{\sim}{X}, \underset{\sim}{H}_n}{E} (b)$$

$$= \underset{\underset{\sim}{X}}{E} [\underset{\underset{\sim}{H}_n}{E} (b \mid \underset{\sim}{X})]$$

$$= \underset{\underset{\sim}{X}}{E} [\underset{\underset{\sim}{\theta}^{n-1}, \underset{\sim}{N}^{n-1}}{E} (b \mid \underset{\sim}{X})]$$

$$= \underset{\underset{\sim}{X}}{E} \{\underset{\underset{\sim}{\theta}^{n-1}}{E} \underset{\underset{\sim}{N}^{n-1}}{L \ E} (b \mid \underset{\sim}{\theta}^{n-1}, \underset{\sim}{X}) \mid \underset{\sim}{X}]\}, \qquad (29)$$

where $\underset{\sim}{N}^{n-1} = \{\underset{\sim}{N}_1, \underset{\sim}{N}_2, \ldots, \underset{\sim}{N}^{n-1}\}$. A non-recursive form of (20) of
the a posteriori mean of $\underset{\sim}{X}$ is given by [7], [8]

$$\underset{\sim}{H}_n = \underset{\sim}{\Phi}(r \ \underset{\sim}{\Phi} + \underset{\sim}{\Sigma})^{-1} \sum_{i=1}^{n-1} \theta_i \ (\underset{\sim}{Y}_i + \underset{\sim}{u}_i) + \underset{\sim}{\Sigma} (r \ \underset{\sim}{\Phi} + \underset{\sim}{\Sigma})^{-1} \ \underset{\sim}{M},$$
$$(30)$$

where $r = \sum_{i=1}^{n-1} \theta i$. When $\theta_i = 1$, we have

$$\underset{\sim}{Y}_i + \underset{\sim}{u}_i = \underset{\sim}{X} + \underset{\sim}{N}_i. \qquad (31)$$

Therefore, let

$$\beta(r) = r \ \underset{\sim}{\Phi} + \underset{\sim}{\Sigma},$$

then

$$\underset{\sim}{H}_n = \underset{\sim}{\Phi}\, \underset{\sim}{\beta}^{-1} (r \underset{\sim}{X} + \sum_{i=1}^{n-1} \theta_i \underset{\sim}{N}_i) + \underset{\sim}{\Sigma}\, \underset{\sim}{\beta}^{-1} \underset{\sim}{M}. \tag{32}$$

Hence

$$E_{\underset{\sim}{N}^{n-1}} (\underset{\sim}{H}_n | \underset{\sim}{\theta}^{n-1}, \underset{\sim}{X}) = \underset{\sim}{\Phi}\, \underset{\sim}{\beta}^{-1} r \underset{\sim}{X} + \underset{\sim}{\Sigma}\, \underset{\sim}{\beta}^{-1} \underset{\sim}{M} \tag{33}$$

$$\stackrel{\Delta}{=} \overline{\underset{\sim}{H}}_n$$

$$E_{\underset{\sim}{N}^{n-1}} (\underset{\sim}{H}_n \underset{\sim}{H}_n^T | \underset{\sim}{\theta}^{n-1}, \underset{\sim}{X}) = \overline{\underset{\sim}{H}}_n \overline{\underset{\sim}{H}}_n^T + r\, \underset{\sim}{\Phi}\, \underset{\sim}{\beta}^{-1} \underset{\sim}{\Sigma}\, \underset{\sim}{\beta}^{-1} \underset{\sim}{\Phi} \tag{34}$$

$$\stackrel{\Delta}{=} \overline{\underset{\sim}{H}_n \underset{\sim}{H}_n^T}\ .$$

Substituting (28) into (29), and using

$$E_{\underset{\sim}{X}}(\underset{\sim}{X} \underset{\sim}{X}^T) = \underset{\sim}{\Delta} + \underset{\sim}{\Phi}\ , \tag{35}$$

where

$$\underset{\sim}{\Delta} = \underset{\sim}{M} \underset{\sim}{M}^T,$$

we obtain the following equation after some manipulations;

$$\overline{J}_n^{*F} = \sum_{r=0}^{n-1} {}_{n-1}C_r p^r (1-p)^{n-1-r}\ \mathrm{tr} \{ \underset{\sim}{A}_n [\ p(\underset{\sim}{\Delta}+\underset{\sim}{\Phi}) - p^2 \{\underset{\sim}{\Delta} + r\ \underset{\sim}{\beta}^{-1} (r)\ \underset{\sim}{\Phi}^2\}]\}. \tag{36}$$

When n is large enough, since we can approximate

$$r \simeq p(n-1),$$

we have

$$\overline{J}_n^{*F} \simeq \mathrm{tr}\ \{ \underset{\sim}{A}_n [p(\underset{\sim}{\Delta}+\underset{\sim}{\Phi}) - p^2 \{\underset{\sim}{\Delta} + p(n-1)\ \underset{\sim}{\beta}^{-1}(p(n-1))\ \underset{\sim}{\Phi}^2\}]\}\ . \tag{37}$$

V. NON-FEEDBACK OPTIMUM METHOD

Let us consider the minimization of the transmission energy for a communication system without feedback link, i.e., a self-bias method in which an appropriate value $\underset{\sim}{\nu}_n$ is subtracted beforehand from the transmitter output. In this method, we have

$$\underset{\sim}{Y}_n = \theta_n \underset{\sim}{X}_n - \underset{\sim}{\nu}_n + \underset{\sim}{N}_n. \tag{38}$$

The expectation of the transmission energy is given by

$$J_n = E_{\underset{\sim}{\theta}_n} [(\theta_n \underset{\sim}{X}_n - \underset{\sim}{\nu}_n)^T \underset{\sim}{A}_n (\theta_n \underset{\sim}{X}_n - \underset{\sim}{\nu}_n)] .$$

The optimum $\underset{\sim}{\nu}_n$ that makes J_n minimum is

$$\underset{\sim}{\nu}_n^* = E_{\underset{\sim}{\theta}_n} (\theta_n \underset{\sim}{X}_n) = p \underset{\sim}{X}_n. \tag{39}$$

Then the optimum J_n for $\underset{\sim}{\nu}_n^*$ is given by

$$J_n^* = p(1-p) \underset{\sim}{X}_n^T \underset{\sim}{A}_n \underset{\sim}{X}_n. \tag{40}$$

We can see that J_n^* is (1-p) times of transmission energy $p \underset{\sim}{X}_n^T \underset{\sim}{A}_n \underset{\sim}{X}_n$ for non-feedback system, i.e., $\underset{\sim}{\nu}_n=0$. In the following, we consider only the case of fixed signal, i.e., $\underset{\sim}{X}_n = \underset{\sim}{X}$, n=1,2,... In this case, the ensemble average of (40) is given by

$$\overline{J}_n^{*NF} = p(1-p) \text{ tr } [\underset{\sim}{A}_N(\underset{\sim}{\Delta}+ \underset{\sim}{\Phi})] . \tag{41}$$

In (25.b), since $\underset{\sim}{\Xi}_n \to \underset{\sim}{0}$, and $\underset{\sim}{H}_n \to \underset{\sim}{X}$ for $n \to \infty$, we see that the transmission energy of the optimum feedback method converges to that of the non-feedback optimum method. Their difference is

$$\delta_n \triangleq \overline{J}_n^{*F} - \overline{J}_n^{*NF}$$

$$= \sum_{r=0}^{n-1} {}_{n-1}C_r \, p^r (1-p)^{n-1-r} p^2 \text{ tr}[\underset{\sim}{A}_n \, \beta^{-1}(r) \underset{\sim}{\Sigma} \underset{\sim}{\Phi}] . \tag{42}$$

As well as (37), when n is large enough,

$$\delta_n \approx p^2 \, \mathrm{tr}[\underset{\sim}{A}_n \, \underset{\sim}{\beta}^{-1}(p(n-1)) \, \underset{\sim}{\Sigma} \, \underset{\sim}{\Phi}] \,. \tag{43}$$

From these results, it may be concluded that the transmission energy of a non-feedback optimum method is less than that of optimum feedback method, and that they coincide with $n \to \infty$.

Let us discuss the estimation error of both methods. The estimation error (a posteriori variance) of the estimate $\underset{\sim}{H}_n$ of $\underset{\sim}{X}$ for the optimum feedback method is given in the non-recursive form of (21) by [7], [8]

$$\begin{aligned} \underset{\sim}{\Xi}_n &= \underset{\sim}{\Sigma}(r \, \underset{\sim}{\Phi} + \underset{\sim}{\Sigma})^{-1} \, \underset{\sim}{\Phi} \\ &= (r \, \underset{\sim}{\Phi} \, \underset{\sim}{\Sigma}^{-1} + \underset{\sim}{I})^{-1} \, \underset{\sim}{\Phi} \,. \end{aligned} \tag{44}$$

Hence, its ensemble average is

$$\begin{aligned} \underset{\sim}{\Xi}_n^F &\triangleq E_{\underset{\sim}{\theta}^{n-1}} (\underset{\sim}{\Xi}_n) \\ &= \sum_{r=0}^{n-1} {}_{n-1}C_r \, p^r \, (1-p)^{n-1-r} \, (r \, \underset{\sim}{\Phi} \, \underset{\sim}{\Sigma}^{-1} + I)^{-1} \, \underset{\sim}{\Phi} \,. \end{aligned} \tag{45}$$

On the other hand, in the non-feedback optimum method, since the output of the transmitter for $\theta_n = 1$ and 0 are $(1-p)\underset{\sim}{X}$ and $-p\underset{\sim}{X}$, respectively, the equivalent measurement for $\underset{\sim}{X}$ is

$$\underset{\sim}{Y}_n/(\theta_n - p) = \underset{\sim}{X} + \underset{\sim}{N}_n/(\theta_n - p). \tag{46}$$

Then, since the equivalent noise is $\underset{\sim}{N}_n/(\theta_n - p)$, its variance is $\underset{\sim}{\Sigma}/(\theta_n - p)^2$.

Generally, observing n_1 intervals with the additive noise $N(\underset{\sim}{0}, \underset{\sim}{\Sigma}_1)$ followed by n_0 intervals with the additive noise $N(\underset{\sim}{0}, \underset{\sim}{\Sigma}_0)$, we have

$$\underset{\sim}{\Xi}_{n_1+1} = \underset{\sim}{\Sigma}^1 \, (n_1 \, \underset{\sim}{\Phi} + \underset{\sim}{\Sigma}^1)^{-1} \underset{\sim}{\Phi} \tag{47}$$

and

$$\underset{\sim}{\Xi}_{n_1+n_0+1} = \underset{\sim}{\Sigma}^0 (n_1 \underset{\sim}{\Xi}_{n_1+1} + \underset{\sim}{\Sigma}^0)^{-1} \underset{\sim}{\Xi}_{n_1+1}$$

$$= [n_1 \underset{\sim}{\Phi} (\underset{\sim}{\Sigma}^1)^{-1} + n_0 \underset{\sim}{\Phi} (\underset{\sim}{\Sigma}^0)^{-1} + \underset{\sim}{I}]^{-1} \underset{\sim}{\Phi} . \quad (48)$$

Therefore, in the non-feedback optimum method, if we make the following correspondences;

$$n_1 = r, \qquad n_0 = n-1-r$$

$$\underset{\sim}{\Sigma}^1 = \underset{\sim}{\Sigma}/(1-p)^2, \qquad \underset{\sim}{\Sigma}^0 = \underset{\sim}{\Sigma}/p^2,$$

the a posteriori variance (48) yields

$$\underset{\sim}{\Xi}_n = [r(1-p)^2 \underset{\sim}{\Phi} \underset{\sim}{\Sigma}^{-1} + (n-1-r)\, p^2 \underset{\sim}{\Phi} \underset{\sim}{\Sigma}^{-1} + \underset{\sim}{I}]^{-1} \underset{\sim}{\Phi} . \quad (49)$$

Let the ensemble average of (49) be

$$\overline{\underset{\sim}{\Xi}}_n^{NF} \triangleq E_{\underset{\sim}{\theta}^{n-1}} (\underset{\sim}{\Xi}_n), \quad (50)$$

then for large n, since $r \simeq p(n-1)$, we have

$$\overline{\underset{\sim}{\Xi}}_n^F \; \overline{\underset{\sim}{\Xi}}_n^{NF\;-1} \simeq [p(n-1) \; \underset{\sim}{\Phi} \; \underset{\sim}{\Sigma}^{-1}]^{-1} \; \underset{\sim}{\Phi} \; \underset{\sim}{\Phi}^{-1} \, [p(1-p)(n-1) \; \underset{\sim}{\Phi} \; \underset{\sim}{\Sigma}^{-1}]$$

$$= (1-p)\, \underset{\sim}{I}. \quad (51)$$

Therefore, for large n,

$$\overline{\underset{\sim}{\Xi}}_n^F \simeq (1-p)\, \overline{\underset{\sim}{\Xi}}_n^{NF} , \quad (52)$$

and it can be said that the estimation error in the optimum feedback method is less than the non-feedback optimum method. This is caused by the fact that besides the measurement of the non-feedback optimum method contains noise for all n, each noise is equivalently large, i.e.; the estimate is disturbed more by its large noise. On the other hand, since both methods coincide with the case for $n \to \infty$ in respect of the transmission energy, we have the same relation as (52) with respect to the estimation error for

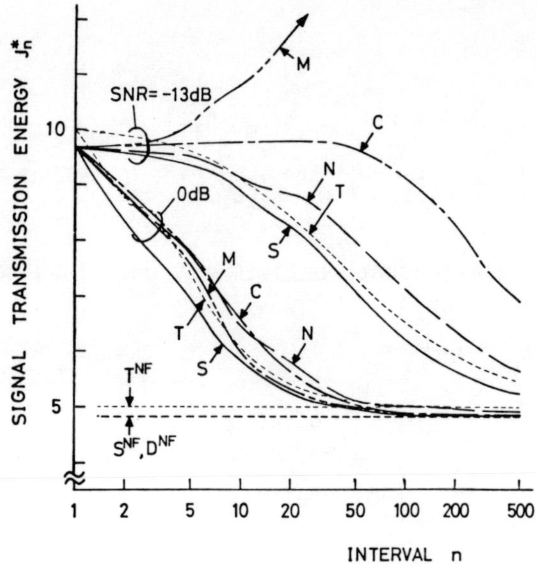

Fig. 2. Signal transmission energy, S - supervised; N - NDD; C - CDD; M - MDD; D - DD; T - theoretical value; superscript NF - non-feedback method; no superscript - feedback method.

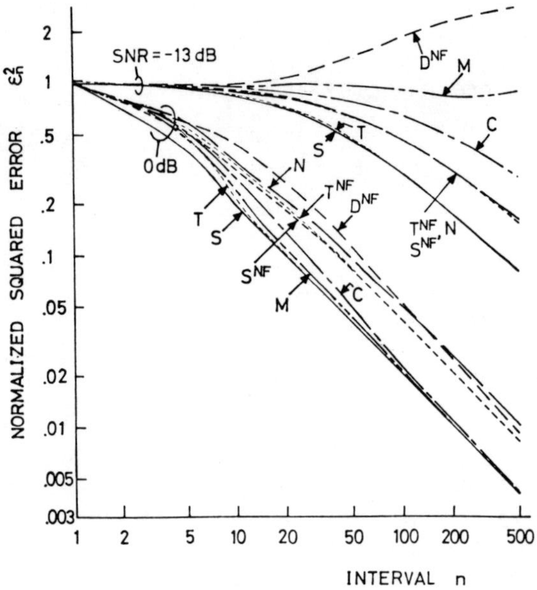

Fig. 3. Normalized squared error of estimate. Abbreviations are the same as Fig. 2.

the same transmission energy.

VI. NONSUPERVISED LEARNING

So far, we have discussed only the supervised learning scheme. However, in an ordinary communication system, since we have no supervisor who teaches the sequence of θ_n, the receiver must learn without supervisor. As is well known, the construction of a non-supervised learning machine leads to the difficulty of exponential growing. Therefore, some practical realizable methods have been proposed [1]. A representative method of them is the decision-directed (DD) method that substitutes its own decision for the direction of the supervisor [7], [8]. Besides this method, we have consistent-estimator-type DD (CDD), non-DD (NDD), and modified-DD (MDD) methods [16], [18], [19]. Each of them is made to improve the shortcomings of the DD method: The CDD is made to obtain a consistent estimator, and the NDD is made to obtain an estimator that does not depend on the decisions. However, these two methods can not be applied to the non-feedback case. The MDD is a method that obtains the estimators for all 2^m message sequences of the last m intervals. The MDD method is applied to the feedback case by sending back the value that is determined by the a posteriori mean that corresponds to a sequence that has maximum a posteriori probability.

A computer simulation of the above methods for the fixed signal under the conditions L=20, p=1/2, $\underset{\sim}{M}=\underset{\sim}{0}$, $\underset{\sim}{\Phi}=\Phi\underset{\sim}{I}$, $\underset{\sim}{\Sigma}=\Sigma\underset{\sim}{I}$, m=5, and $\underset{\sim}{A}_n=\underset{\sim}{I}$ is made. Average results of one hundred runs is shown in Fig. 2-4, where SNR= $10\log_{10}(\Phi/\Sigma)$ [dB]. When $\underset{\sim}{\Phi}$ is large compared with $\underset{\sim}{M}\underset{\sim}{M}^T$ in the non-feedback case, the labeling of the binary massage cannot be known without supervisor; i.e., when the initial knowledge with respect to $\underset{\sim}{X}$ is little, it is essentially difficult to decide whether $(1-\tilde{p})\underset{\sim}{X}$ corresponds to $\theta_n=1$ or $-p\underset{\sim}{X}$ corresponds to $\theta_n=1$. In fact, in our computer simulation, the convergence of DD method is unstable; and we have had two convergence cases $\underset{\sim}{H}_n\to\underset{\sim}{X}$ and $\underset{\sim}{H}_n\to-\underset{\sim}{X}$ with each probability 1/2. Therefore, in this case, the differential encoding [20] that does not need the labeling must be used. In Fig. 2-4, the result that was made the label exchange is shown for the non-feedback DD method. The transmission energy of Fig. 2 is calculated using (28) and (40). Further, the theoretical values of supervised optimum method are calculated by (36) and (41). We see that when the SNR is small (SNR=-13dB), the MDD method that is based on the DD biased estimators [8] seems to diverge. Fig. 3 shows the normalized squared error of estimate

$$\varepsilon_n^2 = (\underset{\sim}{H}_n - \underset{\sim}{X})^T (\underset{\sim}{H}_n - \underset{\sim}{X})/L\,\Phi\ . \tag{53}$$

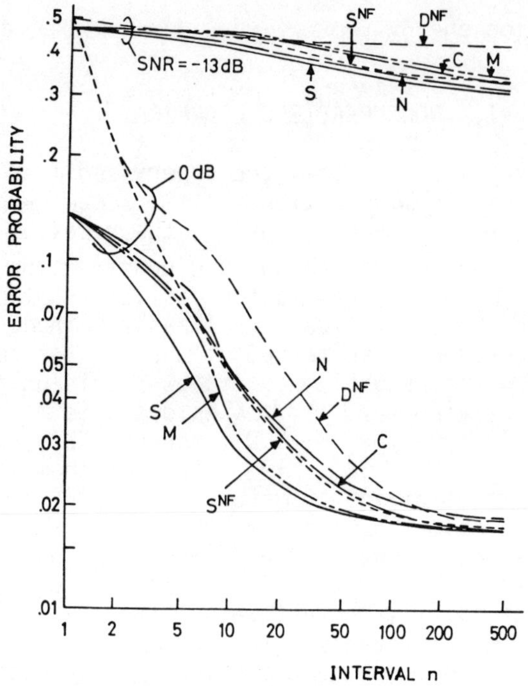

Fig. 4. Probability of dichotomization error. Abbreviations are the same as Fig. 2.

The theoretical values in the supervised case are calculated by (45) and (49). Fig. 4 shows the error probability, which is given by a non-central chi-squared distribution [18], [19]. The squared estimation error and the error probability for MDD method are both approximately obtained using the a posteriori mean corresponding to a sequence with maximum a posteriori probability. Further, when SNR=0dB, the CDD method is almost the same as the DD method.

From these results, it can be said that as to the feedback method, the DD and MDD are both good for large SNR, and the NDD is good for small SNR. Further, in a non-feedback case, the nonsupervised learning is practically impossible for a small SNR. Generally speaking, we can see that the feedback is effective.

VII. CONCLUSION

Kalman filter type optimum supervised learning receiver for unknown signal has been presented. Then, the optimal feedback value that minimizes the forward transmission energy has been given. It is the a posteriori mean of the unknown signal multiplied by a probability of signal occurence. It can be said that this method is a kind of center-of-gravity feedback. Besides deriving the expressions for the required transmission energy of this method, it is shown that the estimation error is less than the optimum non-feedback method. These methods have been extended to the nonsupervised case. The extension to the multi-class signal case is also possible. By computer simulation, it had been made clear that the feedback is effective. However, when the feedback energy is constrained and the feedback link has noise of $N(\underset{\sim}{0}, \underset{\sim}{D}_n)$, the transmission energy increases by $\text{tr}(\underset{\sim}{A}_n \underset{\sim}{D}_n)$ because of the noise. Further, since the variance of the additive noise in the forward channel can be regarded to increase equivalently by $\underset{\sim}{D}_n$, the SNR equivalently decreases. Therefore, the estimation error increases, and the effect of the feedback decreases. In the noisy feedback case, a recursive estimation device of $\underset{\sim}{H}_n$ is required in the transmitter to increase the efficiency of the system [12], [14].

The approach described in this paper may be applied not only to the communication system but also to automatic slice level setting of mark sensors and so on, or enhancement of images.

REFERENCES

(1) J. Spragins, "Learning without a teacher," IEEE Trans. Inform. Theory, vol. IT-12, pp.223-230, Apr. 1966.
(2) C.V. Jakowatz, R.L. Shuey, and G.M. White, "Adaptive waveform recognition," in Proc. 4th Symp. on Information Theory, C. Cherry, Ed., Washington, D.C.: Butterworths, 1961.
(3) M.J. Hinch, "A model for a self-adapting filter," Inform. Contr., vol. 5, pp. 185-203, Sept. 1962.
(4) K. Tanaka, S. Higuchi, and S. Maekawa, "An identification method of system characteristics using a new type of adaptive correlating filter," in Proc. Int. Fed. Automatic Control Tokyo Symp., Aug. 1965, pp.245-254.
(5) E.M. Glaser, "Signal detection by adaptive filters," IRE Trans. Inform. Theory, vol. IT-7, pp.87-97, Apr. 1961.
(6) N. Abramson and D. Braverman, "Learning to recognize patterns in a random environment," IRE Trans. Inform. Theory, vol. IT-8, pp.58-63, Sept. 1962.
(7) H.J. Scudder, III, "Adaptive communication receivers," IEEE Trans. Inform. Theory, vol. IT-11, pp.167-174, Apr. 1965.
(8) H.J. Scudder, "Probability of error of some adaptive pattern-recognition machines," IEEE Trans. Inform. Theory, vol. IT-11, pp.363-371. July 1965.
(9) S. Tamura, S. Higuchi, and K. Tanaka, "On the recognition of time-varying patterns using learning procedures, "IEEE Trans. Inform. Theory, vol. IT-17, pp.445-452, July 1971.

(10) C.E. Shannon, "The zero error capacity of a noisy channel," IRE Trans. Inform. Theory, vol IT-2, pp.3-13, Sept. 1956.

(11) J.P.M. Schalkwijk and T. Kailath, "A coding scheme for additive noise channels with feedback-Part I and II," IEEE Trans. Inform. Theory, vol. IT-12, pp.172-182, pp.183-189, Apr. 1966.

(12) R.L. Kashyap, "Feedback coding schemes for an additive noise channel with a noisy feedback link," IEEE Trans. Inform. Theory, vol. IT-14, pp.471-480, May 1968.

(13) G.L. Turin, "Signal design for sequential detection systems with feeback," IEEE Trans. Inform. Theory, vol. IT-11, pp.401-408, July 1965.

(14) J.K. Omura, "Optimal linear transmission of analogue data for channels with feedback," IEEE Trans. Inform. Theory, vol. IT-14, pp.38-43, January 1968.

(15) J.P.M. Schalkwijk, "Center-of-gravity information feedback," IEEE Trans. Inform. Theory, vol. IT-14, pp.324-331, March 1968.

(16) S. Tamura and K. Tanaka, "Synchronization for unknown signal sequence by learning procedure," IEEE Trans. Commun., vol. COM-20, pp.780-787, August 1972.

(17) J.S. Meditch, Stochastic Optimal Linear Estimation and Control. New York: McGraw-Hill, 1969.

(18) K. Tanaka and S. Tamura,"Some considerations on a type of pattern recognition using nonsupervised learning procedure," in Int. Fed. Automatic Control Int. Symp. Technical and Biological Problems of Control, Yerevan, Armenia, Sept. 1968.

(19) K. Tanaka, "Some studies on pattern recognition with nonsupervised learning procedures," in Pattern Recognition and Machine Learning, K.S. Fu Ed., New York: Plenum, 1971, pp.1-17.

(20) J.C. Hancock and P.A. Wintz, Signal Detection Theory, New York: McGraw-Hill, 1966.

COMPUTATIONAL ALGORITHMS FOR INTERACTIVE PATTERN RECOGNITION*

Y. T. Chien

The University of Connecticut

Storrs, Ct. 06268

I. INTRODUCTION

Until recently, digital computers have been used in the design of pattern recognition systems largely for the purpose of simulating or testing computational algorithms before they are implemented in the way of hardware. The evolving use of computer graphics, however, has changed the computer's role significantly. In the past few years, the need for man-machine interaction in data analysis and algorithm synthesis has become increasingly evident to anyone who has undertaken a serious pattern recognition problem. Like many computer-aided systems, the interactive approach allows the speed and accuracy of a computer to be combined with man's ability and intuition at various stages of the pattern recognition process. This approach is particularly desirable whenever one is faced with a large data base of which analytic and statistical properties must be calculated dynamically and in real time. In the event of minimum prior knowledge regarding the data base, man-machine interplay via some graphical medium often becomes a necessity.

The interactive approach to the design of pattern recognition systems must pay special attention to the problem of putting various computational algorithms in a form suitable for computer display and human observation. This paper deals with some of the computational algorithms that have been proposed for interactive pattern recognition.

---

*Presented at the Second US-Japan Seminar on Learing Controls and intelligent Controls, Gainesville, Florida, October 22-26, 1973.

One important class of such algorithms involves the mapping (transformation) of h-dimensional (h usually large) pattern samples into 2- or 3-dimensional spaces. This transformation must be performed so that the multivariate pattern samples can be easily evaluated by human observations through the use of computer graphics. For example, when very little is known about the pattern samples, one would be interested in the 2- or 3-dimensional configuration of the pattern samples to see if, and perhaps how, clusters are formed. This information could then be utilized to help develop the recognition logic.

Conceptually, mappings of this kind are not entirely new. However, in order to be effective in an interactive environment, they must be able to meet at least two added requirements. First, they must be computationally fast to avoid any excessive idle time on the part of the human observer. The second requirement is that the mappings should be recursively applicable in that the mapping obtained from a set of samples need not be totally recomputed when new samples are introduced.

Several classes of display-oriented mapping algorithms will be discussed and their special properties in relation to an interactive setting will be presented. In particular a modified version of Sammon's nonlinear mapping [6] will be presented to demonstrate the special characteristics and added requirements of a class of computational algorithms. This mapping will then be evaluated using a number of artificial examples and real applications for the interactive design of pattern recognition systems. When possible, a comparison of this mapping with other related algorithms will be made. A possible use of this transformation as a means for interactive feature selection will be discussed.

II. THE INTERACTIVE APPROACH

When we speak of man-machine interaction in pattern recognition, we are referring to some type of pattern recognition systems (e.g., a computer programmed to perform the recognition function) that has a person seated at a terminal, usually with a graphical display, who works interactively with the computer in an attempt to extract the characteristic information from the input patterns or to improve the recognition performance of the system (see Figure 1). In the simplest case, this interaction may involve a person acting as a clerk, being "asked" by the computer to check on the input data if a decision as to its class membership cannot be made. Or, the person, acting as a supervisor, is asked to provide "names" for the input pattern samples the computer has classified in order to determine the correctness of these classifications. On the other extreme, human assistance may be used as part of the feature extraction and recognition procedure.

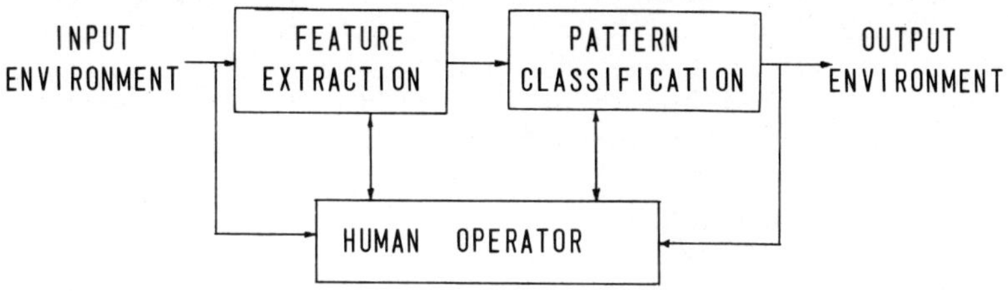

INTERACTIVE PATTERN RECOGNITION

Fig. 1.

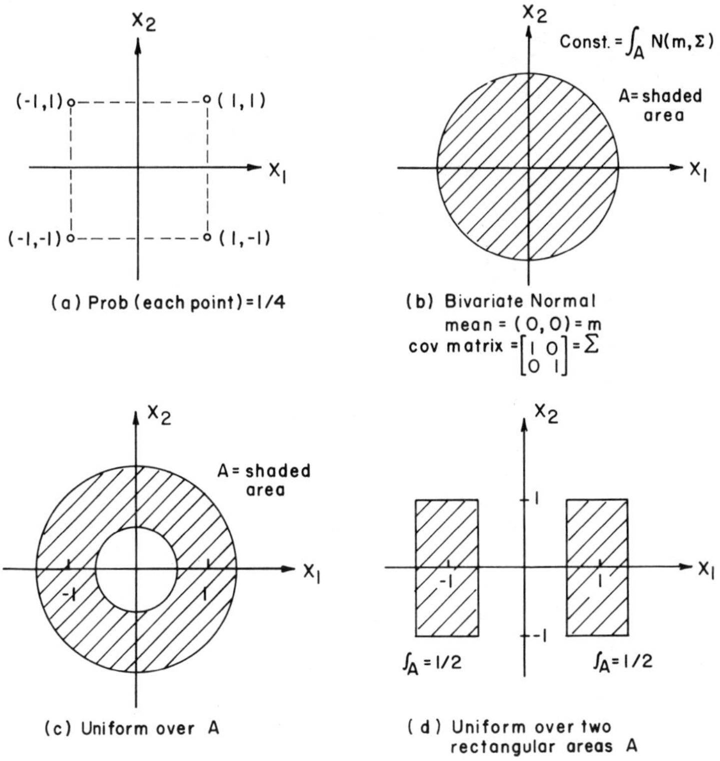

Fig. 2.

That is, the person at the terminal may be helping discover and evaluate the characteristics that are most informative to the classification process. He may also initiate and later modify the classification algorithms to be used by the recognition logic. Using the computer and its associated graphical facilities, he may try various classification techniques as he goes along, calling on the computer when needed to give him a certain two-dimensional display of the information on hand. This information may include the original data input or any intermediate results of partial analysis of the data.

Much of the research work in interactive pattern recognition to date seems to have one or more types of the activities we have just described. Depending upon the goals that a particular system is designed to achieve, each system has proven to be useful in that, in addition to doing old things differently, it can use the new medium to do things better. Applications of the interactive pattern recognition systems have been found in many scientific and engineering areas where automated means for information processing and decision making often requires human assistance. Studies of the various interactive systems that have been developed for pattern analysis and/or recognition can be found in [8, 11].

III. DISPLAY-ORIENTED MAPPING ALGORITHMS

In an interactive environment, the principle concern of pattern recognition system design is how to describe and present information in a form suitable for computer display and human observation. The information in question may take many forms, thus requiring a variety of techniques for its representation and subsequent evaluation. One of the major techniques relates to the mapping (or transforming) of multi-dimensional vector information into 2- or 3-dimensional spaces. Such mappings are often performed to allow pattern samples in higher dimensional space to be evaluated by human observations through the use of computer displays. Algorithms that are designed to implement mappings of this kind are therefore said to be display-oriented.

Conceptually, the need for mapping data from one space to another is not entirely new. However, in order to be effective in a man-machine setting, display-oriented mapping algorithms must be able to meet at least two added requirements. First, they must be computationally fast so as to avoid any excessive idle time on the part of the human observer. This is an essential requirement if human abilities are to be effectively utilized in conjunction with the capabilities of a machine. Secondly, the mapping should be recursively applicable. That is, the mapping obtained from a set of samples need not be totally recomputed whenever new samples are introduced.

In this section, two classes of display-oriented mapping algorithms that have been proposed will be discussed. The discussion will concentrate on the computational properties of these algorithms in relation to the interactive environment and its requirements. Throughout this and the following discussion, we will assume that the main objective of the display-oriented mapping algorithms is to present a 2-dimensional* view of the available pattern samples in order to allow a human observer to analyze if and how clusters** are formed in the high-dimensional space.

To describe the various algorithms in a unified manner, the following notations will be used. Assume that we have a set of N pattern samples, each is an h-dimensional vector in an h-space. Let these vectors be denoted by X_i, $i = 1, 2, \ldots, N$, where

$$X_i = (x_{i1}, x_{i2}, \ldots, x_{ih}) \qquad (1)$$

Let the mapping (or transformation) be denoted by T, and the corresponding N vectors in a d-space (d=2 in this discussion) after the mapping be designated as Y_i, $i = 1, 2, \ldots, N$, where

$$Y_i = T(X_i) = (y_{i1}, y_{i2}, \ldots, y_{id}) \qquad (2)$$

1. Linear Mappings.

One class of display-oriented algorithms for mapping h-dimensional vectors X_i into d-dimensional vectors Y_i is to apply linear operations, such as scaling, translation and rotation to X_i. There are a variety of such algorithms which have been developed to transform the pattern samples into a new space for easier clustering identification, dimensionality reduction, and in which hopefully simpler recognition logic may be designed and evaluated. A good discussion of most of the linear mapping algorithms can be found in Chapter 2 of [1].

---

* Three-dimensional views displayed as a perspective configuration of pattern samples are also possible but will not be discussed here.

** A cluster of pattern samples is considered to be a set of points in a multi-dimensional space in which they share certain common characteristics and tend to exhibit a higher density compared to the density of points in the surrounding region of the space. This is an intuitive description of a cluster and therefore should not be taken as a precise definition of clusters.

Most of the linear algorithms depends on the computation of the covariance matrix or matrices of the N pattern samples upon which the resulting form of the mapping T is determined. Depending on the clustering criterion and the constraints placed upon the criterion in the optimization process, the precise form of T and its properties will be different. For example, if the clustering criterion is to minimize the mean-square distance $\overline{D^2}$ between a set of N h-dimensional vectors [21]

$$\overline{D^2} = \frac{1}{N(N-1)} \sum_{i=1}^{N} \sum_{j=1}^{N} \sum_{k=1}^{h} w_k^2(x_{ik} - x_{jk})^2 \qquad (3)$$

under the constraint that the volume of the space is constant,

$$\prod_{k=1}^{h} w_k = 1 \qquad (4)$$

then the linear mapping T is a rotation R followed by a diagonal transformation W. The rows of the matrix R are eigenvectors of the covariance matrix of the set of N vectors, and the elements of W, w_k, are determined by the standard deviations of the individual components of the N vectors. That is, the relationship between X_i and Y_i is described by

$$Y_i = T(X_i) = X_i \ R \ W \qquad (5)$$

It should be noted that linear mappings of this kind are not immediately applicable to display-oriented pattern recognition problems. The reason is that the dimension of the transformed space is identical to that of the original space. However, a variation of these mappings, called the generalized Karhunen-Loeve expansion, can be easily adapted to allow us to select the two "most informative" dimensions for displaying the data configuration. This is achieved by arranging the rotational matrix R so that the first two eigenvectors are chosen to correspond to the two largest eigenvalues. In [9], it has been shown that the first two components of the pattern samples in the transformed space, compared to any other two components, are optimal in that (1) they can represent the original vectors with minimum error, and (2) they contain the most information in describing the pattern samples in the sense of entropy.

Several other variations of the linear mappings described above exist, and they are also limited to pattern recognition problems where second order statistics and covariance matrices are sufficient descriptors of the data. This is often a severe limitation, for there are many problems whose data cannot be completely described by covariance matrices. Figure 2 shows an example of several different data distributions (2-dimensional) having the same covariance matrix $\sum = \begin{bmatrix} 1 & 0 \\ 0 & 1 \end{bmatrix}$.

2. Nonlinear Mappings.

Sammon [16, 17] has studied a class of mappings that are not based upon the second order statistics of the data. These mappings are nonlinear in that they are not limited to such linear operations as scaling, translation and rotation, which tend to preserve the global structure of the data. The emphasis of nonlinear mappings is to preserve the local information of the data so that the details of relationships among the pattern samples may be analyzed. This criterion is particularly important when the discovery of clusters by a machine is to be assisted by human's ability to concentrate on fine details through graphical displays.

For purposes of later discussions, we shall describe Sammon's nonlinear mapping in some detail although other variations of the mapping also exist [6,7] . Sammon's algorithms is primarily designed to map h-dimensional data X_i into 2- or 3-dimensional space while maintaining the local, intrinsic structure of the data. This is accomplished by attempting to preserve the interpoint distances between data points under the mapping. Although any distance measure could be used, the Euclidean distance will be used here in describing the algorithm.

Let d^*_{ij} denote the distance between the vectors X_i and X_j in the h-space and the distance between the corresponding vectors Y_i and Y_j in the transformed d-space be denoted by d_{ij}. The nonlinear mapping T, where

$$Y_i = T(X_i) \text{ for } i = 1,2,\ldots,N$$

is described by defining the following error function E as a criterion for local structure preservation:

$$E = \text{Error} = \frac{1}{\sum\limits_{i<j} d^*_{ij}} \sum_{i<j}^{N} \frac{(d^*_{ij} - d_{ij})^2}{d^*_{ij}} \qquad (6)$$

We say that a configuration of Y_i, i = 1, 2, ..., N in the d-space preserves the local structure of X_i, i = 1, 2, ..., N, in the h-space if E is minimized. Thus, the approximate structure preservation is maintained by fitting the N points in the d-space such that their interpoint distances approximate the corresponding interpoint distances in the h-space.

Note that the error function E to be minimized is a function in d·N variables, due to the d components of the N vectors, Y_i, i = 1, 2, ..., N to be determined. To find the optimal configuration of Y_i's, Sammon used a steepest decent procedure to search for the minimum of the error function E. This method works as follows. Initially, a set of N arbitrary points is chosen in the d-space. Let this initial configuration be denoted by $Y_i(0)$, i = 1, 2, ..., N, where $Y_i(0) = (y_{i1}(0), y_{i2}(0), \ldots, y_{id}(0))$. The initial configuration is then adjusted or corrected along the direction of a steepest decent. This correction procedure is then applied to the new configuration and the process is repeated until either the minimum value of E is achieved, or one becomes satisfied with the new configuration in the d-space. This procedure can be iteratively described as follows. If we denote the mapping error E after the nth correction by E(n), i.e.,

$$E(n) = \frac{1}{\sum\limits_{i<j}^{N} d^*_{ij}} \sum_{i<j}^{N} \frac{[d^*_{ij} - d_{ij}(n)]^2}{d^*_{ij}} \tag{7}$$

where

$$d_{ij}(n) = \{ \sum_{k=1}^{d} [y_{ik}(n) - y_{jk}(n)]^2 \}^{1/2} \tag{8}$$

and

$$Y_i(n) = (y_{i1}(n), y_{i2}(n), \ldots, y_{id}(n)) \qquad i = 1, 2, \ldots, N \tag{9}$$

is the configuration of Y_i in d-space after the nth correction. The (n+1)th configuration is obtained by computing

$$y_{ik}(n+1) = y_{ik}(n) - (CF) \cdot \Delta_{ik}(n) \tag{10}$$

where CF is a correction factor and

$$\Delta_{ik}(n) = \frac{\partial E(n)}{\partial y_{ik}(n)} \Bigg/ \left| \frac{\partial^2 E(n)}{\partial y_{ik}(n)^2} \right| \tag{11}$$

involves the partial derivatives of E(n) calculated along the kth direction at the nth configuration of $Y_i(n)$.

Figure 4 shows a 2-dimensional display of cluster points corresponding to nine handwritten numbers represented as a

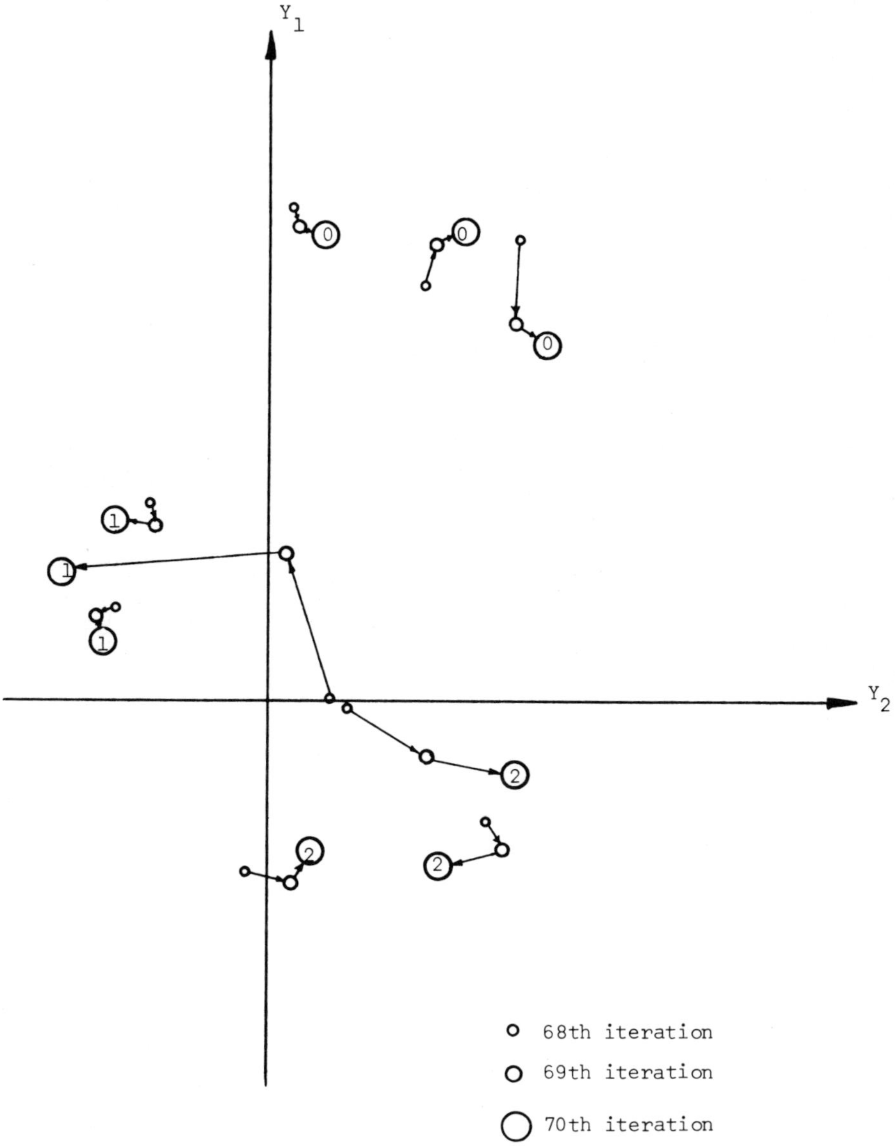

Y1
Y2
0
0
0
1
1
1
2
2
2
68th iteration
69th iteration
70th iteration

Fig. 3.

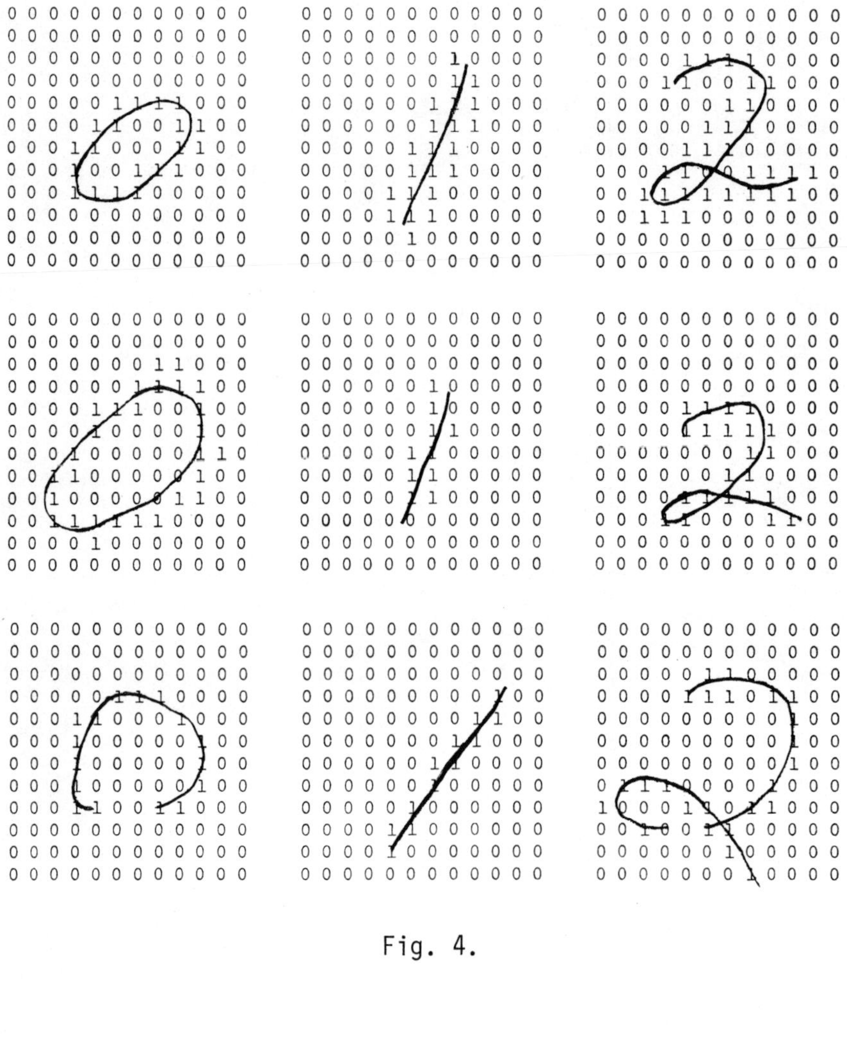

Fig. 4.

144-element vector. It also demonstrates the migration of points during the later stages of the iteration process. The numbers are 0, 1, and 2 (See Figure 3) selected from the Highleyman data, one of the standard pattern recognition data bases made available through the IEEE Computer Society.

The nonlinear mapping algorithm described above has been successively implemented for clustering-seeking purposes, notably in the OLPARS systems [19,20] . The algorithm requires computation of all the N(N-1)/2 interpoint distances. For large N and h, such computational requirements may be prohibitive especially in a man-machine environment. In a variation of this method, Chang and Lee [7] have described a heuristic relaxation method to compute the new configuration of Y_i's in the d-space. Instead of minimizing the error function E by adjusting the N points simultaneously, a pair of points will be adjusted at a time iteratively, minimizing each term of the error E. This variation has been shown to be more efficient for large numbers of data points.

A different approach to construct nonlinear mappings has been proposed by Calvert [6]. The scheme is to randomly perturb the N points Y_i in the d-space under the constraints that ensure the preservation of local structure of data points and linear separation between classes. This method is also iteratively applicable, and the perturbation can be terminated as soon as one is satisfied with the new configuration in the d-space.

IV. FURTHER CONSIDERATIONS AND APPLICATIONS OF NONLINEAR ALGORITHMS

The nonlinear mapping algorithm described by Sammon works satisfactorily when the number of pattern samples N is not very large. If N is large, the amount of computation and storage required to achieve the optimal mapping (in the sense of minimized error function E) could be prohibitively large. Since the effectiveness of any mapping algorithm in an interactive environment depends upon the speedy display of the data configuration in the transformed space, it is desirable to improve the nonlinear algorithms so as to reduce the computation and storage required for large values of N.

Another problem with the nonlinear algorithms we discussed relates to the fact that they do not have an explicit mathematical form. Unlike the linear mappings whose explicit form can be precisely determined, nonlinear mappings (e.g. Sammon's algorithm) are generally described in an iterative manner. Each iteration of the mapping process involves an adjustment of the configuration of all of N data points in the new space. This iteration is repeated until a solution or approximate solution is reached.

Such iterative definition of a mapping algorithm makes it difficult to apply the algorithm recursively when new pattern samples are introduced to the pattern recognition system. This means that the iterative adjustment of the configuration in the new space must be re-initiated and repeated for the new as well as the old samples. No attempt is made in these algorithms to preserve the local details of the configuration for the old data points. Consequently, a great deal of the computation is wasted in re-adjusting the old sample points along with the new data points.

1. A Modified Algorithm

To improve this situation, we describe a modified Sammon's algorithm which will take into account the two problems mentioned above. Two new elements are introduced to arrive at the optimal nonlinear mapping.

The first new element is intended to make the convergence of the iterative process to achieve the minimum-error configuration faster. This is accomplished by letting the correction factor CF in the algorithm (eq. 10) become a linear function of the error. Thus, in general we have

$$CF = a\,E(n) + b \qquad (12)$$

where a and b are constants and E(n) is the error at the nth iteration. The effect of this modification is to allow the amount of adjustment for each point of configuration in the new space to be controlled by the error E(n) at each iteration, rather than by a constant throughout the entire process. When E(n) becomes smaller as n increases, which indicates a closer approximation between the two configurations in the original and the new space, the amount of adjustment is proportionally decreased. Similarly, longer E(n) would allow the configuration in the new space to call for larger adjustments.

The second modification deals with the recursive application of the mapping algorithm when new samples are introduced. We say that a nonlinear mapping algorithm is recursively applicable if the computation of the mapping involving the new sample (or samples) depends upon only the new sample (or samples) and the configuration of the old samples in the transformed d-space. Thus, for example, the Sammon's nonlinear mapping algorithm is not recursively applicable. This is because the computation of a new mapping must involve the adjustment of all the samples - new as well as old - in the transformed space.

To eliminate the re-computations in non-recursive algorithms, we must take advantage of the man-machine environment in determining the new mapping when a new sample or a group of new samples is introduced. A recursively applicable algorithm may be described as follows: Assume that a nonlinear mapping has been applied to a set of samples X_i, i = 1, 2,...,N, and the resulting transformed samples are Y_i, i = 1,2,...,N. The configuration of the Y_i's in the transformed space is then displayed for human observation. Upon observation of the display, the human observer determines what he considers to be the clusters formed within the configuration. Let the centers (e.g. mean vectors) of these clusters be denoted by C_1, C_2,...,C_P, where P is the number of clusters detected, which is usually much smaller than N as each C_i consists of one or more samples.

Now, suppose that a group of M new samples, denoted by X_i, i = N+1, N+2,..., N+M ($M \geq 1$), is introduced to the system. Instead of adjusting the entire N+M samples to arrive at an optimal configuration of Y_i, i = 1, 2,...,N+M, the set of M new samples are adjusted in conjunction with the P cluster centers C_i, i = 1, 2, ...,P. That is, the samples in each C_i will migrate as a group towards a new configuration in the transformation with minimum error. The number of sample points that need to be adjusted for each iteration reduces from N+M to N+P. Furthermore, the algorithm becomes recursively applicable since the configuration of the old samples, in the form of cluster centers rather than individual samples, is now adjusted along with the new samples to achieve the updated optimal mapping.

2. Examples in Imagery Study

An experimental comparison of the two versions of the nonlinear mapping has been made to demonstrate the utility of the modified algorithm. The experiments involved computer-simulated studies of satellite imagery of the New England area taken by the NASA ERTS-1 satellite. Results of these studies have also been found useful to provide the necessary information needed for the design of an interactive system for ERTS imagery processing currently being implemented at the University of Connecticut.

For the purpose of this discussion, digitized satellite photographs of the same location in the New England area in four different spectral bands are shown below.

Band 7		Band 6
C C D E F F D D C C	definitely water	A A A B E D B A A 9
C C D F F E D C C C		A A A D E C A A A A
C C E F F D D C C C	definitely land	A A C E E B A A A A

Band 5	Band 4
D D C D E E D D D D	C C C C D D C C C C
E D C E E D D E E E	D C C C D C C C C C
E D D E E D D D D E	C C C D D C C C C C

The 30-element picture for each of the four bands corresponds to geographically a section of the Cape Cod canal and its adjacent land. Each element of the picture represents one of the sixteen gray levels coded in hexidecimal (0 to 9, A, B, C, D, E, F). A vector X_i can be made from the ith elements of the four bands as follows:

$$X_1 = \begin{bmatrix} \text{ith element from band 7} \\ \text{ith element from band 6} \\ \text{ith element from band 5} \\ \text{ith element from band 4} \end{bmatrix}^t$$

For example,

$$X_1 = (C, A, D, C)$$

The letters in the vectors X_i, i = 1,..., 30 must of course be converted to numeric values before applying the mapping algorithms. The 30-element 4-band picture above yields a total of 30 four-element vectors X_i, i = 1, 2,..., 30. Each vector in reality is either water or land. However, due to the uncertainty in interpreting the pictures by human observation (see band 7), the 30 vectors will be classified into 3 classes: Definitely water, definitely land, and uncertain. Our objective here is to determine if, and how, clusters may be formed in the 2-dimensional space when a nonlinear mapping algorithm is applied.

Figure 5 shows the configuration of the 30 vectors after 20 iterations of Sammon's algorithm (Error $\simeq$ .04). An improved configuration was obtained by removing the vector component which represents the Band 5 information. This configuration is shown in Figure 6 (Error $\simeq$ .01 after 20 iterations). A comparison of these two configurations suggests that the three bands (7, 6, 4) provide a better set of spectural features by which the local details of the photo imagery can be preserved under the nonlinear mapping. Selection of features by elimination of this kind is particularly attractive in an interactive environment, as the human observer, with the aid of a display, can play a direct and immediate role in the selection process.

To apply the recursive algorithm as well as the Sammon's

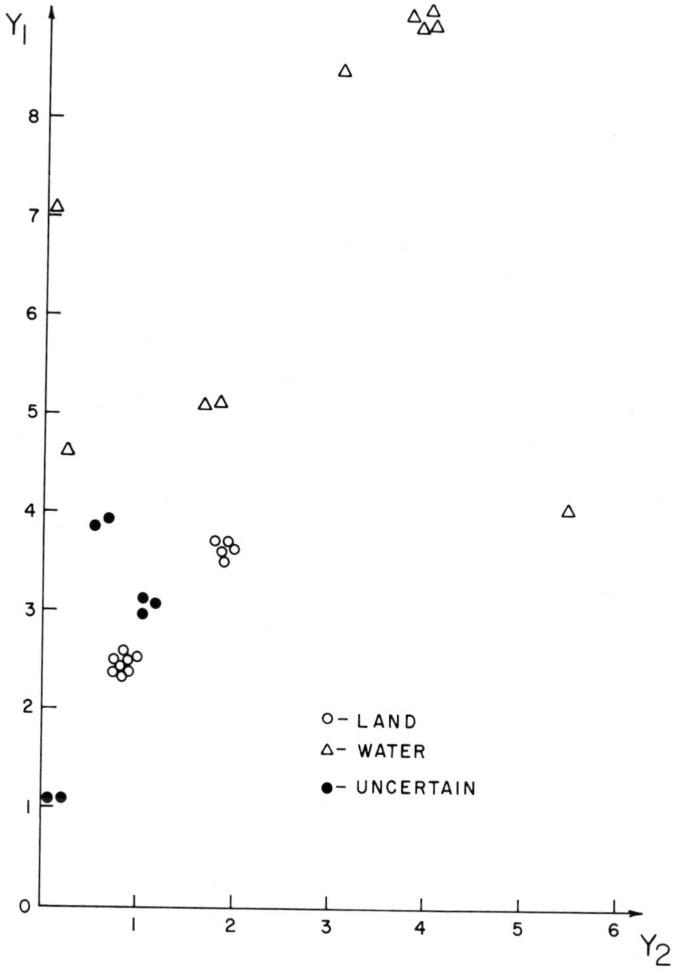

Fig. 5.

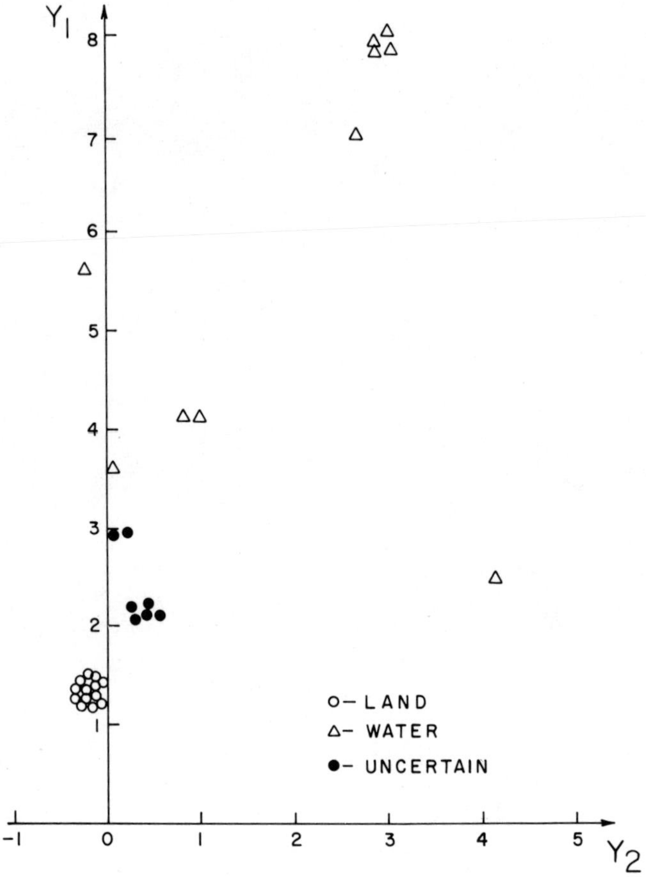

Fig. 6.

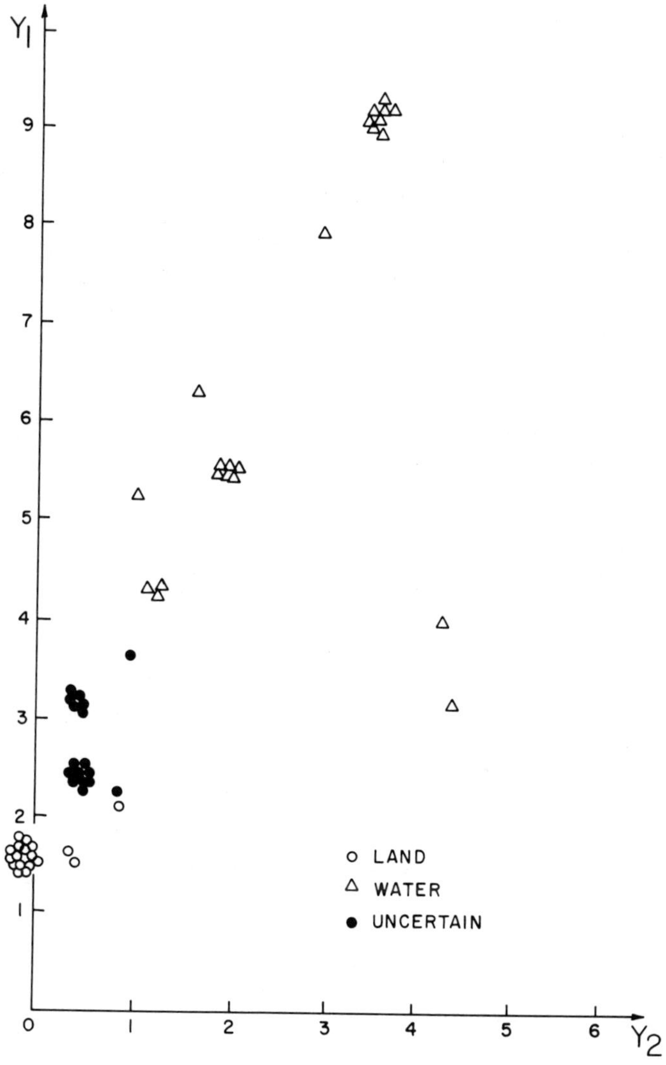

Fig. 7.

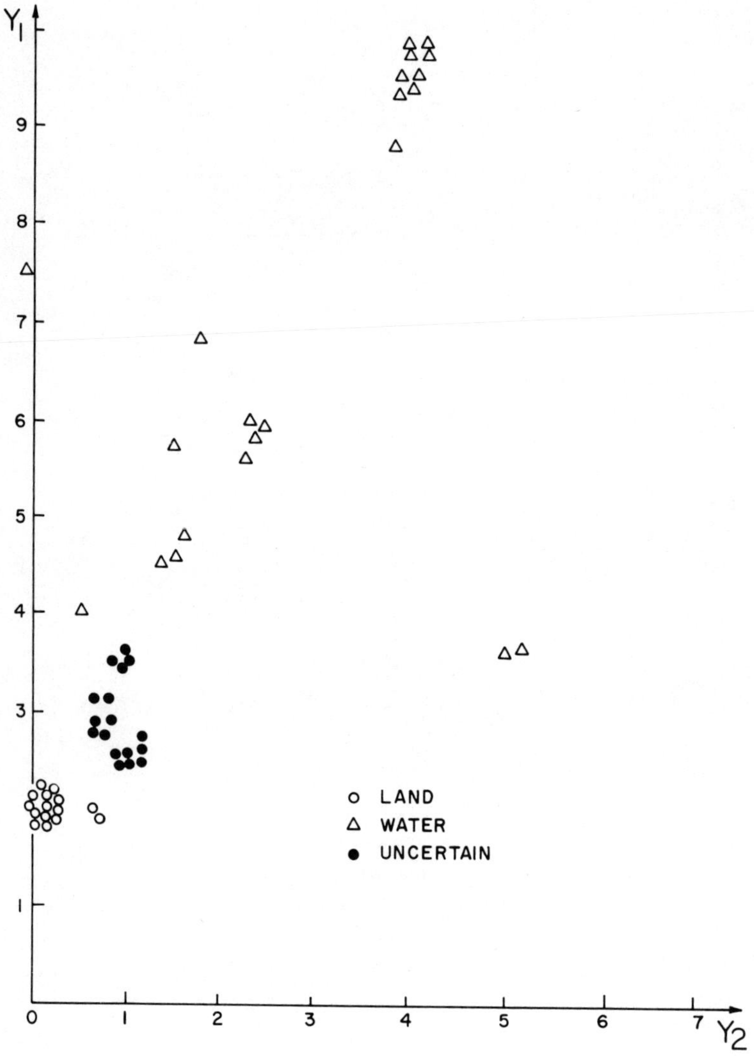

Fig. 8.

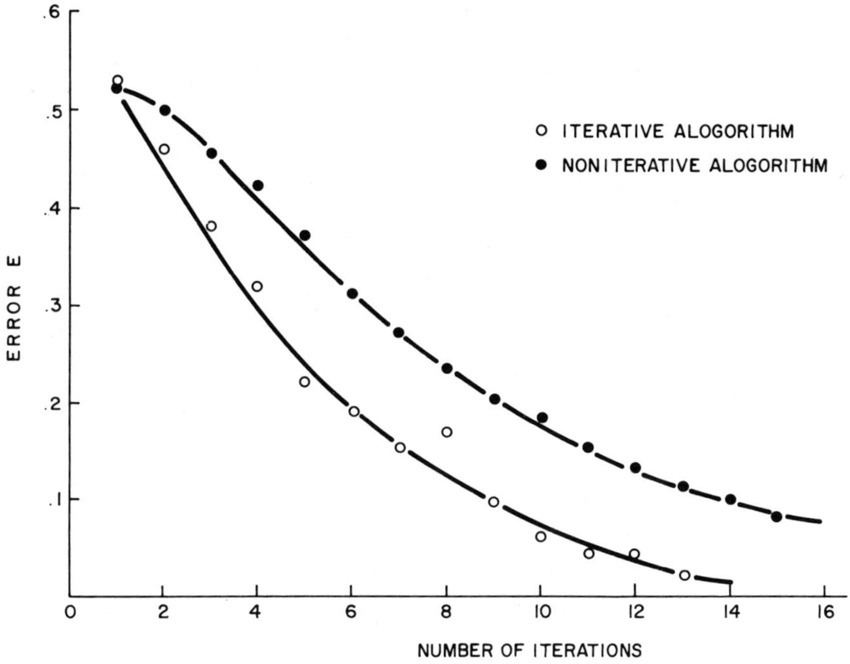
.6
.5
.4
.3
.2
.1
ERROR E
O ITERATIVE ALOGORITHM
● NONITERATIVE ALOGORITHM
0
2
4
6
8
10
12
14
16
NUMBER OF ITERATIONS

Fig. 9.

algorithm, an additional 30-element picture corresponding to an area near the previous location is selected. The satellite photographs of this area for the 3 selected bands (Band 5 removed) are shown below.

Band 7

D C D D D F F F D C definitely water

C C D E F F F D D D

C C E F F F D D C C definitely land

Band 6

A 9 A A B E E D B 9

A 9 A C E E D B A A

A A D E D D A B A A

Band 4

B B C B C C D C C C

C C C C D D C C D C

B C C D C C C C C C

These additional 30-elements are used to form 30 three-component vectors to be adjusted along with the previous 30 vectors to arrive at a new 2-dimensional configuration. Figure 7 shows the configuration (after 15 iterations) of the 60 vectors using the nonrecursive Sammon's algorithm. The final error is approximately .08. Figure 8 shows the configuration (after 13 iterations) of the 60 vectors using the recursive algorithm. The final error is about .02. The recursive algorithm computes the optimal adjustment for the 30 new vectors, along with the three cluster centers (mean vectors) of the previous 30 vectors. A comparison of these two algorithms was made on the basis of convergence rate by plotting the errors as a function of the numbers of iterations. This is shown in Figure 9 which indicates a faster convergence for the recursive algorithm.

V. CONCLUSIONS AND FURTHER REMARKS

The computational algorithms discussed in this paper deal primarily with the mapping of multi-dimensional data to lower-dimensional (2 or 3) spaces convenient for human observation and evaluation. Both linear and nonlinear methods have been discussed and their merits and limitations compared.

The notion of recursively applicable algorithms has been introduced. Such algorithms require less computation and storage to arrive at updated optimal nonlinear mappings when such mappings must be repeatedly applied to allow man-machine interaction in the design of a pattern recognition system.

Acknowledgment-This work was supported in part by the NSF grant GJ-36434 and the U. S. Dept. Interior/UConn Institute of Water Resources Project A-042. The computer programming work was done by Mr. BRUCE SCHACTER.

REFERENCES

1. H.C. ANDREWS, Introduction to Mathematical Techniques in Pattern Recognition, Wiley-Interscience, 1972.
2. G.H. BALL, "A Comparison of Some Cluster-Seeking Techniques", RADC-TR-66-514, November, 1966.
3. G.H. BALL, et al., "Implications of Interactive Graphic Computers for Pattern Recognition Methodology", Methodologies of Pattern Recognition (ed. S. WATARABE), Academic Press, 1969, pp. 23-31.
4. THOMAS W. CALVERT, "Nonorthogonal Projections for Feature Extraction in Pattern Recognition", IEEE Trans. on Computers, C-19, 5, May 1970, pp. 447-452.
5. THOMAS W. CALVERT, "Projections of Multidimensional Data for Use in Man-Computer Graphics", Proc. Fall Joint Computer Conference, 1968, pp. 227-231.
6. THOMAS W. CALVERT and T.Y. YOUNG, "Randomly Generated Nonlinear Transformations for Pattern Recognition", IEEE Trans. on Systems Science and Cybernetics, SSC-5, 4, Oct. 1969, pp. 266-273.
7. C.L. CHANG and R.C. Lee, "A Heuristic Relaxation Method for Nonlinear Mapping in Cluster Analysis", IEEE Transactions on Systems, Man and Cybernetics, March 1973, pp. 197-200.
8. Y.T. CHIEN, "Interactive Pattern Recognition - A Review and Outlook", Proc. 11th IEEE Symposium on Adaptive Processes, December 1972, pp. 106-110.
9. Y.T. CHIEN and K.S. FU., "Selection and Ordering of Feature Observations in a Pattern Recognition System", Information and Control, 12, 1968, pp. 394-415.
10. N.M. HERBST and P.M. WILL, "An Experimental Laboratory for Pattern Recognition and Signal Processing", Communications of the ACM, 15, 4, April 1972, pp. 231-244.
11. L.N. KANAL, "Interactive Pattern Analysis and Classification System: A Survey and Commentary", Proc. IEEE, 60-10, October 1972, pp. 1200-1215.
12. E.A. PATRICK and F.P. FISCHER, II., "Cluster Mapping with Experimental Computer Graphics", IEEE Trans. on Computers, C-18, 11, Nov. 1969, pp. 987-991.
13. E.A. PATRICK and L.Y. SHEN, "Interactive Use of Problem Knowledge for Clustering and Decision-Making", IEEE Trans. on Computers, 20, 2, Feb. 1971, pp. 216-222.
14. E.A. PATRICK, et al., "Mapping Multidimensional Space to One Dimension for Computer Output Display", IEEE Trans. Computers, 17, Oct. 1968, pp. 949-953.

15. KENDALL PRESTON, JR., and P.E. NORGREN, "Interactive Image Processor Speeds Pattern Recognition by Computer", Electronics, Oct. 1972.
16. JOHN W. SAMMON, JR., "A Nonlinear Mapping for Data Structure Analysis", IEEE Trans. on Computers, C-18, 5, May 1969, pp. 401-409.
17. JOHN W. SAMMON, JR., "Interactive Pattern Analysis and Classification", IEEE Trans. on Computers, C-19, 7, July 1970, pp. 594-616.
18. JOHN W. SAMMON, JR., et al., "Programs for On-Line Pattern Analysis", Vol. I, and II, RADC-TR-71-177, September 1971.
19. JOHN W. SAMMON, JR., A.H. PROCTOR and D.F. ROBERTS, "An interactive-Graphic Subsystem for Pattern Analysis", Pattern Recognition, 3, 1971, pp. 37-52.
20. G.S. SUBESTYEN, Decision-Making Processes in Pattern Recognition, The Macmillan Company, 1962.

A METHODOLOGY FOR INTERACTIVE SYSTEMS

K. Hanakata

University of Stuttgart

Stuttgart, W. Germany

I INTRODUCTION

In the recent activities in the fields of applied computer science such as Artificial Intelligence, Information Retrieval, Pattern Recognition etc., a methodology of interactive systems is one of the most important problems in designing a system for the above-mentioned purposes. A basic condition which we have to consider in designing an interactive system is that the dialogue partner of a machine is always a man who uses the interactive system and who evaluates its performance on the basis of his different characteristics of information processing. In this respect, it is obvious that the machine part of such an interactive system should be designed so that the human characteristics are also taken into account in terms of performance criterion. In the following brief note, a basic scheme of interactive systems is discussed focusing on the problems which basically come from human ambiguous or fuzzy expression [1].

---

[1]As expressed by L.A. Zadeh (1), while "mechanistic systems, that is, inanimate systems whose behavior is governed by the law of mechanics, physics,..." and therefore "precise, rigorous and quantitative", humanistic systems whose behavior is strongly "fuzzy unrigorous and qualitative, and have proved to be rather impervious to mathematical analysis and computer simulation". "Indeed, it is entirely possible that only through the use of such more tolerant approaches which are approximate in nature, could computer simulation become truly effective as a tool...".

II INTERACTIVE SYSTEMS

With the development of computer techniques, the capabilities of machine problem solving in its general sense has so increased that not only its problem solving methods, but also a certain amount of information which is necessary for problem solving can be gradually shifted from the human intelligence to the machine (intelligence). This means that the solvable problem space is enlarged by means of an interactive system with its effective use of the advantageous characteristics of both sides. In the general discussion of interactive systems the following three components are to be considered.

(1) Problem: A complex of tasks, instructions or missions which we want to implement or to achieve, or from which we want to draw conclusions with the help of an interactive system. For example, find a necessary and sufficient amount of data from a static or dynamic data bank, or find a sequence of simple tasks which can be executed by a computer-controlled hardware device.

(2) Man: The one side of an interactive system that gives the problem mentioned in (1) to the other side. This side has its own capabilities[4] in relation to the problem solving such as induction, logical deduction, logical expression, abstraction, generalization, its own degree of description precision by means of natural language, reliability of calculation, recalling speed of stored memory, perception, etc. Another important factor is that this side is the user of the system, and so evaluates the system on the basis of his own criterion.

(3) Machine: Similar to the man-side, this side has also its own capabilities in relation to the problem solving but in a different degree.

Fig. 1 illustrates the relation between these three factors and a basic communication model within both sides. The direct man-machine communication can take place only through the common communication media from or in which each side derives or puts information in a prescribed form of expression. Because of the difference of characteristics and degree of capabilities between man and machine, these common communication media, e.g. language, visual pattern, etc. are only the subsets of those which are inherent in each side, for instance, natural language for man-side or high level programming language for machine-side. In this situation the following questions arise in relation to the problem solving by the given interactive system.

(i) Is it possible for man to sufficiently describe a given problem in this restricted common language so that the machine can solve the problem on the base of this problem description?

This question is basically related to the problem of solvability under certain description power of a given language. It is a matter of system capability in terms of solvable problem space.

(ii) Because of the human characteristics of ambiguous or fuzzy expressions, and because of the lower description power of the common communication language, we may not be able to establish the semantic relation of one to one correspondence between man and machine (shown by a broken line in Fig. 1.) Therefore, the necessary information for the machine problem solving is not always transmitted with the meaning the human partner wanted to describe.

If it is allowed for the machine to request the man for additional information trying to turn the problem solvable (otherwise, every user has to be a perfect mathematician), what is the optimal question from machine to man in terms of the problem solving cost on the machine side, answering cost on the man side.

These two questions are major factors related to the performance criterion of the given interactive system.

III PROBLEM SOLVING

As is hinted in the previous chapter, the object of an interactive system can be interpreted as the problem solving in any kind of application, e.g. information retrieval, computer-aided instruction, process on-line control, information inquiring etc.

For the simple question-answering man-machine dialogue a given problem may be solved immediately by checking the well prepared answer lists, as far as the given question contains enough information described in an appropriate form. If a corresponding answer cannot be found on the list, then the problem is considered unsolvable, and a standard response is given to the user, who evaluates the system as, say, "primitive", or "the answering repertory is limited". If an appropriate answer which is not expected by the user is found from a well structured large data bank, the evaluation of the well trained user may be "usable". Even in this case man's partner can hardly be predicated with "intelligent", because of the restriction on the language structure and the description form in which to describe the problem perfectly makes the system difficult to use. Therefore, the next step for

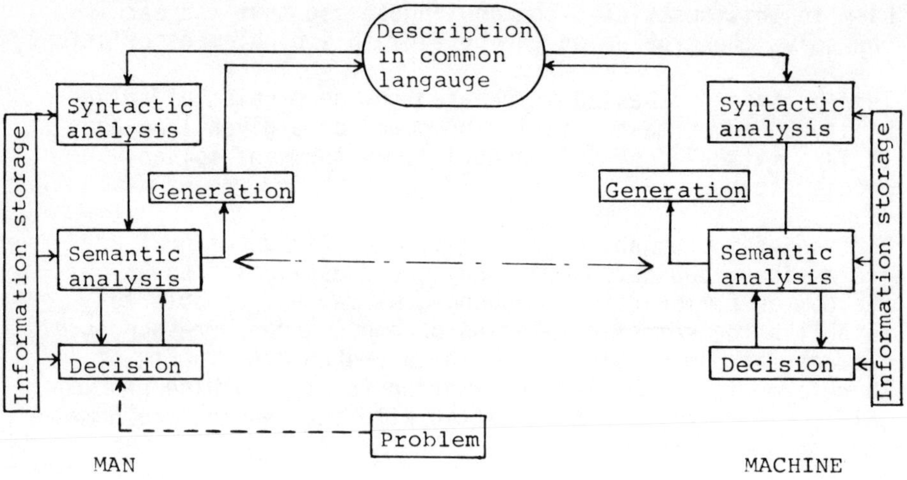

Fig. 1.

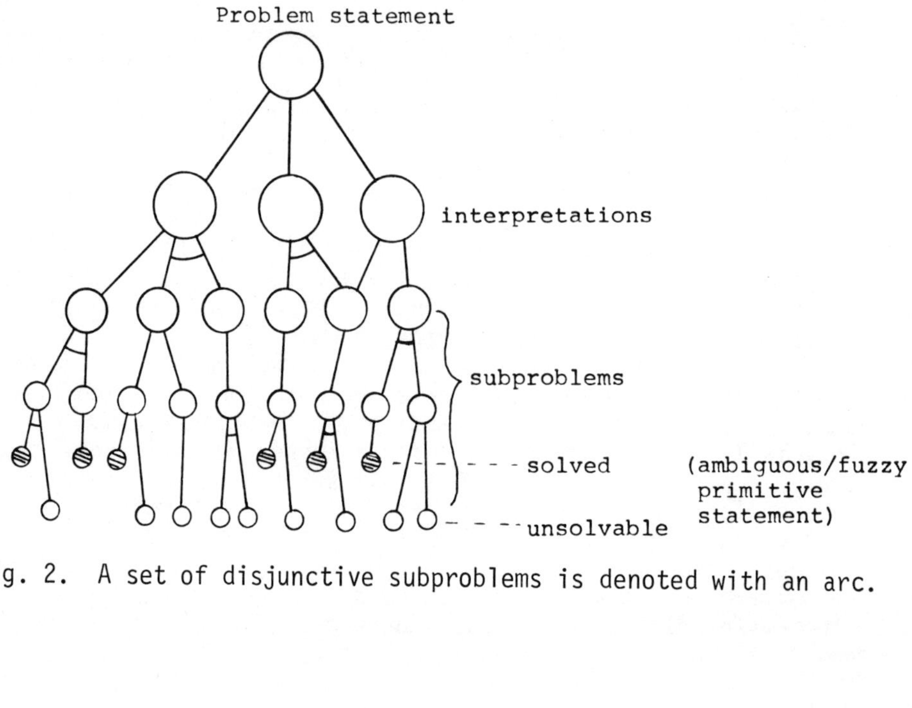

Fig. 2. A set of disjunctive subproblems is denoted with an arc.

the intelligence level -up of the machine side is to expand the common language space covering larger parts of the natural language and a more sophisticated problem analysis method such that it can accept a complex problem compactly expressed in terms of higher level of concepts.

With the increasing description power of the common language for interactive systems a trouble which arises in relation to the description freedom of the man-side is to deal with ambiguous and or fuzzy description which is now, semantically at least, acceptable.

This ambiguity or fuzziness contained in the problem description may be intrinsic because of the lack of knowledge or ostensible because of the convenient expression. In both cases, the ambiguity or fuzziness may be able to be reduced from the relational structure within the problem description, so that it is not essential any more to the problem solving.

Example:

SHE IS QUITE YOUNG, I DON'T KNOW HOW OLD SHE IS. HER YOUNGER BROTHER, A FRIEND OF MINE IS ENROLLED IN PURDUE.

DOES SHE HAVE THE VOTE?

(Such sort of ambiguous or fuzzy expression which is not intrinsic with regard to the problem solving may be defined as not ambiguous or fuzzy.)

Given a set of problem statements, it is important to decide whether the problem is solvable without any additional information or if not, what kind of information is necessary to make the problem solvable, and in this case, what is the appropriate machine question.

An approach to this problem of solvability and optimal machine questioning may be illustrated by the simple scheme of problem-reduction method.

Defining a set of axioms in relation to the given problem which may be described, say, in a natural language containing some kind of ambiguity or fuzziness, we apply an appropriate subset of these axioms to the initial problem statements, generating a set of different interpretations. Some of these interpretations applied by axioms are reduced to a set of subproblems which are all conjunctive with each other. By the successive application of axioms to the present state of problem statements, a subproblem is reduced to a set of either disjunctive or conjunctive successor subproblems, each of them simpler to be solved than its predecessor, and finally there remains either subproblems whose solutions are axioms themselves, or primitive ambiguous or fuzzy statements which cannot be partitioned by means of axioms (Fig. 2).

An intermediate subproblem is defined to be solvable if either one of its disjunctive successor subproblems is solvable or all of its conjunctive successor subproblems are solvable. According to this definition of "solvable", the initial given problem is solvable if and only if there exists a sequence of conjunctive subproblems which are all solvable.

Applying resolution principle we have a set of clauses including those which express ambiguous or fuzzy statements. A resolution of two clauses, either of which is ambiguous or fuzzy, produces either a distinct or unambiguous resolvent, or again, an ambiguous or fuzzy one. In this case, the successive resolution process propogates the ambiguity or fuzziness. If we cannot produce any empty clause in an appropriate time still producing the ambiguous or fuzzy resolvents, we have reason to doubt that it cannot be solved unless the ambiguity or fuzziness of the original problem statement is eliminated.

IV EVALUATION OF MACHINE QUESTIONING

In the last section we discussed the case where a given problem cannot be solved, probably because of its inherent ambiguity or fuzziness within the feasible time for interactive systems.

In terms of the resolution tree (refutation graph) the ambiguous or fuzzy clauses cannot find their partner to resolve into an unambiguous resolvent, or even in this case, the unambiguous resolvent is too weak to produce its empty descendant clause. Here at this point we have an access for the formulation of the machine question whose answer from the man is to reduce the ambiguity so as to make the original problem solvable.

After trying to produce an empty clause in vain for feasible steps of resolution process, we have to find which ambiguous or fuzzy clauses among those to be resolved,including newly generated resolvents, can be easily modified by additional information so as to produce an empty clause.

<u>Example:</u>

		Notation
(i)	Everybody above 18 in Japan has the vote	$J(x) \wedge P_{18}(x) \Rightarrow W(x)$

(ii) Every student is older than 18 $S(x) \Rightarrow P_{18}(x)$

(iii) X is older than Y implies that if Y is above 18, then x is also above 18 $O(x,Y) \Rightarrow [P_{18}(Y) \Rightarrow P_{18}(x)]$

(iv) A is an elder sister of B $O(A,B)$

(v) B is a student $S(B)$

(vi) Does A have the vote? $W(A)?$

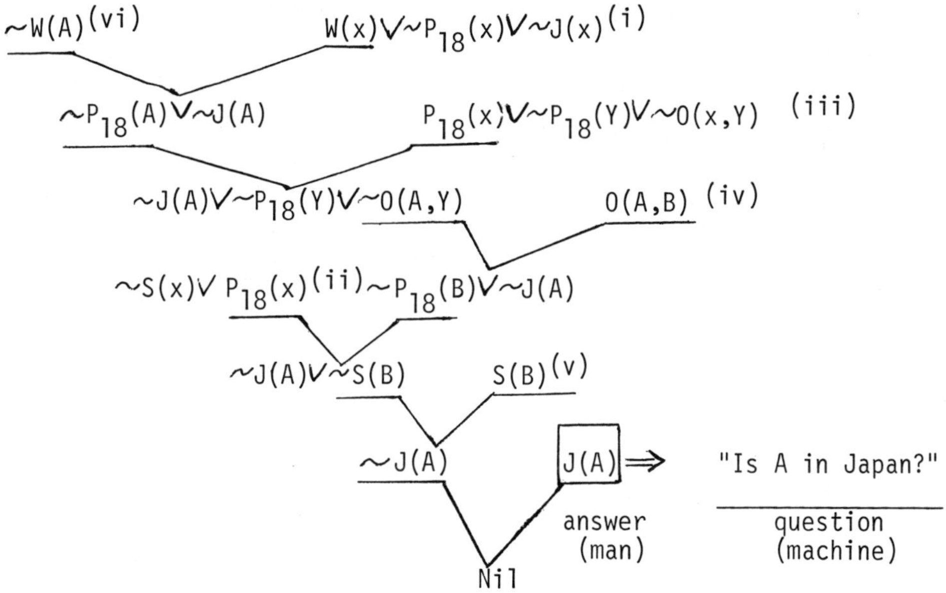

$\sim J(A)$ i.e. "A is not in Japan" is the last level resolvent which cannot find its partner for resolution, and so the empty clause is not generated. We may attribute this unsolvable situation to the incomplete problem description. If we had obtained some additional information of premise such as "A is in Japan", i.e. J(A) (in the box) then we could have produced the empty clause i.e. "Nil". From this conjectural information we can generate such an interactive question from machine to man as "Is A in Japan?"

In the above simple example additional information is requested from the machine in order to produce an empty clause. It is important to note that such a question should be generated whose answer brings additional information putting an end to the resolution process as early as possible but not because of hoping to prove that the conjecture (w(A)) follows from a given set of premises. If the conjecture doesn't intrinsically follow from the set of premises in the truth regardless of the completeness or incompleteness of the problem description, then the machine may keep asking the questions in vain not getting the expected answers which can generate an empty clause. In this case, of course, the answer is supposed to be always consistent to the *a priori* given set of premises. Therefore, if it is not known that the conjectural conclusion does necessarily follow from the premises, then it is not reasonable to generate questions only hoping to solve the given problem. In such a situation, it is better to attack the problem from both sides, that is, parallel to the resolution process trying to prove that the conjectural conclusion follows from the given premises, we try to prove that the reverse of the conjectural conclusion follows from the given premises by using the resolution process[3]. In this case, those questions are at first given from both processes whose answers give the additional information so as to terminate either of two resolution processes as early as possible.

So far we discussed how to generate questions from the resolution processes in order to achieve some conclusion relevant to the problem. In this case, we considered the goodness of machine questioning only from the viewpoint of machine problem solving, but not from the human side where the machine performance of question answering is evaluated. Therefore, in order to evaluate the goodness of the machine questioning from the human side, we must have some performance function not only in terms of the distance of resolution steps to the empty clause, but also in terms of the easiness of human answering. For some questions, it is difficult for the human to answer depending upon his level of intelligence or knowledge and cost related to the information requested. For example, it is quite easy to give an answer such as "A is larger than B" than to give the exact measurements of A and B if that information is enough to convert the problem solvable. This problem of performance criterion of a whole interactive system is fairly complex including the model of human problem solving[4] in psychology.

REFERENCES

(1) Zadeh, L.A., "The concept of a linguistic variable and its application to approximate reasoning", Proc. of 2nd joing U.S.-Japan seminar, Oct. 1973.

(2) Nilsson, N.J., *Problem-solving methods in artificial intelligence*, McGraw-Hill, 1971.

(3) Laubsch J., Private discussion, Univ. of Stuttgart, Dec.6,1973.

(4) Newell, A., etal, *Human Problem solving*, Prentice Hall, N.J. 1972.

AUTOMATIC RECOGNITION OF COMPLEX THREE-DIMENSIONAL OBJECTS FROM OPTICAL IMAGES*

Robert B. McGhee

Ohio State University

Columbus, Ohio 73210

INTRODUCTION

The possibility of duplicating in a machine the ability of animals and man to interpret visual information has intrigued many investigators. While this problem has been treated with considerable success relative to automatic classification of two-dimensional objects, especially with regard to recognition of printed characters, comparatively little has been achieved in automatic identification of three-dimensional objects. Generally speaking, such successes as have been attained in the latter area can be grouped under two broad headings: 1) scene analysis in which various specialized procedures are used to determine spatial relationships between simple objects such as solid polyhedra, cylinders, cones, etc., and 2) statistical pattern recognition in which certain numerical features of an image of an unknown object are used to classify it and perhaps to estimate its position and orientation [1]. This paper is addressed entirely to the second approach and is concerned with recognition of complex man-made objects rather than natural objects or simple solids.

PROBLEM STATEMENT

In what follows, it is assumed that an object to be recognized belongs to one of a set of known classes and that all elements of a

*Research on this problem is supported at Ohio State University, by the United States Air Force Office of Scientific Research under Grant AF-AFOSR-71-2048.

given class are essentially "cast from the same mold." That is, each object is a more or less perfect replica of a given solid model for its class. It is also assumed that classification must be automatically accomplished from the information available in a single optical image obtained with the object in an unknown orientation relative to a camera. With these restrictions, the difficulties associated with object recognition arise from two principal sources: 1) in most cases, an unlimited number of different optical images can be obtained for a given object class depending on object position and orientation, and 2) the images available may in many circumstances be of low quality and may also contain confusing background information (optical clutter.)

Research on the above stated problem was originally motivated by a need for automatic methods for identifying aircraft types in an air traffic control situation [2]. More recently, the importance of this problem in industrial automation has been recognized [3]. However, a survey of the literature has revealed that essentially all published work involving arbitrary object orientation has dealt with the aircraft identification problem. Consequently, the remainder of this paper is devoted exclusively to this aspect of solid object recognition. In the interest of conciseness, only those studies which have been carried through to an evaluation of recognition accuracy with real or simulated images are reported. To the author's knowledge, all published results are summarized in the following.

COMPONENTS OF A RECOGNITION SCHEME

It is generally recognized that the statistical (geometrical) approach to automatic recognition of objects from two-dimensional images usually involves three distinct phases: 1) picture segmentation, 2) feature extraction, and 3) classification [1,4]. The first problem has to do with isolating the object to be recognized from its background. Many different approaches to this problem have been attempted in various applications including detection of gray level edges, spatial filtering, gray level histogram analysis, connectivity analysis, etc. [1,4]. One of the more successful methods in many practical applications has been to simply threshold the picture so that only objects lying above (or below) a given gray level are considered to be of interest [1,5]. Once this has been done, the picture is reduced to what amounts to a silhouette [6] of the object or objects of interest with perhaps additional undesired background elements also appearing. It appears that such thresholding is the only method which has been used to date with respect to aircraft identification schemes. However, other interesting possibilities exist including color analysis [1], comparison of successive frames for moving object detection, and range gating (with active illumination).

One of the more difficult problems in the design of a pictorial pattern recognition system relates to the selection of a set of appropriate numerical attributes or <u>features</u>, to be extracted from the object of interest for purposes of classification. By and large, the success of any practical system depends critically upon this decision. While there is little in the way of a general theory to guide in the selection of features for an arbitrary problem [7], it is possible to state some desirable attributes of features for identification of solid objects. Among these are:

1. The features should be informative. That is, the dimensionality of the vector of measurements (feature vector) should be as low as possible consistent with acceptable recognition accuracy.

2. The features should be invariant with translation of the object normal to the camera optical axis.

3. The features should either be invariant or depend in a known way upon the distance from the camera to the object.

Some examples of features satisfying these criteria to varying degrees are presented later in this paper.

With respect to classification, there is no evidence that aircraft image features possess any special properties. Hence, the full range of statistical pattern recognition theory is applicable [1,8,9]. This theory is far too vast to review here and no attempt will be made to do so. Instead, in what follows, a survey of all successful approaches to the problem under consideration will be presented. Because of the importance and special aspects of the feature selection problem, this component of the overall recognition problem will be emphasized.

FOURIER INVARIANTS

Cosgriff [2] was among the first to study the problem of automatic aircraft recognition. In his work, he defined a function called the "tangent angle vs. arc length" function for a closed planar region as

$$\Theta(s) = \int_0^s k(\zeta)\, d\zeta \qquad (1)$$

where k(s) is the curvature of the boundary of the region as a function of arc length from an arbitrary starting point [2]. In order to make $\Theta(s)$ amenable to Fourier analysis, it is useful to make it periodic by transforming it to the "slope intrinsic function" defined by [1].

$$\hat{\Theta}(s) = \Theta(s) - \frac{2\pi s}{L} - \mu \tag{2}$$

where μ is just the average value of $\Theta(s)$ over the entire boundary contour and L is the length of the boundary. Finally, if

$$\ell = \frac{2\pi s}{L} \tag{3}$$

and $\phi(\ell)$ is the "normalized cumulative angle bend function" [10]

$$\phi(\ell) = \hat{\Theta}(\ell L/2\pi) \tag{4}$$

then

$$\phi(\ell) = \sum_{k=1}^{\infty} A_k \cos(k\ell - \alpha_k) \tag{5}$$

An example of such a function is provided in [11] where it is also shown that the coefficients A_k possess the desired invariance with image size, translation, and rotation in the image plane.

The invariance properties of the coefficients A_k are sufficient to permit fixed-font character recognition from a single prototype for each class and such an application was proposed by Brill [10]. However, for three-dimensional objects, two further degrees of object motion are possible. Specifically, if the direction of the camera optical axis is taken as the object elevation axis, then complete specification of the object angular position depends upon the definition of two more rotation axes not colinear with the optical axis. These two axes will be called the "roll" and "azimuth" axes [12]. Together with the elevation axis they form a basis for an <u>Euler angle</u> set of rotations [13]. The exact definition of these angles is not important and will not be pursued further here. All that matters is that any pair of values for the azimuth angle ψ, and the roll angle ϕ, uniquely describes the <u>object viewing aspect</u> [6] and hence determines the appearance of its silhouette. Of course the silhouette of any airplane varies considerably with aspect so instead of being described by a vector of <u>constants</u>

$$\rho = [A_1, A_2, \ldots, A_N] \tag{6}$$

each aircraft type gives rise to a vector of <u>functions</u>

$$\rho(\phi,\psi) = [A_1(\phi,\psi),\ A_2(\phi,\psi),\ldots,A_N(\phi,\psi)] \tag{7}$$

These functions are in general unknown and must be approximated from values obtained by viewing a member of a given object class from a finite number of aspects. This is a basic problem associated with all but one of the methods to follow.

Cosgriff attempted to overcome the difficulty associated with the functional dependence of the A_k on viewing aspect by segregating the features associated with the various classes by means of separating hyperplanes [1]. In a five class problem making use of a six-dimensional feature vector consisting of A_1 through A_6, he reported almost perfect results with respect to a training set consisting of 65 essentially noiseless photographs for each object class [14]. However, no evaluation of this system with an independent test set was reported and no means for obtaining images in real-time was considered. It must be stated, therefore, that the practical value of these invariants in aircraft recognition remains to be determined.

In work closely related to that of Cosgriff et al, Richard and Hemami described the boundary of a silhouette in the parametric form [15]

$$z(t) = (x(t),\ y(t)),\quad t \in [0,1] \tag{8}$$

They then expanded z(t) in complex Fourier series as

$$z(t) = \sum_{k=-\infty}^{\infty} c_k\, e^{i2\pi kt} \tag{9}$$

If the coefficients c_k are normalized by the relation

$$c_k^* = c_k / [(\sum_{\ell=-\infty}^{\infty} |c_\ell|^2) - |c_0|^2]^{1/2} \tag{10}$$

then it can be shown that $|c_k^*|$ are size and rotation invariants[15].

Using a "wire-frame" simulation model [16] for each of four aircraft, Richard and Hemami produced a total of 666 synthetic images for each aircraft type by varying the two aspect angles in small increments, resulting in a total training set of 2664 noiseless images. Following this, they generated a test sample consisting of one hundred images at random aspects for each aircraft type. Utilizing a simple nearest neighbor decision rule [1], they found that no identification errors occurred. When image points were displaced by the addition of Gaussian noise, a few errors were observed. For example, in a typical experiment involving 100 samples of a given aircraft type in which each sample point was displaced by noise acting independently in the x and y directions with a standard deviation of about 20 percent of the image radius, the observed error rate was 8 percent [15].

As in the earlier work of Cosgriff, in [15] no consideration is given to a real-time implementation of the recognition scheme developed. Instead, all results were obtained either by computer simulation or by processing of digitized photographs. However, the effect of noise was examined and the ability of a nearest neighbor decision rule to achieve perfect discrimination with noiseless data was demonstrated. It must be presumed that this would also be true for the invariants A_k defined by Eq. (5), however, so no comparison of the relative value of these invariants and those defined by Eq. (10) can be made at this time.

SEQUENTIAL IMAGE MATCHING

One of the most fundamental ways of accomplishing optical character recognition is to simply directly compare a set of reference images or "templates" [1] to a test image and to then identify the object as belonging to the class yielding the best match. If this approach is attempted in aircraft recognition, it is found that a very large number of such templates must be provided for each class because of the variability of the silhouette with viewing aspect. In an attempt to overcome this difficulty, Advani [17] utilized a sequential approach in which a series of synthetic images was generated and compared to a test image. In this approach, suggested by earlier work of Hemami et al [18], the degree of match between images is measured by the sum-squared error function

$$\Phi(x,y,z,\Theta,\phi,\psi,n) = \sum_{i=1}^{N} (u_i - u_i'(n))^2 + (v_i - v_i'(n))^2 \tag{11}$$

where n is an integer denoting aircraft type and the other six independent variables are the position coordinates and Euler angles used to generate the synthetic image points (u_i', v_i') for comparison to the test image points (u_i, v_i). By making use of a nonlinear programming algorithm, Advani was able to force Φ to a global minimum, thereby determining the best match among all possible orientations over all types of aircraft for which a silhouette generating subroutine was provided to the computer.

Advani's synthetic silhouettes were also obtained by computing the projection of a wire-frame model for a given aircraft type onto an image plane and then deleting all interior lines. He was able to demonstrate global convergence and excellent immunity to uncorrelated Gaussian distortion noise. Unfortunately, no real images were used in this work so no evaluation of error rate is available.

The main appeal in Advani's approach is that since all reference image points are used and the reference images are computed on-line, no approximations are required in the evaluation of Eq. (11). Thus, one might expect a maximum degree of recognition accuracy to be ultimately available from such a procedure. The primary obstacle to the use of this approach seems to lie in the time required for synthetic silhouette generation. Using a medium-sized general purpose computer programmed in Fortran, Advani reported recognition times ranging from several minutes to one hour. It remains to be seen whether or not further improvements in computer hardware and software can yield acceptable computation times for this method.

SLOPE DENSITY

In all of the work reported above, no attention was given to the problem of obtaining images in real-time. However, this is of course an essential aspect of any practical system. Sklansky et al [6,19] have addressed this problem and report on a parallel mechanism for describing silhouettes which makes use of a nutating array of circular photocells to scan a binary image. This array computes in real-time an approximation to the slope density function defined by

$$f(\alpha) = \frac{d}{d\alpha} F(\alpha) \qquad (12)$$

where $F(\alpha)$ is the probability that a randomly chosen point on the boundary of a silhouette possesses a tangent with the slope less than or equal to α. That is, referring to Eq. (1)

$$F(\alpha) = P(\Theta(s) \leq \alpha) \qquad (13)$$

The main advantage of using slope density to characterize a closed contour seems to reside in the availability of a special piece of apparatus for its measurement in real time. With this equipment, it has been shown that image rotation merely introduces a time shift in the output waveform, which is periodic for a fixed scene [20]. This means that either matched filtering [1] or Fourier analysis can be used for identification providing that a sufficient number of prototypes is available for each aircraft class. That is, the slope density of course depends on viewing aspect and thus it too suffers from the function approximation problem described in conjunction with Eq. (7).

In a limited laboratory experiment using back-lighted three-dimensional models of five different aircraft types, Davidson et al, reported approximately fifty percent correct recognition based on harmonic analysis of the slope density function [20]. While this result cannot be said to be very impressive, it appears that a very small training sample was used and that better results might be obtained in a larger experiment. Again, as for all previous methods, it must be admitted that no statistically significant evaluation of recognition accuracy has been reported for this method of identification.

MOMENT INVARIANTS

In work paralleling that of Sklansky et al., Dudani also sought image features which would be particularly easy to compute [12]. Building on earlier work by Hu on character recognition [21], he chose to use the central moments of a binary image as the basis for defining a set of rotation and translation invariants. Specifically, considering either a solid silhouette or its boundary to be represented by a matrix of ones and zeros in a discretized two-dimensional image plane, the central moments of the image are given by

$$\mu_{pq} = \frac{1}{N} \sum_{i=1}^{N} (u_i - \bar{u})^p \quad (v_i - \bar{v})^q \tag{14}$$

where $\bar{u}$ and $\bar{v}$ are the mean values of the image coordinates u and v, respectively, and the summation is over all image points. Dudani modified Hu's invariants to obtain size invariance in addition to translation and image rotation invariance with the following results [22].

$$M_1 = (\mu_{20} + \mu_{02})^{1/2} \cdot B = r \cdot B \tag{15}$$

$$M_2 = \frac{1}{r^4} [(\mu_{20} - \mu_{02})^2 + 4\mu_{11}{}^2] \tag{16}$$

$$M_3 = \frac{1}{r^6} [(\mu_{30} - 3\mu_{12})^2 + (3\mu_{12} - \mu_{03})^2] \tag{17}$$

$$M_4 = \frac{1}{r^6} [(\mu_{30} + \mu_{12})^2 + (\mu_{21} + \mu_{03})^2] \tag{18}$$

$$M_5 = \frac{1}{r^{12}} [(\mu_{30} - 3\mu_{12})(\mu_{30} + \mu_{12}) \{(\mu_{30} + \mu_{12})^2 + 3(\mu_{21} + \mu_{03})^2\}$$
$$+ (3\mu_{21} - \mu_{03})(\mu_{21} + \mu_{03}) \{3(\mu_{30} + \mu_{12})^2 - (\mu_{21} + \mu_{03})^2\}] \tag{19}$$

$$M_6 = \frac{1}{r^8} [(\mu_{20} - \mu_{02})\{(\mu_{30} + \mu_{12})^2 - (\mu_{21} + \mu_{03})^2\} + 4\mu_{11}(\mu_{30} + \mu_{12})(\mu_{21} + \mu_{03})] \quad (20)$$

$$M_7 = \frac{1}{r^{12}} [(3\mu_{21} - \mu_{03})(\mu_{30} + \mu_{12})\{(\mu_{30} + \mu_{12})^2 - 3(\mu_{21} + \mu_{03})^2\} - (\mu_{30} - 3\mu_{12})(\mu_{21} + \mu_{03})\{3(\mu_{30} + \mu_{12})^2 - (\mu_{21} + \mu_{03})^2\}] \quad (21)$$

In Eq. (15) above, B is an estimate of the distance to the object from the camera. If such an estimate is not available, M_1 cannot be used as an invariant, but all other M_i are unaffected.

In order to investigate the value of moment invariants in aircraft identification, Dudani carried out a lengthy experiment using live television images. The system used for this purpose was organized as shown in Figure 1. Briefly, in the operation of this system, a conventional closed-circuit television camera was used to acquire images of model airplanes. The models were painted white and were supported in front of a black background so that image segmentation was obtained by simply thresholding the video signal. The signal coming out of the threshold circuit was presented to an operator on a cathode ray tube monitor and was also passed on to an interface circuit where each horizontal line was quantized into 180 discrete resolution cells. Since the television camera in this system made use of an interlaced raster scan, only one half-frame (256 lines) of video data was entered into the computer by the interface system. The image used for identification was thus in the form of a 180 x 256 binary matrix with the desired image being represented by ones and the background by zeros.

Typically, when digitized images were examined on the CRT display unit, it was found that they contained not only the desired silhouette, but also a number of small "blobs" of ones caused by background objects, noise in the video signal, camera imperfections, etc. The desired image also usually contained a few holes consisting of zeros produced by the same effects. Both of these defects were easily remedied by a simple picture cleaning algorithm which removed all clusters of ones or zeros with a perimeter less than a specified length. After this operation, the cleaned image was again presented on the CRT display and was simultaneously used to evaluate the invariants defined by Eq. (15) through (21) These invariants were then stored in the memory of the computer and the picture was again processed to obtain its outline. Subsequently, Eq. (15) through (21) were evaluated once more to get the corresponding invariants of the boundary points. Thus, the feature

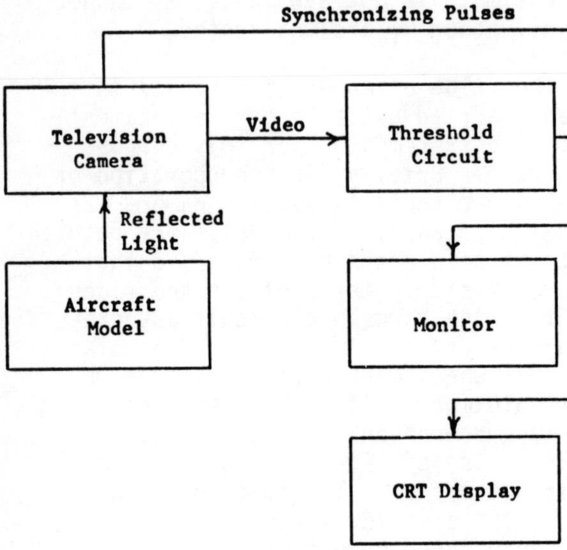

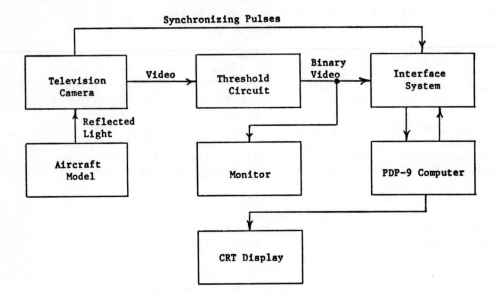

Fig. 1: Image Acquisition and Processing System for Moment Invariant Experiments

vector used for classification was of dimension fourteen, consisting of seven invariants derived from the entire silhouette and seven derived from its boundary.

In order to obtain a large sample for the design of a recognition algorithm, Dudani used the system shown on Figure 1 to collect 551 images for each of six aircraft classes. A typical set of such images is shown on Figure 2 which also indicates the six classes. These images are direct negative copies taken from the CRT display on the PDP-9 computer shown on Figure 1 and are representative of the training set used in this experiment.

Dudani used two distinct decision rules in his classification experiments: a Bayes decision rule and a weighted k-nearest neighbor rule. Considering first the Bayes rule, if Ω is the Cartesian product space defined by

$$\Omega = \{i=1,2,3,\ldots,6\} \times \{-\frac{\pi}{2} \leq \psi \leq \frac{\pi}{2}\} \times \{0 \leq \phi \leq \frac{\pi}{2}\} \quad (22)$$

then Dudani showed that all distinct noiseless images of any airplane in any of the six classes contained in the training set can be associated with some $\omega = \{i,\psi,\phi\} \in \Omega$. However, as mentioned previously, and as can clearly be seen from Figure 2, the real images obtained from television were in fact quite noisy so that the 14-dimensional feature vector ρ is also noisy. In such circumstances it is necessary to deal with the <u>conditional probability density function</u> for ρ, namely $p(\rho/\omega)$, rather than simply considering ω to be a function of ρ as in Eq. (7). Dudani assumed that ρ was normally distributed with a mean ρ_0 depending on ω and a covariance matrix Σ_n independent of ω; i.e.,

$$p(\rho/\omega) = N(\rho_0(\omega), \Sigma_n) \quad (23)$$

He then proceeded to estimate ρ_0 and Σ_n from the training sample. Having obtained these estimates, he also assumed a particular form for the <u>a priori density function</u> for ω as follows:

$$p(\omega) = P(i)\ p(\psi/\psi_0)\ p(\phi/\phi_0) \quad (24)$$

where ϕ_0 and ψ_0 are roll and azimuth estimates obtained from some other source such as radar, flight plans, etc. From Eq. (23) and (24), it is possible to obtain the <u>posterior density function</u> for ω by means of the relationship

$$p(\omega/\rho) = \frac{p(\rho/\omega)\ p(\omega)}{p(\rho)} \quad (25)$$

where $p(\rho)$ is the <u>marginal density function</u>

$$p(\rho) = \int p(\rho/\omega)\ p(\omega)d\omega \quad (26)$$

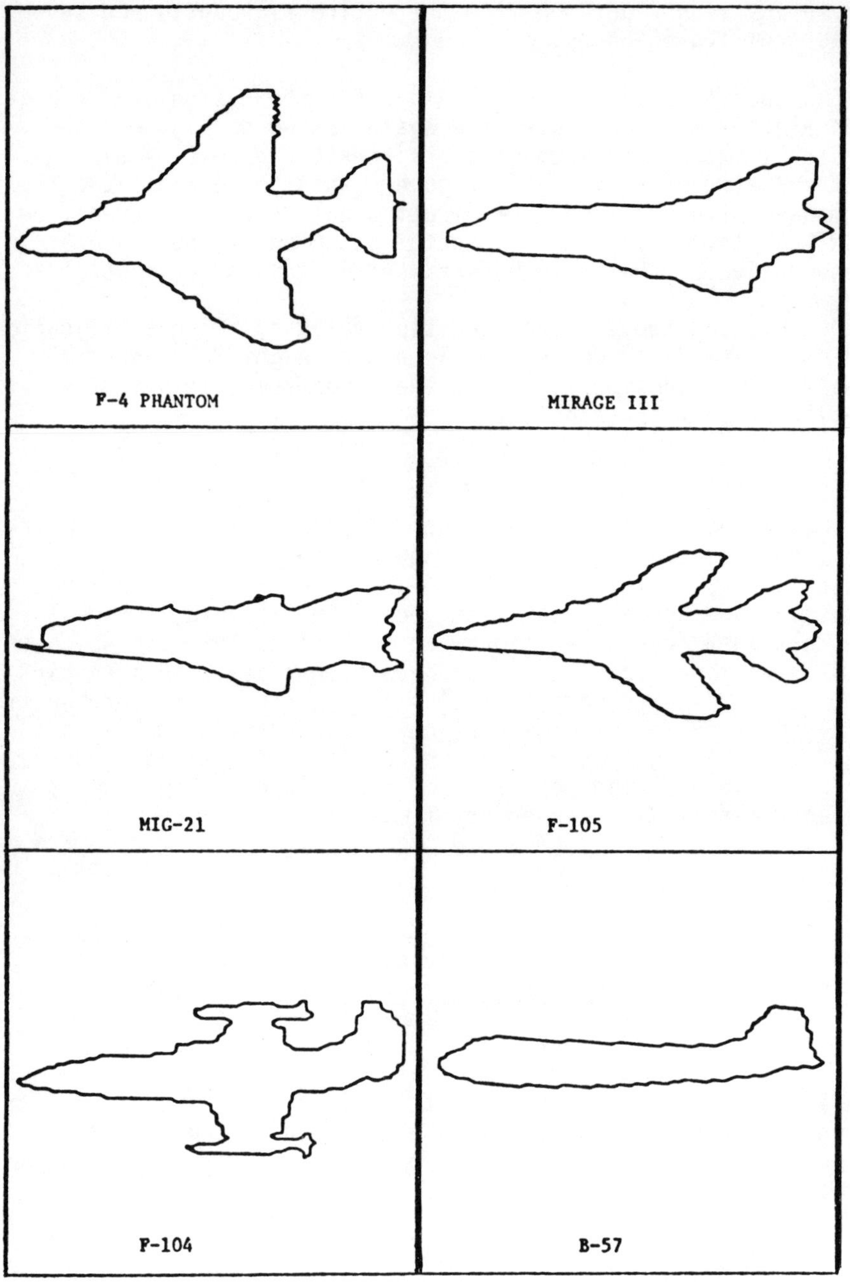

Fig. 2: Typical Television Images Used In Moment Invariant Experiments.

Eq. (25) is central to Bayes estimation. That is, for any given loss function, this equation can be used to make a decision which minimizes average loss [1]. In Particular, if a 1,0 loss function is used [1], then the best decision is the one which maximizes Eq. (25). This loss function was adopted by Dudani and used in an experimental evaluation of recognition accuracy to be presented later in this section of this paper.

To the extent that the assumed models for the underlying probability density functions are correct, Bayes estimates are, by definition, optimal with respect to average loss or risk. However, if such models are erroneous, simpler procedures may give better results. One particularly appealing possibility is to not model the data statistics at all, but to merely use all training samples as the basis for comparison with an unidentified test sample to obtain its classification. Dudani did this with all 551 x 6 = 3306 training samples being used as an "authority file". To compensate for scaling effects, all of the training samples were subjected to a linear transformation to obtain zero mean and unit variance for each component when averaged over the entire test set. The feature vector for an image to be classified was first subjected to the same transformation and then compared to every element in the training sample by means of the Euclidian metric

$$d_{ij} = \sum_{i=1}^{14} (\rho_i - \rho_{ij})^2 \tag{27}$$

where ρ_i denotes the i-th component of the unclassified feature vector, ρ, and ρ_{ij} stands for the same component of the j-th training sample vector, j=1,2,,,3306.

Let

$$\alpha = \{\alpha_1, \alpha_2, \ldots, \alpha_k\} \tag{28}$$

be the class index for each of the k-nearest neighbors of ρ arranged in ascending order of distance. Then a simple nearest-neighbor rule selects α_1 as the class of ρ. A k-nearest neighbor rule [1] selects the class appearing most often in α. A weighted nearest neighbor rule assigns a weight vector

$$w = \{w_1, w_2, \ldots, w_k\} \tag{29}$$

to the ordered neighbors according to some predetermined scheme and then computes the class weights

$$Q_j = \sum_{\alpha_i = j} w_i \tag{30}$$

The vector ρ is then assigned to the class with the largest weight.

Table I: Results of Six Class Aircraft Identification Experiment with 132 Test Images

Observer	Number of Errors
Technical, A	10
Technical, B	12
Technical, C	14
Non-Technical	28
Computer, Bayes Rule	8
Computer, Distance Wtd. Rule	6

Dudani tried a number of different nearest neighbor rules and obtained good results for all of them. Among his weighted rules was a distance weighted rule based on the distances associated with the k nearest neighbors. Specifically, if

$$d = \{d_1, d_2, \ldots, d_k\} \qquad (31)$$

are the ordered distances associated with the elements of α, then the weights he used were defined as follows:

$$w_i = \frac{d_k - d_i}{d_k - d_1} \qquad (32)$$

This weight function assigns weights ranging from one to zero for any set α and is of a form which permits a few very close neighbors to overcome the effects of a larger number of more distant neighbors. This tends to make the choice of k less critical than in an unweighted rule. Dudani found that for his data the value k=10 represented a good choice for the evaluation of class weights.

As the above description shows, the data analysis and investigation of decision rules carried out by Dudani were somewhat more extensive than in any earlier work. Having settled on a Bayes rule, and the particular distance weighted k-nearest neighbor rule given by Eq. (32) (with k=10), he prepared a test sample of 132 new images consisting of 22 images of each of the 6 classes in the training sample. These images were obtained at random viewing aspects and were stored in digital form for subsequent processing. In an attempt to provide a more meaningful evaluation of his classification techniques, Dudani asked four human observers to view the same binary images and to decide upon a classification for each. The observers were given as much time as they liked to view each image and were also provided with all six plastic models if they wished. The same images were also classified automatically by the two selected decision rules. The results are summarized in Table I. As can be seen, the computer outperformed all four observers to an impressive degree. This was an unexpected result since it had been thought that, in this problem, human performance represented an implicit upper bound on the classification accuracy attainable in an automatic system.

The research program which included Dudani's work is continuing at Ohio State University. Part of the continuing activity is aimed at a more thorough evaluation of the system which has been described above. However, the results presented in Table I are so encouraging that the major effort is being made in the direction of speeding up the computations At present, approximately 30 seconds are required from the beginning of image acquisition until object identification is completed.

A reduction of this time to a fraction of a second is believed possible by relatively straightforward means, while it appears that special purpose parallel processors can yield identification in one television frame time [23].

SUMMARY AND CONCLUSIONS

Serious work on automatic recognition of complex three-dimensional objects from optical images has gone on for a little more than one decade, mainly with regard to the problem of identification of aircraft types. The present state of the art is that a computer has significantly outperformed a number of human observers in a laboratory experiment. This is a benchmark result which seems to justify a larger effort aimed at the realization of practical systems. It is hoped that this paper will stimulate others to undertake research relating to this important objective.

REFERENCES

1. Duda, R.O., and Hart, P.E., Pattern Classification and Scene Analysis, John Wiley & Sons, New York, 1973.
2. Cosgriff, R.L., Identification of Shape, Report No. 820-11, The Ohio State University Research Foundation, Columbus, Ohio, ASTIA AD-254 792, December 1960.
3. Pugh, A., Heginbotham, W.B., and Kitchin, P.W., "Visual Feedback Applied to Programmable Assembly Machines," Proceedings of 2nd International Symposium on Industrial Robots, Chicago, Ill., May, 1972, pp. 77-88.
4. Rosenfeld, A., Picture Processing by Computer, Academic Press, New York, 1969.
5. Hall, E.L., et al., "A Survey of Preprocessing and Feature Extraction Techniques for Radiographic Images," IEEE Trans. on Computers, Vol. C-20, No. 9, pp. 1032-1044, September 1971.
6. Sklansky, J., and Davidson, G.A., "Recognizing Three-Dimensional Objects by Their Silhouettes," J. Soc. Photo-Opt. Instrum. Eng., Vol. 10, pp. 10-17, October 1971.
7. Levine, M.D., "Feature Extraction: A Survey," Proc. of the IEEE, Vol. 50, No. 8, pp. 1391-1407, August 1969.
8. Nagy, G., "State of the Art in Pattern Recognition," Proc. of IEEE, Vol. 26, No. 5, May 1968.
9. Watanabe, S., Frontiers of Pattern Recognition, Academic Press, New York, 1972.
10. Brill, E.L., "Character Recognition via Fourier Descriptors," in Qualitative Pattern Recognition Through Image Shaping, preprints of WESCON Session 25, Los Angeles, August 1968.
11. Zahn, C.T., and Roskies, R.Z., "Fourier Descriptors for Plane Closed Curves," IEEE Trans. on Computers, Vol. C-21, No. 3, pp. 269-281, March 1972.

12. Dudani, S.A., Moment Methods for the Identification of Three-Dimensional Objects from Optical Images, M.S. thesis, The Ohio State University, Columbus, Ohio, August 1971.
13. Pio, R.L., "Euler Angle Transformations," IEEE Trans. on Automatic Control, Vol. AC-11, No. 4, pp. 707-715, Oct. 1966.
14. Cosgriff, R.L., An Investigation of Techniques for Recognition of Script Numeric Characters, Research Proposal No. 67-14, The Ohio State University Research Foundation, Columbus, Ohio, July 1966.
15. Richard, C.W., and Hemami, H., "Identification of Three-Dimensional Objects using Fourier Descriptors of the Boundary Curve," Proc. of 1972 Symposium on Computer Image Processing and Recognition, Vol. 2, University of Missouri at Columbia, August 1972, pp. 15-2-1 to 15-2-13.
16. Sutherland, I.E., "Computer Displays," Scientific American, June 1970, pp. 57-81.
17. Advani, J.G., Computer Recognition of Three-Dimensional Objects from Optical Images, Ph.D. dissertation, The Ohio State University, Columbus, Ohio, August 1971.
18. Hemami, H., McGhee, R.B., and Gardner, S.R., "Towards a Generalized Template Matching Algorithm for Pictorial Pattern Recognition," Proc. of the 1970 IEEE Symposium on Adaptive Processes, Dec. 1970.
19. Sklansky, J., and Nahin, P.J., "A Parallel Mechanism for Describing Silhouettes," IEEE Trans. on Computers, Vol. C-21, No. 11, pp. 1233-1239, November 1972.
20. Davidson, G.A., Johnson, V.M., and Sklansky, J., Optical Feature Extraction Technique, Report AFAL-TR-70-133, Aerojet-General Corporation, Azusa, Calif., July 1970.
21. Hu, M.K., "Visual Pattern Recognition by Moment Invariants," IRE Trans. on Information Theory, Vol. IT-8, pp. 179-187, February 1962.
22. Dudani, S.A., An Experimental Study of Moment Methods for Automatic Identification of Three-Dimensional Objects from Television Images, Ph.D. dissertation, The Ohio State University, Columbus, Ohio, August 1973.
23. Breeding, K.J., and Miller, P.E., A Symmetrical Rectangular Array Processor and Its Simulation, Technical Note No. 15, Communication and Control Systems Laboratory, The Ohio State University, Columbus, Ohio, October 1973.

EYES OF THE WABOT

S. Ohteru, H. Kobayashi and T. Kato

Waseda University

Tokyo, Japan

This paper is a report on the artificial eyes of the robot from different approaches. The paper is divided into two parts - Part I describes the eyes of WABOT I, the project (1) of which was finished in June 1973, and Part II explains one of the analytical considerations for the intelligent eyes of robots which are now being studied.

PART I - EYES OF WABOT I

The Bio-Engineering Group in Waseda University has been carrying out a robot project for the last five years. The objectives of the project are the following:

1. Move on a human-type biped walking machine.
2. Work with bilateral artificial hands of a human type.
3. Provide hands and feet with joint angle sensors for each joint as proprioceptive means.
4. Provide a sense of equilibrium for a biped move.
5. Provide hands with contact sense for work.
6. Provide artificial eyes as remote receptors.
7. Use a mini-computer as the brain.
8. Speak and listen to Japanese sentences.

The robot is named "WABOT I" (WASEDA ROBOT I). The entire system is presented in a 30-minute movie, while this paper aims to explain mainly the construction of the eyes.

Functions of the Wabot's Eyes

Wabot's eyes are required to have the following functions:

1) After receiving a start command from Wabot's mouth, the eyes detect an object about five meters in distance and measure the exact distance between them.
2) The directions of the trunk and the two legs must be coincided before walking, although the trunk on which the eyes are set rotates in order to search for the object.
3) The setting of the robot in the right direction facing the object makes the direction between the legs and the object cause some errors, because the changing direction is not done continuously around one axis but abruptly by two legs. Therefore, checking of the angle is necessary again before starting.
4) Wabot approaches the object from a long distance and stops at a medium distance in order to check the errors caused by two legs walking.
5) The trunk stops by the eyes' judgment in front of the object at a distance which is capable of handling and the position data of the object is sent to the hands.
6) The focus control and camera position control are to be done with the movement of Wabot in a long, a medium and a near distance, respectively. A remarkable property required of the eyes is closely related to the change of the distance between the eyes and the object during the time of observation. HITAC-10 (Memory 8k Wards, Cycle Time 1.2 μ sec, Interrupt level I) is used as the Central Processing Unit in the whole system. It is difficult for the computer to process the whole frame of the TV camera for the shortage of its memory capacity and cycle time. Visual data are processed by hardware as much as possible.
7) Owing to limitations in the size, the structure (the eyes and the hands must be contained in a trunk house the size of which is 50 cm in front and 40 cm in depth), the weight (total load must be less than 35 kgs.), and the position of the center of gravity, only the TV cameras and the sensors are allowed to be placed in the lower part of the trunk, and the other pre-processors are separated from Wabot.

Wabot's eyes are constructed by two TV cameras installed in its trunk, whose detector scanning lines and focusing are set by the C.P.U. control. When searching for an object, Wabot's trunk rotates to scan the space in front with its TV cameras. When the scanning camera detects an object, it emits a signal which stops the trunk's rotating. All data on distance and angle are then measured and processed at the same time, as an 8-bit digital value. If the eyes do not detect an object, scanning lines are shifted and

Fig. 1.1. Broadening of the visual angle for a short distance.

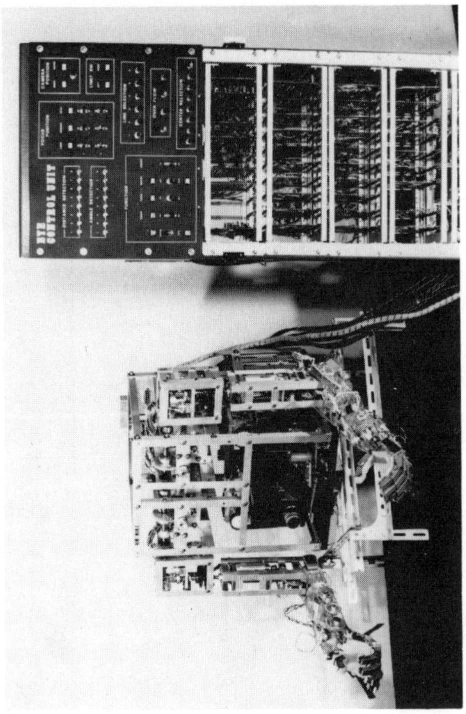

Fig. 1.2. Wabot's eyes and pre-processors.

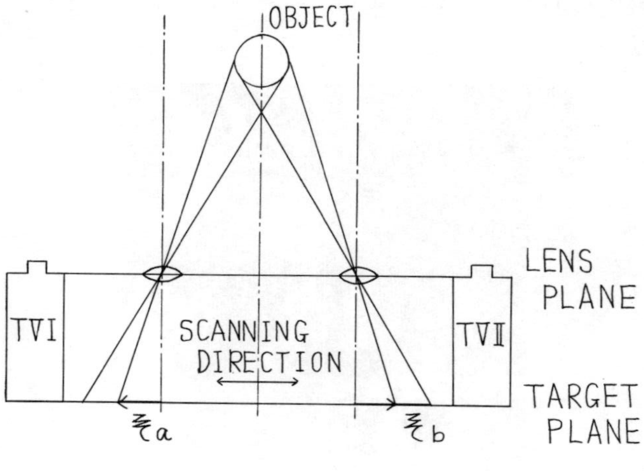

Fig. 1.3. Object detection.

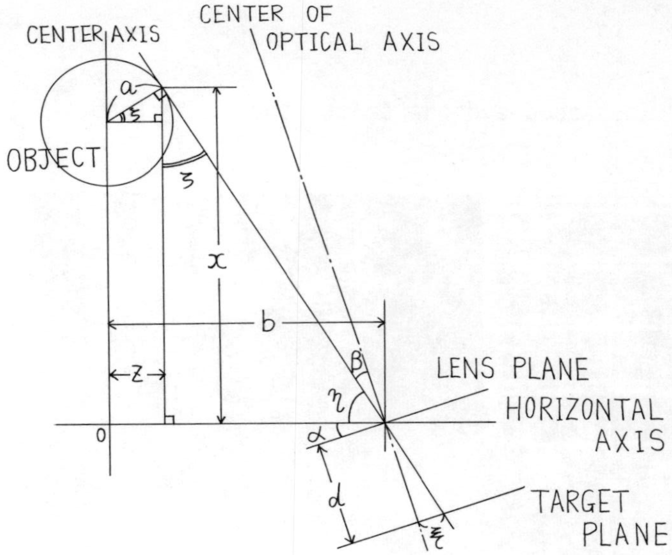

Fig. 1.4. Principle of distance measurement.

a: Radius of the object (cylinder)
2b: Distance between two cameras
d: Interval between lens and target
x: Measured distance
z: Measured radius of the object
α: Setting angle of TV camera
β: Angle from optical axis to edge of the object
ζ: Angle from center axis to edge of the object
η: Angle from horizontal axis to edge of the object
ξ: Object length on target plane

the detecting procedure is repeated following a software algorithm. In order to improve the accuracy of measurement, the zone focusing with a 3-step on-off control is added in the mechanism of the TV camera. In short distances, the focus of the TV camera exceeds the minimum controlled position, and the clear visual data cannot be obtained from TV cameras. For this reason, the lenses are moved by 10 cm backward from the trunk's front edge and the possible distance of handling is shortened to 20 cm from Wabot's shoulder. Furthermore, visual angle of the TV camera equipped with 25 mm standard lens is 30^{0} in the horizontal direction and 25^{0} in the vertical direction, so that an object within 50 cm from the lens is in the blind angle of the TV cameras on the occasion of setting two TV cameras in parallel. Thus, the visual angle for a short distance is changed inwards by 8^{0} by the command of C.P.U. Fig. 1.1 shows the broadening mechanism of the visual angle for a short distance. Fig. 1.2 shows Wabot's eyes and their pre-processors.

The two TV cameras are in the robot's trunk which is rotating scan in inverse directions of each other. When the cameras face to just in front of an object (Fig. 1.3), the trunk stops to rotate and at the same time the distance to the object is calculated.

Wabot was designed so as to keep the slant angle of the trunk within $\pm 2^{0}$ by a posture protection mechanism while not walking; however in practice, it moves by $\pm 2^{0}$-5^{0} in the vertical direction and by $\pm 2^{0}$ in the horizontal direction. Therefore, even when Wabot stands still, the detector scanning lines of the TV cameras cannot be fixed even to a stationary object, because of the slight movements produced in both vertical and horizontal axes. In the trial system, 64 different scanning lines were made available for selection, and one of them was selected by the software.

Fig. 1.4 shows the principle of distance measurement. As an object, we select a cylinder which is seized by a manipulator; therefore, the radius of a cylinder is small enough as compared with camera interval 2b. If the optical axis of the camera is parallel to the floor, we obtain the distance x between camera and object under the condition of:

$$b \gg a \tag{1.1}$$

$$x \approx b \cot(\alpha+\beta) \tag{1.2}$$

From the result of the experiment, the measuring error of the distance is within 7 cm at 5m, and 4 cm at the shortest possible working distance of 0.6m. In order to decrease the measuring error owing to the after-image effect of vidicon, we make the velocity of the rotating trunk for detecting an object under 30^{0}/sec.

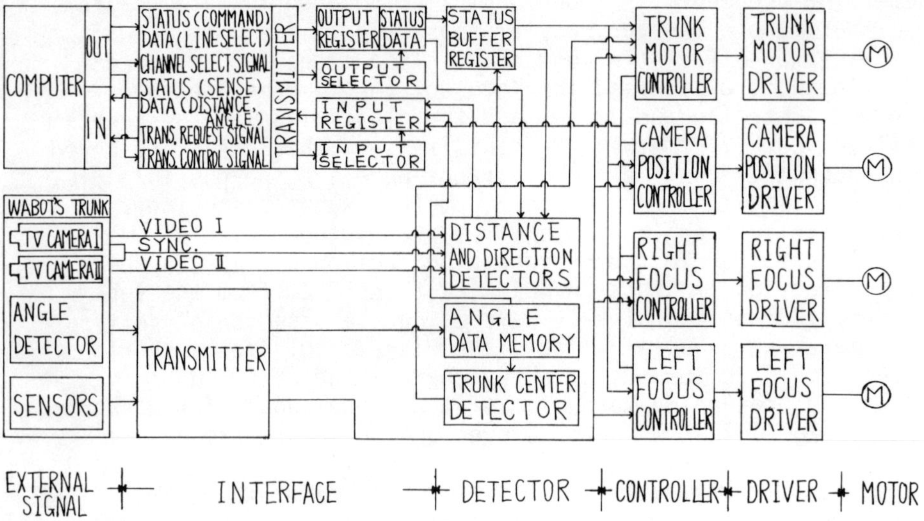

Fig. 1.5. Block diagram of Wabot's eyes and their control system.

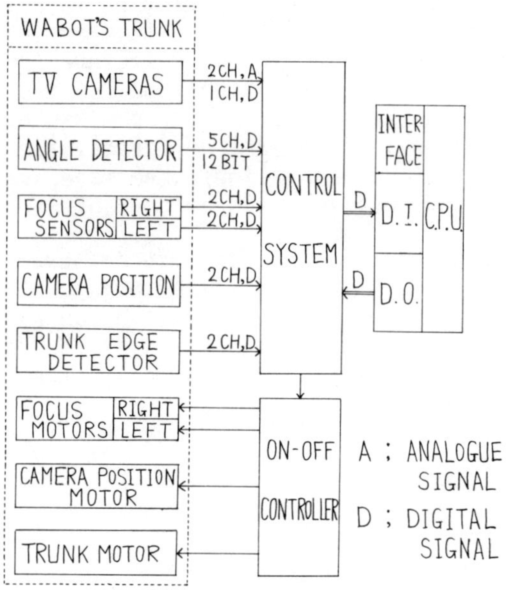

Fig. 1.6. Interface of Wabot's eyes and trunk.

Fig. 1.7. Focus control mechanism.

On-line Operation of Wabot's Eyes

Fig. 1.5 shows a block diagram of the Wabot's eyes and their control system. The whole signals between the eyes and the computer shown in the upper left corner of Fig. 1.5 are digital. Since the whole signal of Wabot interface is linked together by bus-line, the hardwares of channel and transfer control are necessary for the eyes' interface. Transfer control signals of the eyes consist of 6 bits. The status command consists of 4 bits and is the starting signal for each operation of the eyes. The eyes' interface mechanism can judge 15 kinds of messages by decoding the output from the C.P.U. The 16 bits are prepared for the interface considering the future expansion in the eyes' system. The line-select command, which designates one of the detector scanning lines of the TV camera, consists of 6 bits. The input signal consists of 8 bits for distance to the object as well as for angle data with the object, and 15 bits for status sense. Every operation of the eyes should be reported to the C.P.U. for it is constantly controlled on-line. Whenever the eyes' action has stopped, the C.P.U. is interrupted by a transfer request signal. Wabot's trunk is shown in the lower left part of Fig. 1.5 and in the left side of Fig. 1.2. Fig. 1.6 shows the block diagram of the interface of Wabot's eyes and trunk in detail. The TV cameras (CTC-1007B, IKEGAMI LTD.) are externally synchronized, so that the synchronous signal is sent together with video signals (TV I, TV II). For the detection of the angle between Wabot's trunk and legs, a rotary encoder is used (1000 pulse/360° type, RIE-AS, NIKON LTD.). The 10-bit digital data for the angle is sent from Wabot's trunk to the eyes. Wabot's trunk has focus sensors (each of the TV cameras has a focus sensor, respectively), camera position sensors and trunk edge detectors which send data by digital signals through the control system of the eyes to the C.P.U. so as to monitor the operation. Fig. 1.7 shows the focus control mechanism. The block diagram of distance and direction detectors is shown in Fig. 1.8. In order to first catch the object within visual range, the video signals from the object are picked up by both TV cameras (camera I, camera II) using the line selector for the vertical direction and the trunk rotation for the horizontal direction. The gray level of the object signal is sliced and transformed into a binary value. The jumping position of this binary signal is counted by a 4MHz oscillator, and the gate is closed through the hold logic. When, during the rotation of the trunk, the signal sent from each camera is detected at the equal distance from the edge of its sight, the rotation is stopped. At this time, since the trunk is facing the object rightly, the data to be sent to the legs for the rotating angle, between the legs and the trunk is obtained; and thus the walking direction is determined.

The hardware is controlled by 15 kinds of software command.

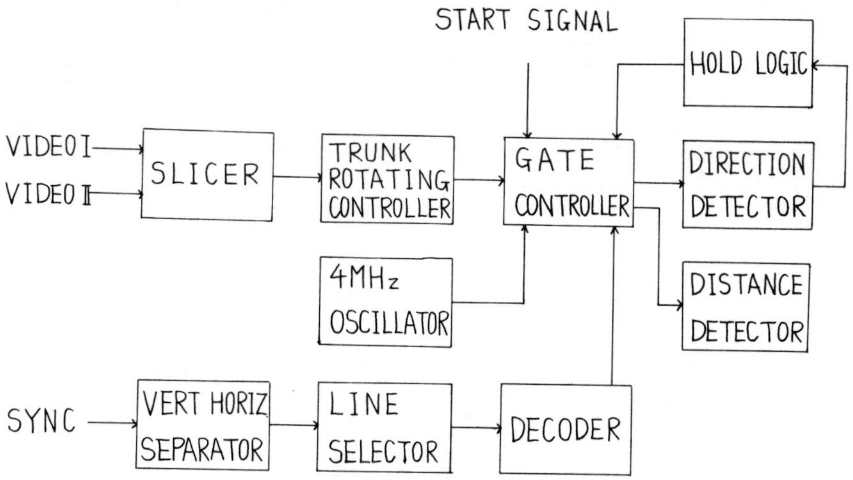

Fig. 1.8. Block diagram of distance and direction detectors.

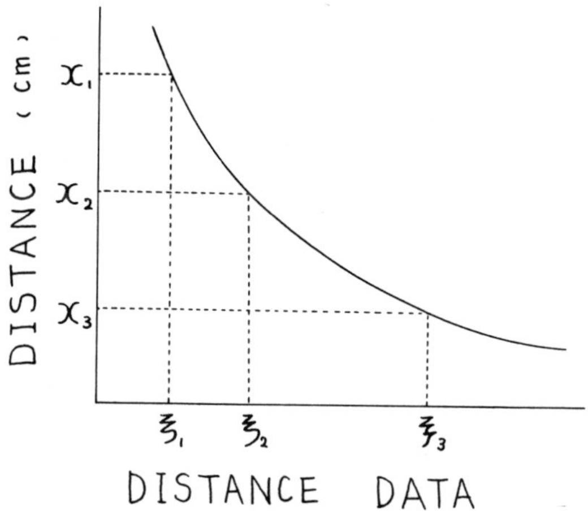

Fig. 1.9. Relation between x and ξ.

The software consists of interrupt analysis, settlement of the focus, searching of the object, detection of the center of the trunk, measurement and calculations of distance and angle, and connection of software in each sub-system. While the system is moving, status of the hardware is always monitored and printed out. The required time of operation is monitored by the software so that the whole system will stop when it exceeds the time limit for operation of any hardware. Unexpected troubles of the system are also checked by this function. A hardware checking program is likewise included in the software. The program for distance detecting uses the correction curves shown in Fig. 1.9 which correspond to the long, medium and near distance so as to minimize the capacity of memory used for the calculation ensuring the accuracy comparable to that of measurement.

Defining x_i, ξ_i and k_i as follows:

x_i: the ordinate value corresponding to the upper edge of the straight line i
ξ_i: the abscissa value corresponding to the upper edge of the straight line i
k_i: the slope of the straight line
distance x can be approximated as

$$x = x_i + K_i(\xi_i - \xi) \qquad 1 \leq i \leq 3 \qquad (1.3)$$

Before the calculation of the distance, the software checks whether or not the measured data falls into the present range, for example, 3-6m, or 1-4m, or 1.5-0.3m. If the data falls into the range, the calculation program can operate; however, if it does not, the detecting procedure is repeated again because there must be some mistakes in the measurement. Thus in a long and medium distance, the result of the measured distance is transmitted to the legs, and in near distance, it is transmitted to the hands in the form of $2^M \times N$(cm) (M; integer, $0 \leq N \leq 1$). Memory saving and rapid processing are realized by the use of machine language.

The miniaturization and simplification of the eye construction are now being tried by the use of self-scanning photodiode arrays.

PART II - AUTOMATIC FOCUSING WITH PATTERN RECOGNITION

The theoretical consideration (2) in a defocused optical system and a focus-assist device (3) have never been reported for an indefinite object. However, it will be useful for a robot to focus automatically on a desired special object. In this case, automatic focusing requires the function of pattern recognition

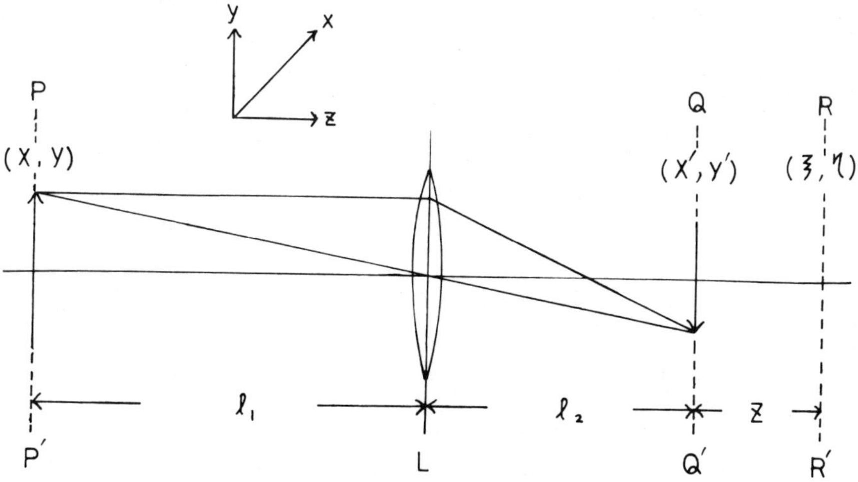

PP' : object plane

QQ' : in-focus plane

RR' : defocused image plane

Fig. 2.1. Optical system.

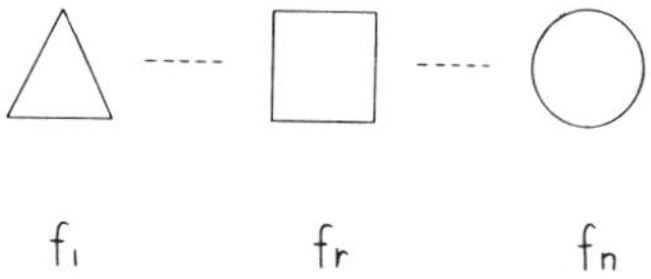

Fig. 2.2. Example of set of standard patterns.

at the same time. Therefore, only the standard patterns which the robot can recognize will be considered. One of the methods of fulfilling both functions is proposed.

Principle of Automatic Focusing with Pattern Recognition

Consider that the object $f(x,y)$ is located at a distance l_1 from an ideal lens L of focal length l_f as shown in Fig. 2.1. The clear image $f(x',y')$ of $f(x,y)$ will be obtained on the in-focus plane QQ', assuming the optical system is ideal. The relation between them is given by (Appendix)

$$f(x,y) \rightarrow \gamma f(c(x'-\alpha),\ c(y'-\beta)) \qquad (2.1)$$

where γ and c are parameters related to light intensity and size of the object, and α and β are phase shift parameters. For simplicity, the rotation of object is neglected.

First, the target plane RR' of robot's eye will be set at a distance z from the in-focus plane QQ', since locations of the object and in-focus plane are still unknown to the robot.

Under the appropriate assumptions, the observed image (defocused image) $F(\xi,\eta)$ of $f(x',y')$ on the target plane RR' is expressed as:

$$F(\xi,\eta) = \iint_{-\infty}^{+\infty} K(\xi-x',\eta-y',z)\gamma\, f(c(x'-\alpha),\ c(y'-\beta))\, dx'dy' \cdot \qquad (2.2)$$

where $f(x',y')$ is an unknown function and the Kernel K is a known function with unknown parameter z. In order to discuss the problem on the object plane, we rewrite

$$F(\xi,\eta) = \frac{\gamma}{c^2}\iint_{-\infty}^{+\infty} K\{\xi-(\frac{x}{c}+\alpha),\eta-(\frac{y}{c}+\beta),\ Z\} f(x,y)dxdy \ldots \qquad (2.3)$$

Next, let a set of standard objects be $f_1(x,y)\cdots f_r(x,y)$ $f_n(x,y)$ as shown in Fig. 2.2. Select $f_r(x,y)$, as a desired special object. If the desired object $f_r(x,y)$ is located on the plane PP', a defocused image $F_r(\xi,\eta)$ would be observed on the plane RR'

$$F_r(\xi,\eta) = \frac{\gamma}{c^2}\iint_{-\infty}^{+\infty} K\{\xi-(\frac{x}{c}+\alpha),\ \eta-(\frac{y}{c}+\beta),\ z\}\ f_r(x,y)dxdy \cdot\cdot \qquad (2.4)$$

The integrand of the above equation is a known function having the unknown parameters c,α,β and z. These four unknown parameters will be determined from four sets of values of ξ_i,η_i from

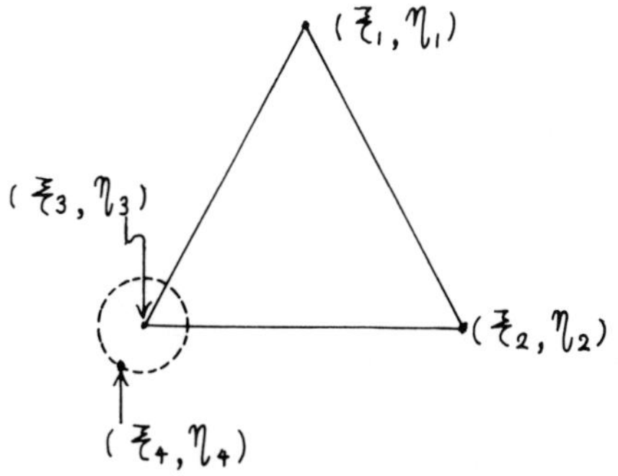

Fig. 2.3. Sampling points of triangle pattern

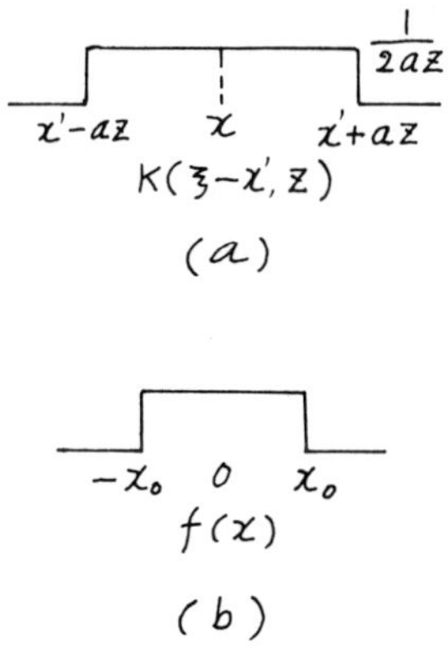

Fig. 2.4. One dimensional model
(a) triangle and square with equal sides
(b) squares with different sides

$$F_r^*(\xi_i,\eta_i) = \frac{\gamma}{c^2} \iint_{-\infty}^{+\infty} K\{\xi_i - (\frac{x}{c}+\alpha),\ \eta_i - (\frac{y}{c}+\beta),\ z\}\ f_r(x,y)dxdy$$

$$i = 1,2,3,4 \qquad (2.5)$$

where $F_r^*(\xi_i,\eta_i)$ is equal to $F(\xi,\eta)$ for $i = 1,2,3,4$

If

$$\iint_{-\infty}^{+\infty} |F(\xi,\eta) - F_r^*(\xi,\eta)|d\xi d\eta = \iint_{-\infty}^{+\infty} |\frac{\gamma}{c^2} \iint K\{\xi-(\frac{x}{c}+\alpha),\ \eta-(\frac{y}{c}+\beta),z\}$$

$$\times \{f(x,y) - f_r(x,y)\}\ dxdy\,|d\xi d\eta < \varepsilon \qquad (2.6)$$

and

$$\iint_{-\infty}^{+\infty} |\ F_k(\xi,\eta) - F_r(\xi,\eta)|d\xi d\eta > \varepsilon \qquad \text{for } k \neq \gamma \qquad (2.7)$$

$$k = 1,2,\ldots,n$$

then the robot makes the decision

$$f(x,y) = f_r(x,y) \qquad \ldots. \qquad (2.8)$$

and the eye can focus to the desired object $f_r(x,y)$ by reducing z to zero. As for the ε in eqs. (2.6) and (2.7), the accuracy of the measuring system of $F(\xi,\eta)$ will give its value, for which a set of standard patterns must be prepared so as to satisfy the relation (2.7).

In actual problems, the effective measuring points to classify the standard patterns can be predicted. Fig. 2.3 shows them in a triangle pattern as an example. Therefore, the decision of eq. (2.8) will be possible during the process of solving eq.(2.5) and the checking procedure of eq. (2.7) may be omitted.

<u>Example I</u>

For simplicity, we consider the simple one-dimensional problem with the rectangular contrast pattern and rectangular response function as shown in Fig. 2.4.

$$K(\xi,x') = \frac{1}{2az}[\ U(\xi- x' + az) - U(\xi- x' - az)]$$

$$= \frac{1}{2az}[\ U\{\xi- (\frac{x}{c}+\alpha) + az\} - U\{\ \xi- (\frac{x}{c}+\alpha) - az\}]$$

$$(2.9)$$

$$f(x) = U(x + x_0) - U(x - x_0) \qquad (2.10)$$

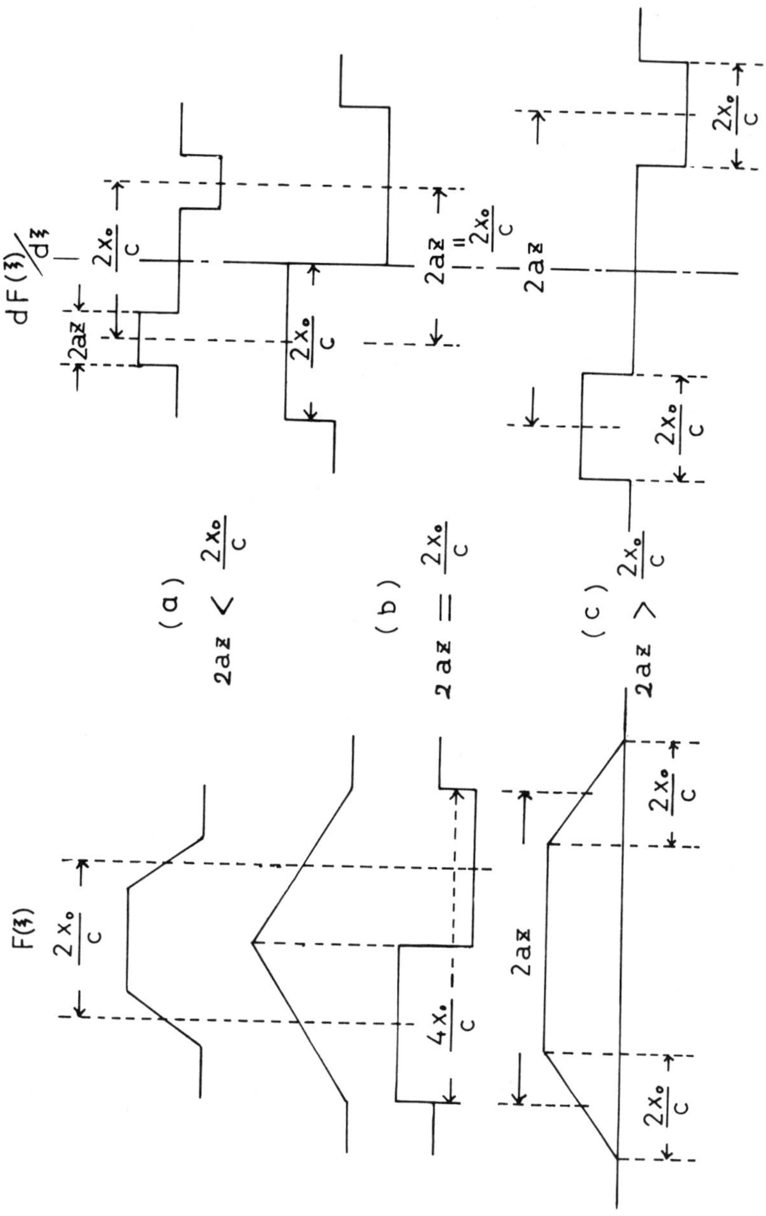

Fig. 2.5. F(ξ) and dF(ξ)/dξ in three cases.

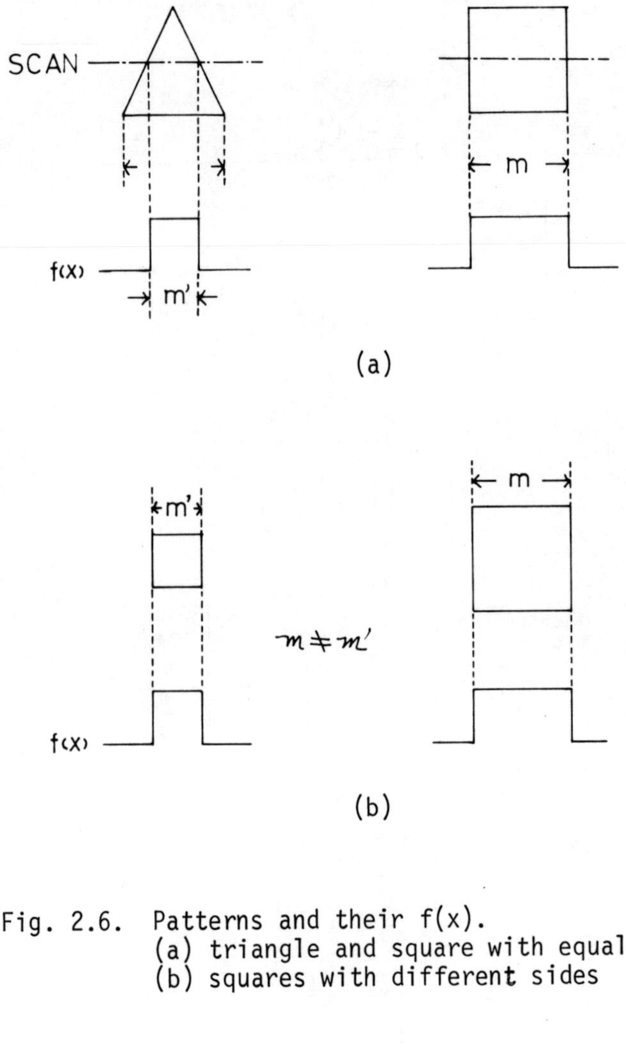

Fig. 2.6. Patterns and their f(x).
(a) triangle and square with equal sides
(b) squares with different sides

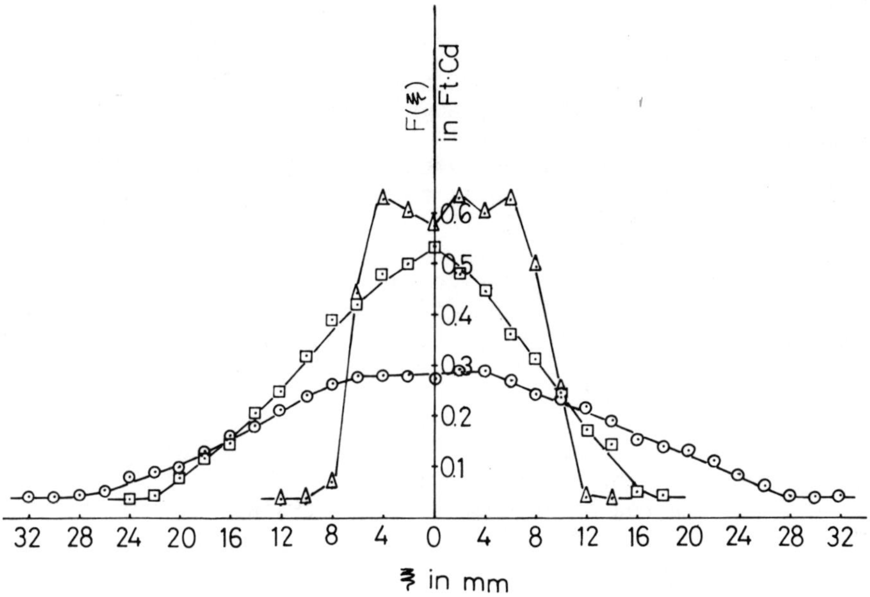

Fig. 2.7. Experimental result of F(ξ).

where function U(x) denotes a step function. Equation (2.3) reduces to

$$F(\xi) = \frac{\gamma}{c} \int_{-x_0}^{x_0} \frac{1}{2az} \{U(\xi - (\frac{x}{c} + \alpha) + az) - U(\xi - (\frac{x}{c} + \alpha) - az\} dx \tag{2.11}$$

From the above equation, we can obtain $F(\xi)$ and $dF(\xi)/d\xi$ as shown in Fig. 2.5 (a), (b) and (c). It must be noted that the two identical $F(\xi)$ will be obtained at equal distances $\pm z$ on the two sides of the in-focus plane.

i) In the case of $2az<2ax_0/c$

According to $az \to 0$, the image $F(\xi)$ will be focused, as shown in Fig. 2.5(a). This effect is the same as the ordinary auto-focusing method without pattern recognition. However, in our case, we can know automatically the length of the original pattern from the observed image which remains defocused. As a result, the identification of the patterns will be possible. For example, consider the f(x) in eq. (2.10), the wave form obtained as a scaning signal from a T.V. camera. A triangle and a square with equal sides are recognized by the comparison of two scanning signals as shown in Fig. 2.6(a), and different size squares may also be classified.

ii) In the Case of $2az \geq 2x_0/c$

Let us follow the change of the form of $F(\xi)$ occuring with increasing the amount of defocusing in Fig. 2.5. The case of $2az \geq 2x_0/c$ defines a region in which the defocusing is remarkable and in which an interesting phenomenon arises. If we try to focus the image, the pulse width of $dF(\xi)/d\xi$ does not decrease; but the interval between two pulses decreases as shown in Fig. 2.5(c). During the process of $z \to 0$, the pulse width begins to decrease after the two pulses just meet (that is $2x_0/c=2az$) as shown in Fig. 2.5(b) and (c). At this point, the state enters the region $2az<2x_0/c$. Figure 2.7 shows the experimental results of $F(\xi)$. Close agreement is obtained with the result to be expected analytically from Fig. 2.5. The fluctuations of the data at the top of the wave shown in Fig. 2.7(a) depends on the diffraction phenomenon, because a narrow optical slit is used in our experiment in order to retain the assumed one dimensional nature of the problem.

Example II

Next, let us consider a continuous contrast pattern

$$f(x) = 1 + \cos c(x-\alpha) \tag{2.12}$$

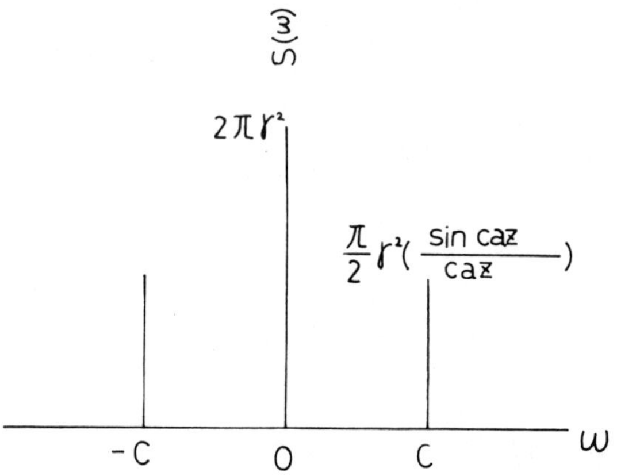

Fig. 2.8. Power spectrum of observed image.

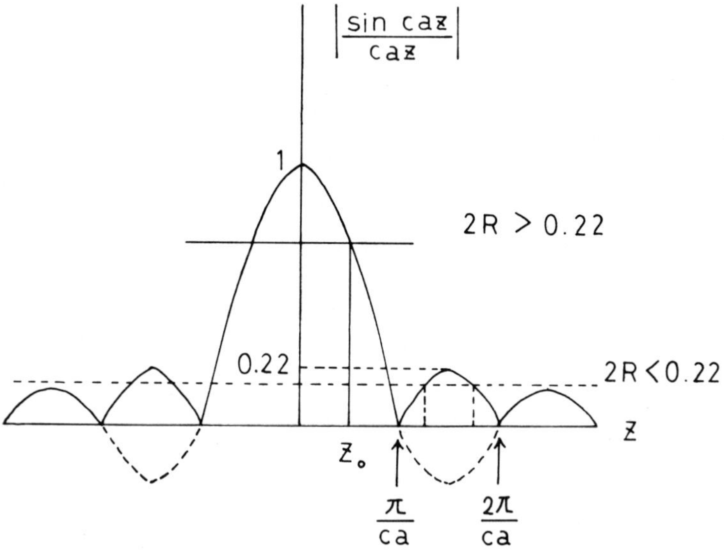

Fig. 2.9. Determination of Z.

$F(\xi)$ is obtained by the same procedure as eq. (2.11)

$$F(\xi) = \frac{\gamma}{c} \int_{-\infty}^{+\infty} \frac{1}{2az}\{ U(\xi-(\frac{x}{c} + \alpha) + az) - U(\xi - (\frac{x}{c} + \alpha) - az\} (1+\cos x)dx$$

$$= \gamma\{1 + \frac{\sin c\ a\ z}{caz} \cdot \cos c\ (\xi-\alpha)\} \qquad (2.13)$$

However, in this case, the edge of the pattern is not clear. Consideration of the spatial frequency domain of the image $F(\xi)$ may be useful for focusing and pattern recognition. The Fourier transform of $F(\xi)$ is obtained as

$$\mathcal{F}[F(\xi)] = \pi\gamma\{2\delta(\omega) + \frac{\sin c\ a\ z}{caz} (\delta(\omega + c) + \delta(\omega - c))e^{-i\alpha\omega}\}$$

$$(2.14)$$

The power spectrum $S(\omega)$ is given by

$$S(\omega) = 2\pi\gamma^2\delta(\omega) + \frac{\pi}{2}\ \gamma^2(\frac{\sin c\ a\ z}{caz})^2\ (\delta(\omega+c) + \delta(\omega-c)) \qquad (2.15)$$

The interval between two impulses gives the value c. (Fig. 2.8)

In order to determine the value z, γ must be eliminated, which means, physically, normalization of the contrast. Let us consider the ratio R^2 of D.C. and A.C. components of $S(\omega)$,

$$R^2 = \frac{\frac{\pi}{2}\ \gamma^2\ (\frac{\sin c\ a\ z}{caz})^2}{2\pi\ \gamma^2} = (\frac{\sin c\ a\ z}{2\ caz})$$

$$R = \frac{1}{2}\ |\frac{\sin c\ a\ z}{c\ a\ z}| \qquad (2.16)$$

The value R is obtained from the observed image. Thus the solution of eq. (2.16) gives the value z. It should be noted, in this case, that the value z will not be determined uniquely when 2R is less than the value (approximately 0.22) of (Sin caz)/caz giving the second peak shown in Fig. 2.9. This means that the contrast of the object falls below 22% of that in the in-focus plane by the increase of a,z, or c.

In Part II, it is explained that the identification of patterns is possible as their images remain defocused and the automatic focusing to a given object is realized.

However, this theory will be applicable only for the lens system with a small depth of focus and for a rather simple form of patterns. Accurate measurement of the amount of defocusing becomes difficult with the increased depth of focus. It is caused by, in the real domain, the decrease of the pulse width 2caz on $dF(\xi)/d\xi$ shown in Fig. 2.5 (a) and, in the spatial frequency domain, the increase of period π/ca on (Sin caz)/caz in Fig. 2.9. For the complex form of patterns, their high frequency components will be filtered by the defocusing action, and the identification becomes difficult. This is easily understandable if one recalls that the Kernel K acts as a low pass filter.

The frequency domain is useful from the theoretical standpoint. However, in many practical problems, the real domain may be more useful, because of the simplicity of the apparatus and the form of patterns to be focused. Examples of two dimensional problems are now being studied.

The authors wish to thank Messrs. S. Hashimoto, M. Nishi, S. Mizuno, K. Hirose, M. Shimamura, M. Shoroji and K. Toyota for their cooperation.

APPENDIX

We consider the following transformation in order to compare an unknown image with the standard patterns in arbitrary pattern size and spatial phase shift.

In the arrangement of Fig. 2.1 the general formula of object $f_0(x,y)$ on the plane PP' can be denoted by

$$f_0(x,y) = \gamma f(c'(x-\alpha'),\ c'(y-\beta')) \qquad (A.1)$$

From geometrical optics, the well-known relations are

$$\frac{x}{\ell_1} = -\frac{x'}{\ell_2}\ ,\quad \frac{y}{y_1} = -\frac{y'}{\ell_2} \qquad (A.2)$$

The clear image $f_0'(x',y')$ on the in-focus plane QQ' from the object $f_0(x,y)$ is given by

$$f_0'(x',y') = \gamma f\{-\frac{\ell_1}{\ell_2}\ c'(x' + \frac{\ell_2}{\ell_1}\alpha'),\ -\frac{\ell_1 c'}{\ell_2}(y' + \frac{\ell_2}{\ell_1}\beta')\} \qquad (A.3)$$

put

$$\alpha = -\frac{\ell_2}{\ell_1}\alpha'\ ,\quad \beta = -\frac{\ell_2}{\ell_1}\beta'\ ,\quad c = -\frac{\ell_1}{\ell_2}c' \qquad (A.4)$$

According to eq. (A.3)

$$f_0'(x',y') = \gamma f(c(x'-\alpha),\ c(y'-\beta)) \qquad (A.5)$$

we obtain the relation (2.1)

$$f_0(x,y) \rightarrow \gamma\, f(c(x'-\alpha),\ c(y'-\beta))$$

REFERENCES

(1) "Special Issue on WABOT", Bulletin of Science and Engineering Research Laboratory, Waseda University. No. 62, 1973

(2) H.H. Hopkins, "The frequency response of a defocused optical system". Proc. Roy. Soc. A. Mathematical and Physical Sciences, Vol. 231, pp. 91-103, 1955

(3) M.A. Kujoory, B.H. Mayall and M.L. Mendelsohn, "Focus-assist device for a flying-spot Microscope, IEEE Trans. on Biomedical Engineering, Vol. BME-20, pp. 126-132, March, 1973

THE "RUBBER-MASK" TECHNIQUE-I

PATTERN MEASUREMENT AND ANALYSIS

Bernard Widrow

Department of Electrical Engineering
Stanford Electronics Laboratories
Stanford, California 94305, U.S.A.

ABSTRACT

Template matching is a fundamental technique of pattern recognition. Although this technique is very general, its applicability has been limited because of the difficulty often encountered when fitting templates to natural data. Natural patterns are often distorted, misshapen, stretched in size, fuzzy, rotated, translated, observed at an unusual perspective, etc. Flexible templates (rubber masks) have been devised which, when fitted to natural data, can be used for measurement, data reduction, data smoothing, and classification of highly irregular waveforms and image shapes. These problems had been largely unsolved by existing template matching methods.

Specific applications to the analysis of human chromosome images, chromatographic recordings, electrocardiogram waveforms, and electroencephalogram waveforms are illustrated. The rubber-mask technique will probably be usable in a wide variety of scientific applications.

1. INTRODUCTION

Much work has been done during the past twenty years in the field of pattern recognition, both theoretical and practical. Several of the schemes that have evolved with matched filters, linear threshold classifiers, and nearest neighbor rules are optimal in some sense; [1-5] but most are ad hoc and have been developed to solve specific problems. On the practical side, the work in optical character recognition has been the most successful.[6-8]

It is hard to typify work in the field because of the profusion of pattern-recognition schemes. However, it is possible to divide the field into two broad schools: one that classifies by comparing feature observations with pattern property lists; and another that uses the pattern information directly and classifies by means of some form of template matching. Other schools may mix these approaches, but the two basic approaches remain.

Pattern features are often found to be invariant to translation, rotation, scale, etc. It is desirable to select such features as a basis for classification. However, there is no general method for choosing or designating features. A feature-detection system, unless very specially tailored to the particular problem, could miss important attributes. Nevertheless, feature detection will retain a permanent place in the methodology of pattern recognition.

The emphasis of this paper will be on template matching. The advantage of template matching comes from being able to use patterns directly without the need for devising, detecting, or measuring special features. The pattern image itself contains all the required information. We are concerned, however, with pattern matching in the presence of rotation, translation, scale change, differing shadow and lighting effects, gross shape distortion, and random noise which makes the pattern fuzzy. Because of the recurrence of such distortions in most data patterns, template matching generally does not work very well.

It is the purpose of this paper to show how template matching can in many cases be used in pattern measurement and analysis, even when patterns are highly irregular in shape. Computer-implemented or computer-simulated flexible templates are proposed. We call these "rubber masks".

The human eye can determine that two non-exactly-matching signatures were made by the same hand. The eye/brain can determine that a face being viewed has been seen before, only not with the same perspective or with the same light and shadows. The eye/brain is remarkably able to ignore pattern variations that for the most part should be ignored. The ability of the eye/brain system to match patterns in the presence of distortion would be useful for incorporation into practical automatic pattern-recognition systems. It is possible that such an ability may be realized in a general way through the use of the flexible masking techniques to be described in this paper.

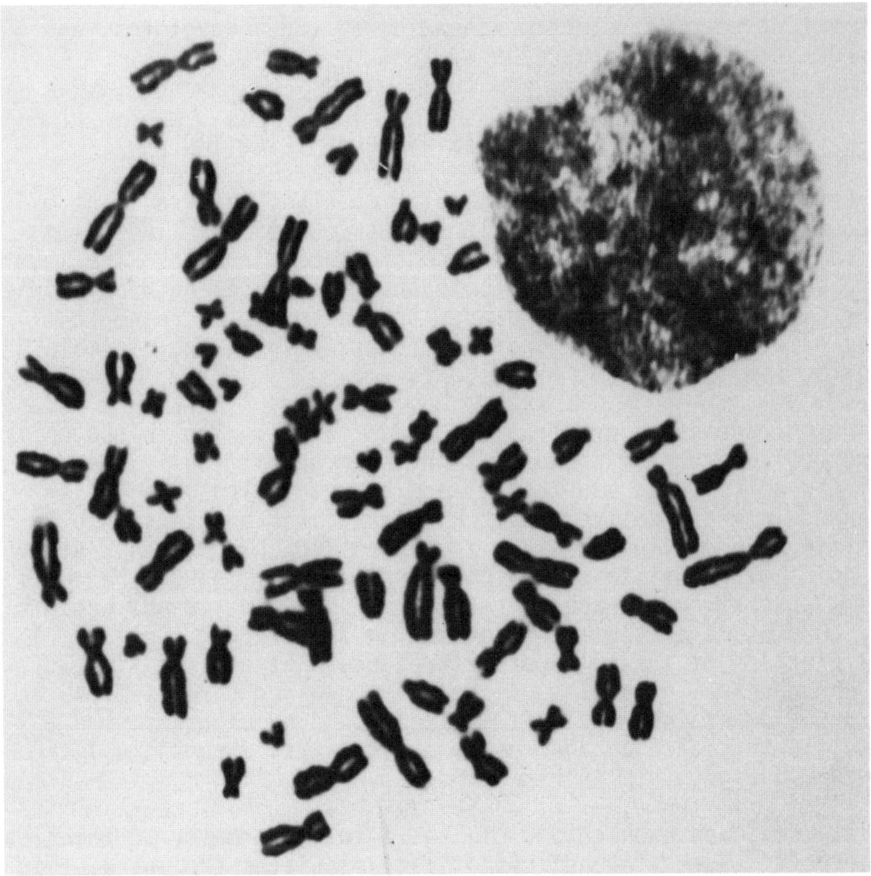

Fig. 1. A Human Chromosome Metaphase Spread.

The flexible template, the rubber mask, is in effect an unusual form of adaptive multi-dimensional matched filter. The basic template could be formed from a mathematical pattern model or from natural pattern data. Rubber masks have been parameterized and adapted as illustrated here with a series of applications to the analysis of chromosomal images, liquid and gas chromatograph output records, electroencephalogram (EEG) waveforms, and electrocardiogram (EKG) waveforms.

2. CHROMOSOME ANALYSIS AND CLASSIFICATION

Chromosome patterns are generally observed under a light microscope that magnifies image size about 2000 times. The chromosomes of a single human cell vary considerably in size. The length of the smallest corresponds to only about ten wavelengths of light in the center of the visible band. Consequently, microscopic images of chromosomes are generally fuzzy, pushing the ultimate resolving limits of light microscopy.

Human chromosome preparations are often made from blood cells or from cells grown in culture. A hypotonic solution is used to swell the cells which are then fixed in acetic alcohol. Cells suspended in the fixative are then dropped from a height onto a glass microscope slide in order to spread the chromosomes. At this point, various stains may be chosen and applied to the cells. Cells caught at "metaphase," i.e. when they are dividing, exhibit their chromosome complement separated in a "spread." Only a small fraction of the cells are at metaphase at a given time. Thus one observes on a glass slide a few stained chromosome spreads amid a sea of stained non-dividing cells. A human chromosome spread (in this case an abnormal one with extra chromosomes) is pictured in Fig. 1.

At metaphase each chromosome has a twin, normally an identical counterpart. The cytogeneticist first associates the pairs, then karyotypes the chromosomes, i.e. classifies them into groups and orders the individuals within the groups. This is a time-consuming and not always error-free process. The problem is difficult, not only because the patterns are fuzzy, but also because the chromosomes sometimes touch one another, or even overlay one another. Furthermore, the "arms" of the chromosomes are often twisted and misshapen due to impact resulting from the samples being dropped onto the glass slide during chromosome preparation.

Chromosomes are karyotyped by their arm lengths and by their shapes. Existing automatic techniques have classified chromosomes on the basis of measurements such as centromeric index (ratio of arm lengths to total body length), ratios of body lengths from one chromosome to another, chromosome areas, etc. Unfortunately,

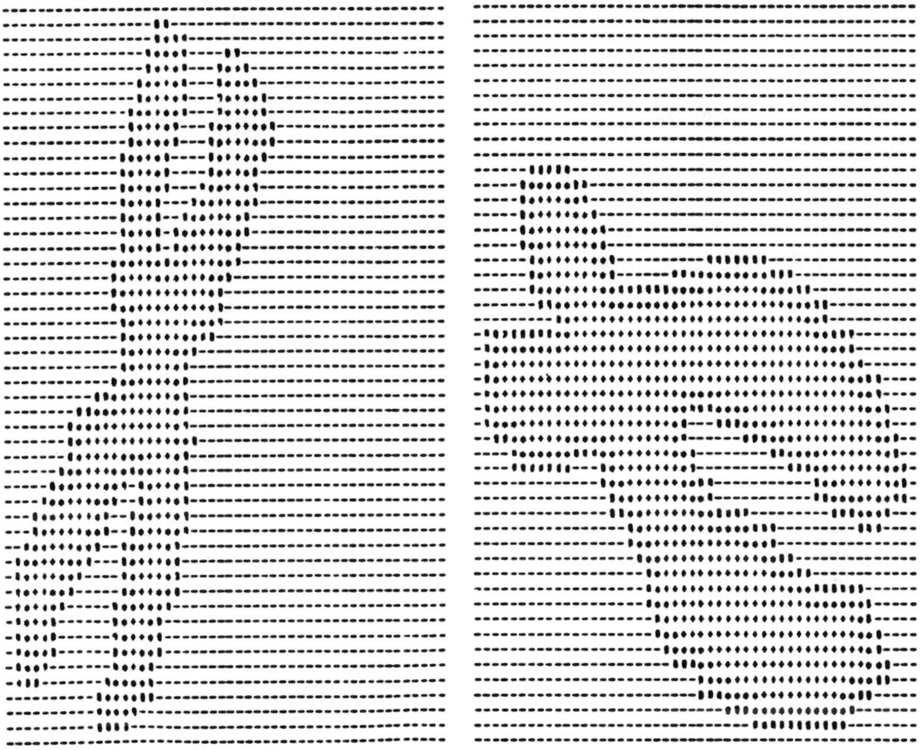

Fig. 2. Two-level Digitization of Chromosome Images.

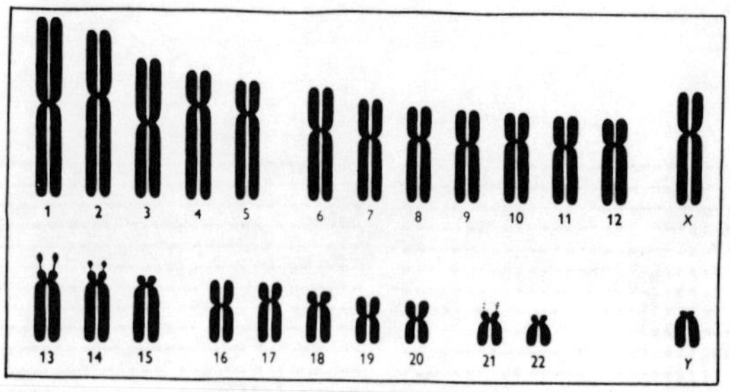

(a) "DENVER STANDARDS"

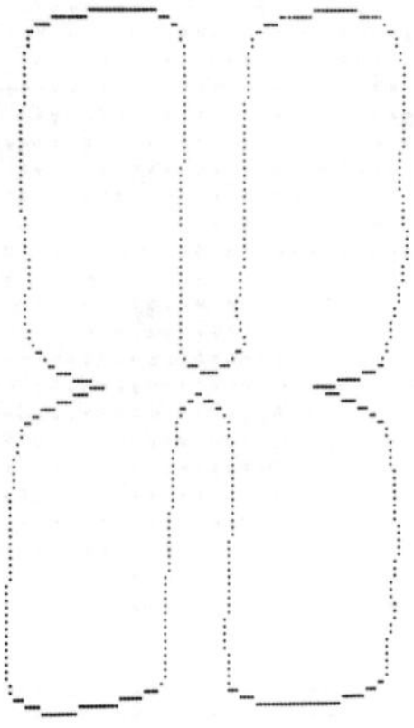

(b) A DIGITIZED "DENVER STANDARD"

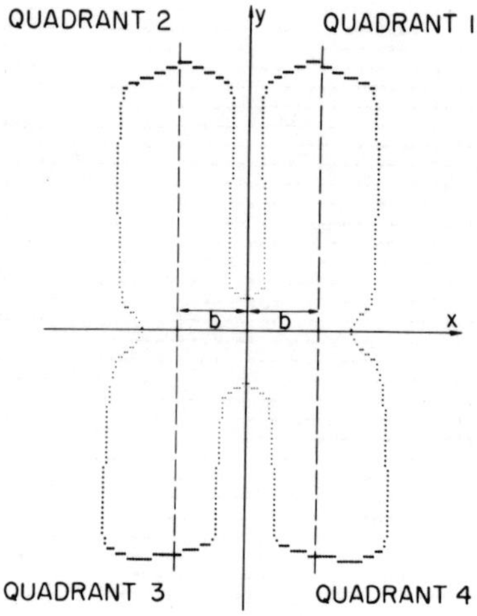

(c) MODIFIED "DENVER STANDARD"

Fig. 3. Human Chromosome Stereotypes.

these parameters cannot generally be measured accurately. It is possible to measure the length of a table top quite accurately and consistently, but it is not easy to measure the length of a chromosome arm which may be twisted and fuzzy. Thus it is almost impossible to determine where the arm begins and where it ends (see Fig. 1).

The rubber mask technique is being developed for the purpose of accurately measuring chromosome geometrical properties. Computer implementation is essential, requiring digitized input patterns. Examples of digitized chromosome images are presented in Fig. 2. Two levels of gray scale were established by reference to an adjustable black/white intensity threshold. Altering the threshold level would have some effect on this form of pattern; it might even be desirable to take several digitizations of the original image with different threshold settings. In any event, the sample patterns in Fig. 2 are typical of those that have been subjected to analysis by rubber masking.

The rubber-mask technique is applied by comparing the shapes of digitized data-sample patterns with those of stored standard patterns. The stored stereotype (idealized pattern) is progressively distorted until the data pattern is fitted to the desired degree of accuracy. The successive distortions, or iterations, to which the stereotype is subjected constitute, in effect, the evolution of a stretched template (rubber mask). By means of a system of coordinates and parameters, the shape finally assumed by the rubber mask can be numerically described.

A rubber-mask chromosome analysis has been made using various versions of the "Denver Standard" chromosomes as stereotypes. These are shown in Fig. 3a. These standard shapes, arranged in a karyotype, were established at a meeting of cytogeneticists organized by T. T. PUCK in Denver in 1960.[9] They were designed to be chromosome-like in shape and to have ratios of upper arm length to lower arm length and arm length to total length which are average for normal human chromosomes. One of these standards (No. 3) was chosen as a stereotype for the present study (illustrated in Figs. 4 and 7) and is shown magnified and in digitized form in Fig. 3b. The form was modified somewhat by adding girth at the center (centromere) in order to create a shape somewhat closer to that of an actual human chromosome. The modified standard chromosome is shown in Fig. 3c. Note the addition of a coordinate system and of dotted lines which are median lines through the chromosome arms. These are essential in keeping track of the distortion process necessary to fit the template to the data sample.

An illustration of template fitting is shown in Fig. 4. The stereotype (the modified Denver Standard No. 3) is shown in Fig. 4a.

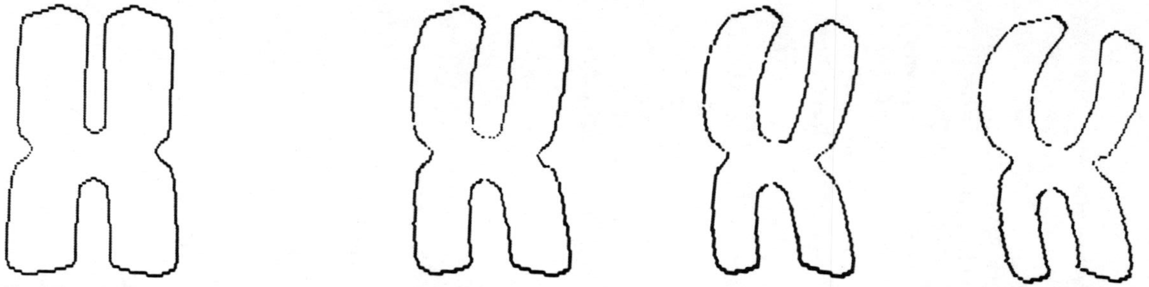

(a) STEREOTYPE MODIFIED (b) ITERATION 1 (c) ITERATION 2 (d) ITERATION 3

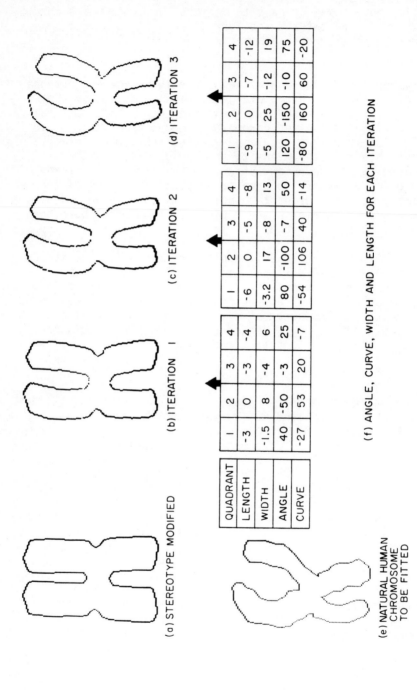

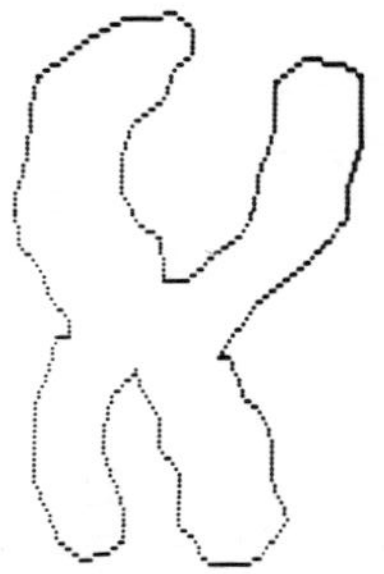

(e) NATURAL HUMAN CHROMOSOME TO BE FITTED

QUADRANT	1	2	3	4
LENGTH	-3	0	-3	-4
WIDTH	-1.5	8	-4	6
ANGLE	40	-50	-3	25
CURVE	-27	53	20	-7

QUADRANT	1	2	3	4
LENGTH	-6	0	-5	-8
WIDTH	-3.2	17	-8	13
ANGLE	80	-100	-7	50
CURVE	-54	106	40	-14

QUADRANT	1	2	3	4
LENGTH	-9	0	-7	-12
WIDTH	-5	25	-12	19
ANGLE	120	-150	-10	75
CURVE	-80	160	60	-20

(f) ANGLE, CURVE, WIDTH AND LENGTH FOR EACH ITERATION

Fig. 4. Stretching a Chromosome Stereotype to Fit Natural Data.

LENGTH $$y' = y + y \cdot \frac{\text{LENGTH}}{100}$$, where LENGTH = % increase in length

WIDTH $$x' = x + (x-b) \cdot \frac{\text{WIDTH}}{100}$$, where WIDTH = % increase in width

ANGLE $$x' = x + y \cdot \text{ANGLE} \cdot k_1$$

where k_1 is a constant chosen so that an ANGLE value of 100 will bend the arm so that the distance from its tip to the y-axis is doubled.

CURVE $$x' = x + y^2 \cdot \text{CURVE} \cdot k_2$$

where k_2 is a constant chosen so that a CURVE value of 100 will bend the arm so that the distance from its tip to the y-axis is doubled.

Fig. 5. Chromosome Distortion Parameters.

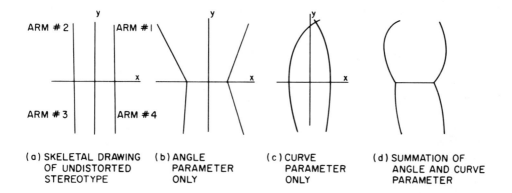

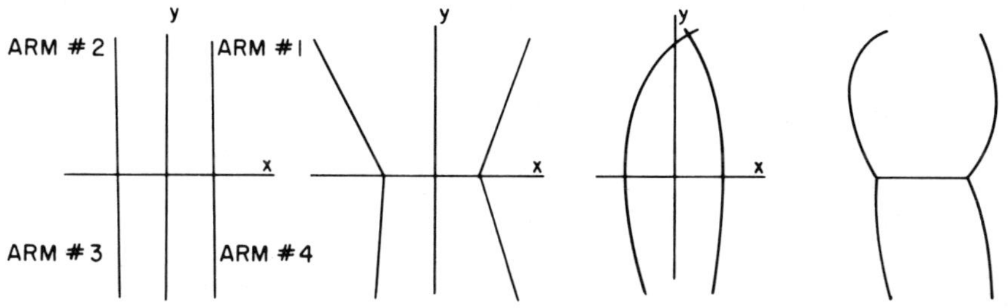

(a) SKELETAL DRAWING OF UNDISTORTED STEREOTYPE (b) ANGLE PARAMETER ONLY (c) CURVE PARAMETER ONLY (d) SUMMATION OF ANGLE AND CURVE PARAMETER

Fig. 6. Skeletal Illustration of Chromosome Distortion Parameters.

The digitized sample of a human chromosome to be fitted is shown in Fig. 4e. Distorted versions of the stereotype at different iterative stages are shown in Figs. 4b-d. Corresponding sets of distortion parameters are listed in Fig. 4f. Through succeeding iterations, the distorted stereotype (stretched template) is seen to experience an evolution of its shape toward that of the chromosome to be fitted.

An explanation of the distortion parameters that have been used in this study, namely LENGTH, WIDTH, ANGLE, and CURVE, is presented in Fig. 5. These parameters are adjusted individually and independently in each of the four quadrants of the stereotype as defined in Fig. 3c. Each arm can be lengthened or shortened, thickened or thinned, offset at an angle along its median line, or curved with a second degree function along its median line; or it can be distorted with a combination of these effects.

The distortion undergone by the stereotype is indicated in skeletal form in Fig. 6. The undistorted stereotype is represented by the "H-pattern" in Fig. 6a. The vertical lines are the median lines through the stereotype arms (refer to Fig. 3c). Figure 6b illustrates the effect of ANGLE only applied to the median lines, while Fig. 6c shows the effect of CURVE only. Figure 6d is a skeletal view of the distorted stereotype incorporating the summed effects of ANGLE and CURVE. Although the effects of LENGTH and WIDTH are not included in this figure, the general shape of the distorted stereotype is quite evident.

A blow-up of the distorted stereotype superposed upon a blow-up of the sample pattern of a human chromosome is shown in Fig. 7. This figure illustrates many aspects of the rubber-mask idea. The "rubberized" template (distorted stereotype) is equivalent to a smoothed version of the actual data, the digitized human chromosome. The "fit" is optimized with the aid of the computer, which counts the number of units of area in the portions of the patterns which are not congruent. The distorted stereotype (rubber mask) is changed by an iterative process until this error count is minimized.

The fitted stereotype (solid outline) could conceivably be used in many circumstances in place of the actual digitized human chromosome (dotted outline). Such use would comprise a form of data reduction. The data image would be represented approximately by the combination of the known undistorted stereotype and by a small number of numerical distortion parameters.

Another application for the fitted stereotype could result in feature measurement. Arm lengths, widths, areas, and centromeric index of the human chromosome could be estimated in terms of the lengths and widths and areas of the distorted stereotype by including the length and width factors derived from the fitting process.

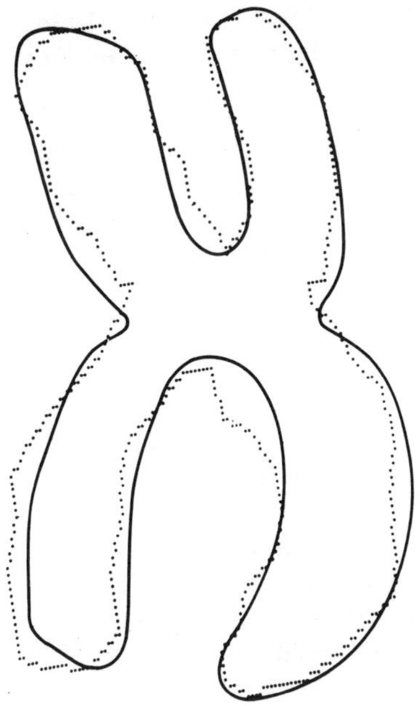

Fig. 7. Rubber Mask at Iteration #3 (Solid) Compared With Natural Chromosome (Dotted).

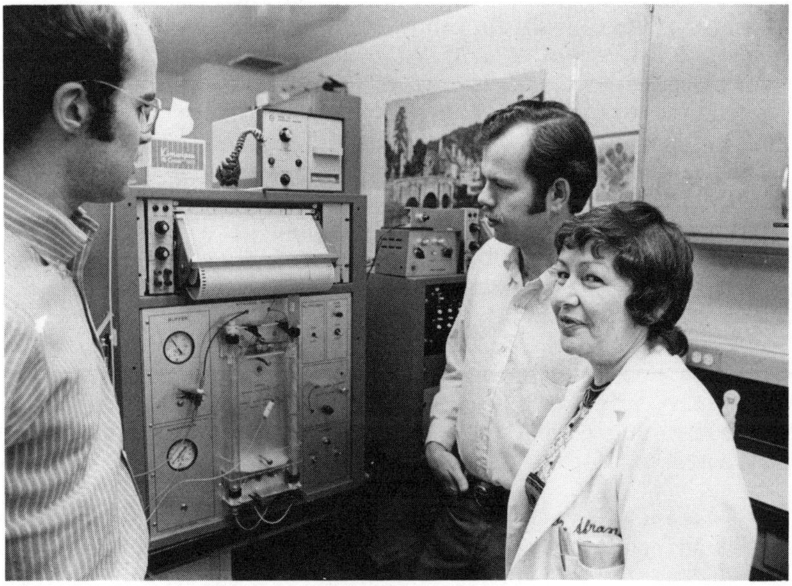

Fig. 8. 'Liquid Chromatograph in the Laboratory.

(The parameters ANGLE and CURVE are merely used in the fitting process and have no value in measuring and classifying chromosomes.)

By placing narrow limits on the variability of the LENGTH and WIDTH parameters, the chromosomes of the spread (Fig. 1) could be karyotyped by associating each data chromosome with the individual modified Denver Standard chromosome that it best fitted. This would require the use of 24 different stereotypes. A more practical approach might be to select a representative stereotype for each of the chromosome groups, comprising Nos. 1-5, 6-12, 13-20, 21-22, X, and Y. A data chromosome would then be grouped by determining to which representative stereotype it could best be fitted. Its position or rank within the group would then be determined by the LENGTH and WIDTH parameters.

The template-fitting experiments illustrated here have been done semiautomatically using an interactive computer terminal. The position and orientation of the axes of the rubber mask relative to those of the data chromosome have been chosen and modified by man/machine interaction. The same is true of the four individual parameters in each quadrant. The number of adjustable parameters is sixteen, plus two for X-Y placing of the mask axes, plus one more for axis rotation, giving a total of nineteen adjustables.

A scheme for fully automatic adjustment is currently under development. A method of initially guessing the centromeric position is being tested which finds four points on the data chromosome located where the arms join in indentations. Diagonal lines through these points cross at a point which seems to be a reasonable initial estimate of the centromere position. The initial angular orientation for the rubber mask is determined by rotating the data pattern so that when squeezed in a software-implemented "vise," a minimal "jaw opening" is obtained. The axes of the rubber mask are initially aligned parallel to the simulated vise jaws. The initial centromere is chosen as above. An iterative process then commences to vary all nineteen parameters in search of a best fit.

Gradient methods are being developed to reduce computation time. Preliminary results show that, using the initial conditions of centromere and angle described above, global optima are attained after small numbers of iterative computational cycles.

Recent research into new staining techniques has resulted in chromosome "banding patterns".[10,11] Under suitable preparation of samples, dark and light bands appear across the chromosome arms. These bands can be photographed or viewed under the microscope. The widths, spacings, and numbers of bands are characteristic of the individual chromosomes in the karyotype. Now, difficult-to-resolve chromosomes are easily separated, greatly simplifying the karyotyping problem.

Work is just beginning on the application of the rubber-masking principle to the problem of measuring geometric properties of banding patterns. Stereotypes of human-chromosome banding patterns have appeared in the literature (12) and this is of great assistance to the development of rubber masks for banding patterns. The banding stain preparations are at present in relatively early stages of development and difficulty is encountered with high variability in the banding patterns. However improvements in preparation techniques are continually appearing in the literature and the usefulness of the banding technique in accurate karyotyping and in the detection of chromosomal abnormalities is being demonstrated in the laboratory.

Two-dimensional rubber masks are illustrated in Figs. 4 and 7. Three-dimensional rubber masks for chromosome banding patterns are being developed adding optical density as another dimension. Actual banding patterns have very wide variations in optical density, (13) and fitting to these variations may turn out to be very advantageous.

The goal of this work is to be able to measure automatically a number of useful geometrical and banding-pattern parameters on individual human chromosomes. If this can be done cheaply enough, the process could be applied in making measurements on the chromosomes of hundreds of cells from the same patient sample. Averaging could provide unprecedented accuracy, effectively placing a new research instrument into the hands of the cytogeneticist to study normal and abnormal measurements. Accuracies of the order of a few per cent is the long-term goal.

The work on chromosome measurement and analysis reported in this section is being developed in collaboration with Dr. Leonard Hayflick of the Stanford University Department of Medical Microbiology. His research interest is in reproducibly growing human cells in nutrient solution in sufficient quantity to make them available to workers all over the world who require standardized sample cells for their experiments. Chromosomal measurement is useful for quality control of the experimental tissue.

3. CHROMATOGRAM ANALYSIS

Figure 7 illustrates a technique for measuring highly irregular shapes, i.e. the digitized shapes of human chromosomes. The same principle could be applied in the analysis of chromatograms outputted by liquid or gas chromatographic systems.

Liquid chromatography is gaining increased importance in the clinical determination of amino-acid content of biological samples, typically taken from blood serum or urine. Certain forms of physical disorders, birth defects, mental disorders, etc., can be dianosed or predicted from the analysis of amino-acid chromatograms. These are graphical outputs, generally recorded on a strip chart, exhibiting a series of peaks. Each amino-acid corresponds to an individual peak. The amount of amino-acid corresponding to a given peak is proportional to the area under the peak. For analysis of amino-acids in blood or urine in clinical application, the chromatograph equipment does not have the resolution to separate all the peaks for quantitative measurement. Some of the peaks tend to overlap and linearly add in output amplitude. The problem is to find the areas under the individual peaks in spite of overlap and possible baseline drift. Similar problems in peak picking and area analysis also exist for measuring amino-acids by gas chromatography. In both chromatographic techniques, it is desirable to locate the individual peaks accurately in time in order to properly identify the associated amino-acid.

Figure 8 shows the amino-acid analyzer, a liquid chromatographic instrument that has been the source of data for this research. Dr. Klara Efron, a pediatrician at the Stanford University Medical Center, has worked with an Electrical Engineering Department student team in preparing the chromatograms for analysis using rubber-mask techniques. The analytical results are checked by testing against known mixtures of amino-acids that have been prepared by Dr. Efron. Each sample, whether it is from a human patient or whether it is a known artificial sample, takes 6 hr to run on the amino-acid analyzer. Thus at the present time, chromatography is a slow and expensive process.

Attempts to speed the process and/or to cut cost generally lead to increased peak overlap (poorer spectrographic resolution). The purpose of the rubber-mask experimentation has therefore been to develop a means, by data processing, to resolve peaks in spite of heavy overlapping and to find their respective areas with great accuracy.

In Fig. 9, a portion of a chromatogram of a typical sample is shown. Eleven peaks are present in this portion. A five-peak section of the same data is shown on an expanded scale in Fig. 10. The approach taken is to evolve a rubber mask (stretchable template) and fit it to the data. The rubber mask consists of a sum of adjustable gaussian peaks. The peaks are not fitted one at a time, but <u>all together</u> in the same process. Each peak has three parameters: position in time t_0, amplitude A, and narrowness $1/2\sigma^2$, where σ is the standard deviation. Thus the fit illustrated in Fig. 10 involves these three parameters tailored for each of the five peaks, plus an additive constant for baseline adjustment. The fit

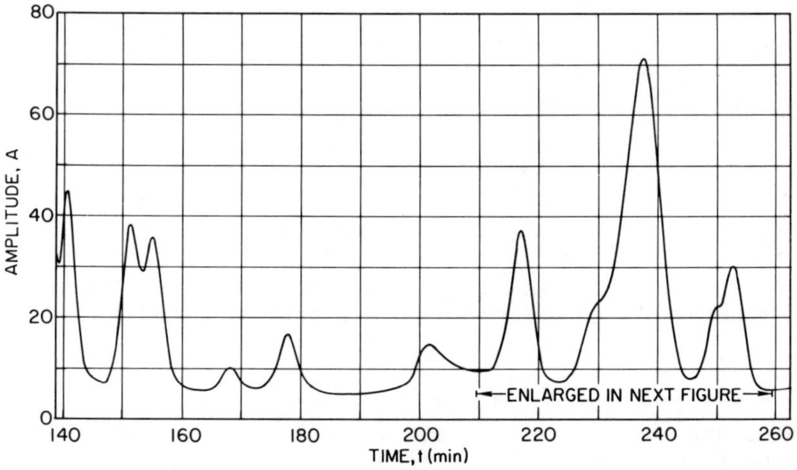

Fig. 9. Portion of Chromatogram From Amino Acid Analyzer.

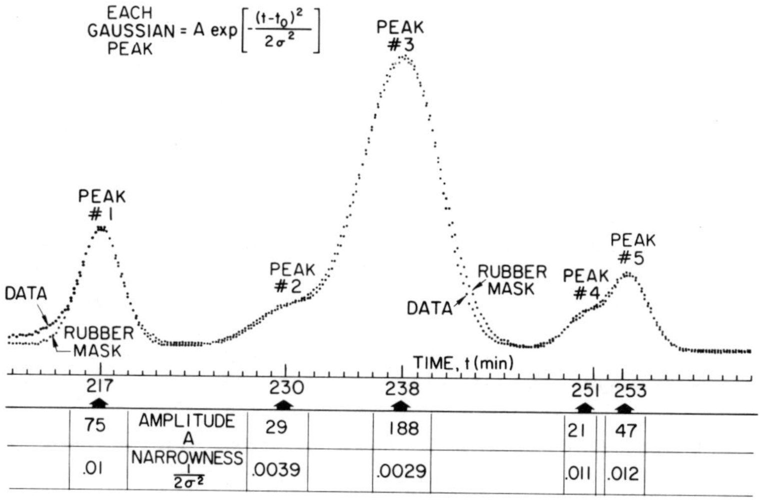

Fig. 10. Portion of Chromatogram With Fitted Template.

shown is very close and is typical of the results obtained with the rubber-mask process. The formula for the fitted rubber mask of Fig. 10 is

$$R = \text{constant} + 75e^{-(0.01)(t-217)^2} + 29e^{-(0.0039)(t-230)^2} + 188e^{-(0.0029)(t-238)^2} + 21e^{-(0.011)(5-251)^2} + 47e^{-(0.012)(t-253)^2}.$$

The suggestion of using a gaussian shape to represent such peaks comes from the chemical literature.(14-16) Diffusion processes take place in the chromatograph equipment, and the gaussian shape is the theoretical solution of the appropriate diffusion equation.

Current research problems involve the development of automatic fitting algorithms using steepest descent. Present methods are man/machine interactive. The problem of segmentation is being studied: (a) How many peaks should be included in each fitting process? (b) What happens when one hypothesizes a given number of peaks in a given portion of output record, whereas the actual number is different? (c) Will multiple-hypothesis testing be required, i.e. trial of various numbers of gaussian peaks? (d) How shall suitable initial parameters for the fitting process be determined? These questions are fairly typical of all the applications of rubber masks and are being studied in particular practical cases.

The chromatogram may be regarded as a string of impulses, one for each amino-acid present, having corresponding time positions and amplitudes, convolved with the gaussian pulse shape by the diffusion process in the chromatograph equipment. The impulse string has thus gone through a linear filter having a two-sided gaussian impulse response. Deconvolution could be done with an inverse filter, which in this case would be the reciprocal of another gaussian shape in the frequency domain. This inverse shape goes to infinity in both directions, and must therefore be cut off by low-pass filtering. In the time domain, this introduces "ringing," causing false peaks. The net results of deconvolution are generally not good, and by no means comparable to the resolving ability of the rubber-mask method.

Illustrated by the example in this section is a problem which is common to many scientific instruments, such as gas chromatographs, mass spectrometers, air pollution monitoring instruments, and other chemical and electronic resolving equipments. The problem is that of resolving overlapping spectral peaks. A good solution to this problem thus has wide practical applicability.

The rubber-mask technique has been used here in data smoothing, measurement, and analysis. Rubber masking can also be used in data reduction. For example, the entire data waveform shown in Fig. 10 could be well represented by the small number of gaussian parameters illustrated there.

4. ANALYSIS OF K-COMPLEXES IN EEG WAVEFORMS

Rubber-mask principles have been utilized in the analysis and classification of complex waveforms that occur in human electroencephalogram (EEG) recordings made during sleep. The signals are obtained from electrodes externally applied to the patient's scalp. This work is being done in collaboration with Dr. Vincent Zarcone, a psychiatrist in the Stanford University School of Medicine who is doing research on sleep and dreams, and with Dr. William Dement, director of the Sleep Research Laboratory, Stanford University.

The "K-wave" or "K-complex"[17-19] is of particular interest to sleep researchers. It is an electrical phenomenon having a characteristic shape which appears superposed upon the EEG background activity. Certain stages of non-dream sleep can be identified by measurement of the frequency of incidence of these waves. Detection of the K-wave in practice is uncertain because of the interference due to the background EEG activity. Also, timing and amplitude vary from K-wave to K-wave of the same patient. The result is that typically, two expert sleep-record readers might agree on the designation of K-waves only 60-80 per cent of the time.

In Fig. 11, the topmost EEG recordings are raw data exhibiting possible K-waves. These non-contiguous portions of a single EEG record were pre-selected by an automatic process as containing "candidate" K-waves. The pre-screening process, based on previous work done at the University of Florida,[20-22] utilizes a set of criteria such as the following: the signal should swing in the negative direction and exceed an amplitude threshold; then within a certain time interval, should swing in the positive direction and fall within a set of range of positive amplitudes; then should swing back toward zero within a certain time range. If such a criterion is met, then the EEG is considered to contain a candidate K-wave. (Note: tradition in electroencephalography is to plot negative signals upward.)

Since there appears to be no precise definition of a K-wave, it can only be defined by showing examples of K-waves judged "good." In order to establish a stereotype of a "good" K-wave, a mask for a given patient is formed by taking portions of the EEG sleep record that occur early in the morning, about 4-5 a.m., when there is generally a substantial quieting in the EEG. About

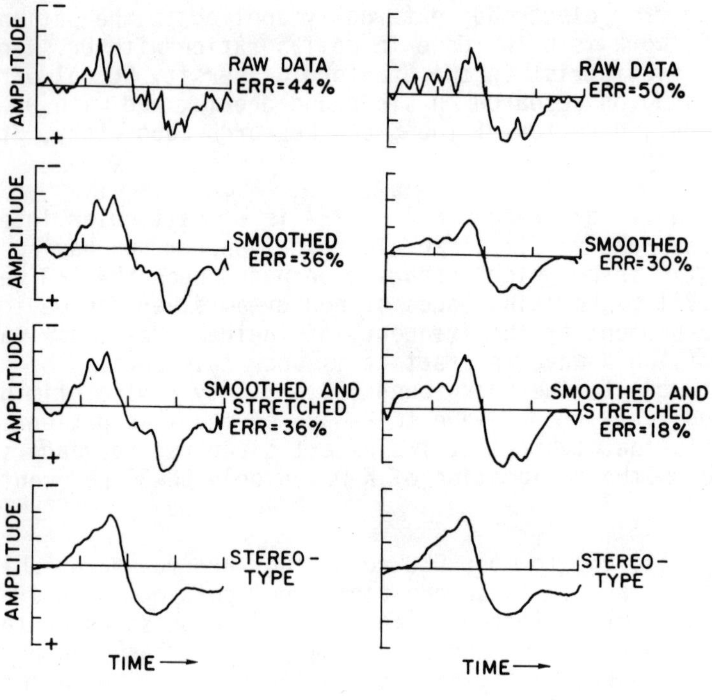

Fig. 11. Candidate K-Waves From Human EEG Sleep Recordings Compared With K-Wave Template.

five to ten good examples of K-waves are selected by an experienced investigator. These are averaged to make the mask according to the following procedure: Two K-waves are selected initially; the first is kept fixed while the second is best aligned as to baseline (dc level) and position along the time axis, and then stretched (first in amplitude, then in time, within about $\pm$ 15 per cent) to achieve best least-squares fit. After several iteration cycles, if the fit is satisfactory, the two K-waves are averaged. Then a third K-wave is taken and best-fitted to the average; it too is averaged in, and so forth until a useful stereotype is formed, as shown at the bottom of each column in Fig. 11.

We are concerned with evaluating candidate K-waves selected from the sleep record. Comparison between each candidate K-wave and the stereotype is done by summing the squares of sample-value differences and normalizing with respect to the sum of the squares of the sample values of the stereotype. All processing is digital. The candidate waveform is translated left-right in time and up-down in baseline level (dc level) to get a best alignment relative to the stereotype giving a minimum sum of squares of error.

In each column of Fig. 11, all of the waveforms have been compared against the stereotype when best-aligned with it, and the corresponding per cent sum-squared errors are indicated in the figure. As stated above, the topmost waveforms are two separate raw EEG events from the same patient, containing candidate waveforms. These raw data waveforms compared to the stereotype with 44 per cent and 50 per cent errors respectively. The next-to-the-top waveforms are similar to the raw-data waveforms but are filtered or smoothed versions. Smoothing is done by digital filtering using a non-recursive moving-average (tapped delay line) filter having symmetrical weights. No phase shift or phase dispersion is introduced by such a filter; only high-frequency components are altered. The raw data and the smoothed waveforms are in exact time registration. The improvement in fit from noise smoothing was significant. Reducing high-frequency noise generally cuts the sum of the squares of the errors at best fit by up to 50 per cent. In the cases shown, the smoothed data waveforms compared to the stereotype with errors of 36 per cent and 30 per cent respectively.

The next-to-the-bottom waveforms in Fig. 11 have been smoothed, best-aligned, and best-stretched (within $\pm$ 15 per cent limits in amplitude and time), and now the errors are down to 36 per cent and 18 per cent respectively. If one looks back at the raw data, the upper left-hand waveform could be rated as a 36 per cent error K-wave candidate, while the upper right-hand waveform could be rated as an 18 per cent error candidate, based on the final best-fitted comparison. The fitting procedure allows candidate K-waves to be ranked in terms of how closely their "intrinsic shapes" match that of the stereotype.

There is considerable interest among the Stanford Sleep Research group in the use of such ranking system. With a moderate amount of experience, this type of ranking seems to agree quite well with that obtained from skilled human scoring of the original sleep EEG record.

It should be noted that in this work, the stereotype remains fixed once it is formed. The data itself is stretched (rubberized) rather than the stereotype.

In the examples previously presented, sums of gaussian shapes (for chromatographs), or geometrically uniform artificial stereotypes (for chromosomes) were used as basic templates for the rubber mask process. In this EEG study, natural data selected by an experienced eye have been used entirely in the formation of stereotypes. The resulting stereotypes have no mathematical formula, geometric symmetry or regularity of any sort.

5. ELECTROCARDIOGRAPHIC WAVEFORM ANALYSIS

The work reported here on EKG waveform analysis has made extensive use of data obtained from Dr. J. von der Groeben of the Stanford University Medical School. From more than a thousand patients having normal heart function (as well as can be determined by clinical workup and EKG analysis), he and his associates have recorded six simultaneous EKG channels plus a seventh channel whose signal came from a strain gauge stretched across the chest to measure respiration.

The application of rubber-mask techniques to the analysis of EKG waveforms has focused on the QRS complex, defined as the main pulse of the electrocardiagram that occurs when the heart muscle "fires."

An EKG "lead" in medical terminology means an output signal of a differential amplifier whose inputs come from a pair of electrodes affixed to the patient. The use of six simultaneous anatomically affixed leads gives six different spatial projections of the complicated electrical phenomena developed during cardiac functioning.[23-25] The cardiac electrical activity is sometimes described in terms of a point electric dipole, the amplitude and direction of whose moment vector change with time during the heart beat. This description is a very rough, first order point of view, however. A more general and more adequate model consists in assuming that the electrical activity is generated by a large number of point dipoles located at different places within the heart muscle. The amplitudes and directions of these dipoles are assumed to be individual and to change with time during the beat. The actual situation is in fact more complicated, since such parameters as the

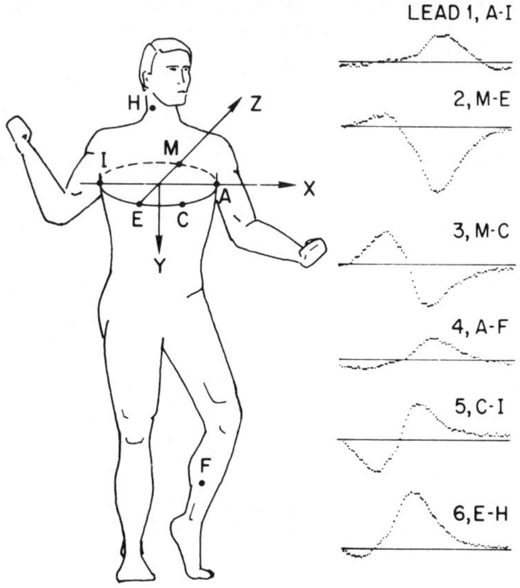

Fig. 12. Electrode Placements for Six-Lead Data.

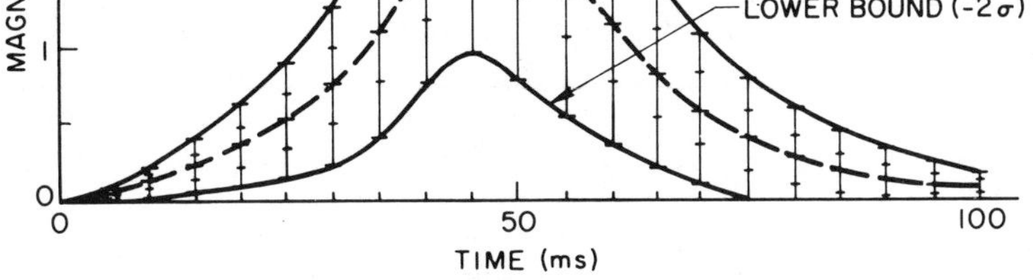

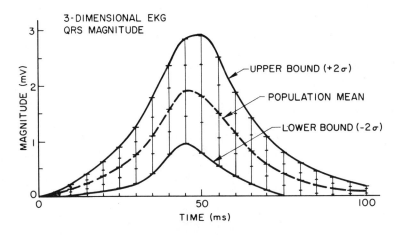

Fig. 13. Bounds of QRS Magnitude for Healthy Males Age 20 and Younger Observed by von der Groeben.

size, shape, position and orientation of the heart undergo substantial mechanical changes during the heart beat. Respiration also has a marked effect on these parameters, further complicating the observed electrical phenomena.

If the heart behaved electrically like a single point dipole whose position remained stationary in space and whose amplitude and orientation changed during the beat, three linearly-independent lead signals, orthogonal or non-orthogonal, would be sufficient to precisely derive or synthesize any other projection (signal from any other lead) during the QRS. It has been found experimentally that a sixth lead can be almost perfectly synthesized by a linear combination of five other leads. Furthermore, it has been found that the QRS of a fifth lead can generally be quite closely synthesized by a linear combination of QRS's of only four other leads.

For the EKG data used here, the placements of the six electrode leads, and typical QRS waveforms from these leads, are illustrated in Fig. 12. This lead system used by Dr. von der Groeben is similar to one proposed by FRANK.[24-26] It has been found possible to linearly combine these six leads to derive three synthetic leads which are essentially spatially orthogonal for an "average" person. Large numbers of normals, i.e. individuals having normal heartbeats, have been studied by Dr. von der Groeben using the synthesized orthogonal three-lead Frank system. In this synthetic three-space, bounds for normals during the QRS have been established. These bounds apply to magnitude in mV and to two angles in a polar coordinate system. Within a normal group for the same sex and age bracket, the QRS magnitudes in 3-space could easily vary over a 3 to 1 (or greater) range. The polar angles show a correspondingly wide range. QRS magnitude data for healthy males age 20 and younger is shown in Fig. 13.[27,28]

These findings raise certain questions. The variations from patient to patient are evidently very high, which is unfortunate. Such variations greatly complicate the problems of automatic EKG analysis. The question is, does high variability really exist from heart to heart, or is much of the evident variability in the QRS measurements caused by variations in chest shape, heart shape, location and orientation of the heart within the chest cavity, respiration, etc.? How consistent is the human heart as a pulse generator? How consistent is it from pulse to pulse, comparing pulses from the same patient, and how consistent is it from patient to patient within a normal group?

In order to analyse or classify human EKG waveforms, it is ultimately necessary to be able to compare one EKG with another. The rubber-mask idea was generalized to do this, including changes in coordinate projections along with the usual time and amplitude stretching and baseline adjustments.

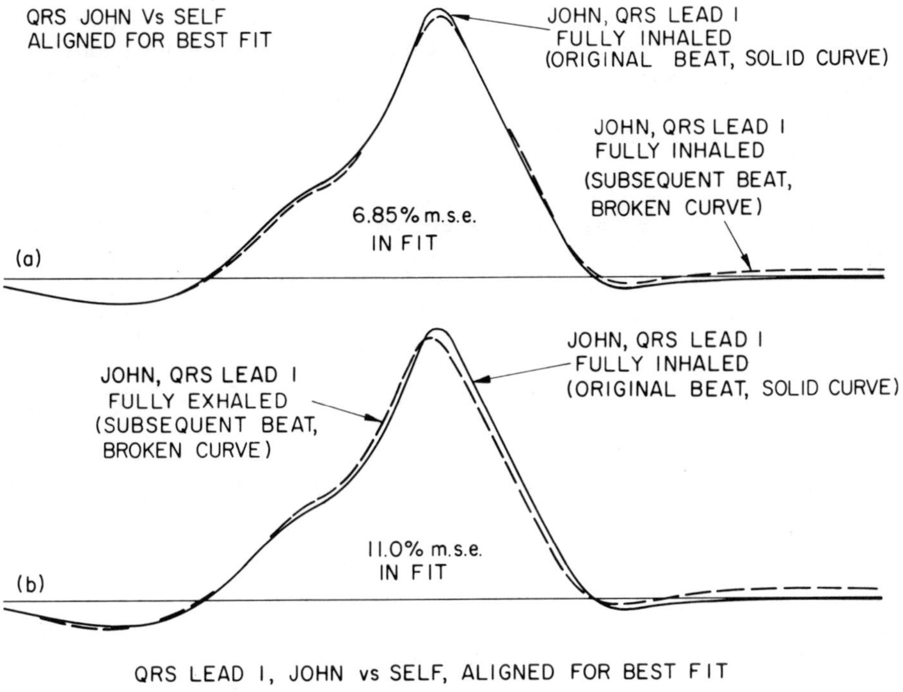

Fig. 14. Beat-to-Beat Comparison for John, QRS, Lead 1.

As a first step, a QRS pulse from a normal patient has been compared with other QRS pulses from subsequent beats of the same patient. Beat-to-beat comparisons for a typical normal patient ("John") are shown in Fig. 14. Taking a best (in the least-square sense) relative alignment in time and in baseline (dc level), a QRS from a fully-inhaled beat is compared with a QRS from a subsequent fully-inhaled beat in Fig. 14a. The mean-square error (m.s.e.) at best fit was 6.85 per cent for the beats shown. For most normals, this figure will vary from about 1 to 10 per cent, and will usually be of the order of a few per cent. In Fig. 14b, a QRS pulse taken at full inspiration is compared with a QRS at full expiration. Here, the per cent mean-square error is much higher; for the pulses shown the m.s.e. is 11.0 per cent. Usually, the error will run from 5 to 15 per cent, generally being about 10 per cent. It is clear from these experiments that the effects of breathing upon the shape of the QRS is highly significant, particularly when the EKG is analysed by computer rather than the "eyeball."

In a recent conversation between the author and Dr. Otto H. Schmitt of the University of Minnesota, it was found that both the Stanford group and a group led by Dr. Schmitt were observing substantial beat-to-beat QRS variation due to respiration, and both groups were taking steps to remove the effects of respiration from the EKG analysis process. Dr. Schmitt has developed a simple, practical means for training the patient within a few minutes to breathe in synchronism with the R-wave (the peak of the QRS). By electronically counting R-waves, every fourth, or every fifth R-wave could be selected to light a lamp, indicating to the patient when to breathe. The patient soon develops a comfortable breathing rhythm synchronized with the EKG. Heart beats are then associated with one another in terms of time count from the R-waves which signal breathing commands. Dr. Schmitt indicated that this procedure gives excellent beat-to-beat consistency, since beats can be chosen at corresponding phases of the respiratory cycle.

The procedure described in the present paper stretches EKG waveforms in amplitude and time and by coordinate transformation in order to remove the effects of respiration. The latter evidently causes mechanical variations in the coordinate projections of the EKG leads.

A method to provide compensation for such variations is illustrated in Fig. 15. There, lead 1 is shown being synthesized as a linear combination of the remaining five leads. The weights of the linear combination (indicated by circles with arrows through them) can be adapted to achieve a best least-squares fit. This figure illustrates a general method of synthesizing desired waveforms by coordinate projection.

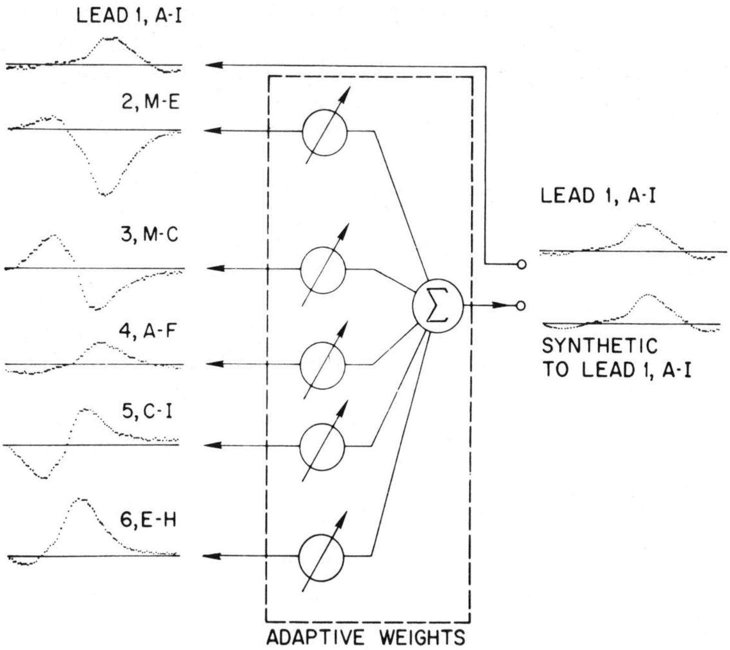

Fig. 15. Synthesis of QRS by Coordinate Projection.

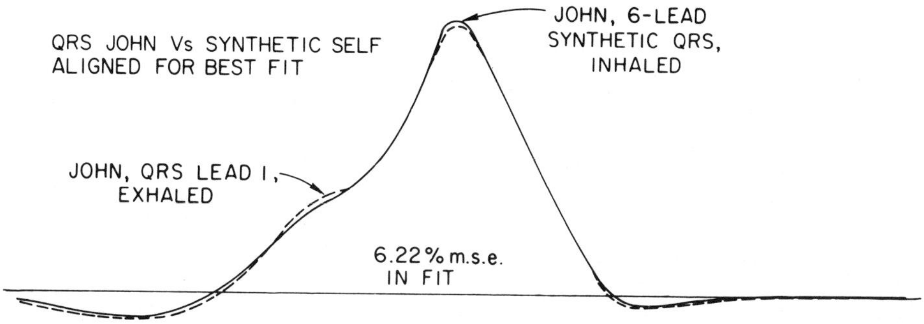

Fig. 16. Synthetic QRS John (inhaled) vs. John Lead 1 (exhaled).

By adapting the synthetic projection to create a best least-squares fit between two QRS's, it appears that one is able to compensate for waveform distortions introduced not only by anatomical effects of respiration, but also by imperfect electrode placement, by variations between individuals with respect to chest shape, heart shape, location and orientation of the heart within the chest cavity, etc.

Synthetic QRS's based on linear combinations of six leads have been adapted to and compared with single-lead QRS's from the same individual. A typical result is presented in Fig. 16. For the comparison shown, best time alignment, best baseline alignment, and best weights were chosen to minimize mean-square error. Synthesizing from data of an original fully-inhaled beat, a fit is made to lead 1 of a subsequent fully-exhaled beat, resulting in 6.22 per cent mean-square error. Figure 16 compares directly with Fig. 14b, where lead 1 of the fully-inhaled beat is compared with lead 1 of the subsequent fully-exhaled beat, resulting in 11.0 per cent error. Synthetic projection has reduced the error, removing that due to respiration, and it has brought the error into the range which is typical of beat-to-beat comparison when maintaining approximately the same phase of respiration. The latter situation is represented in Fig. 14a, where the mean-square error is 6.85 per cent.

The next question to be considered is, why do EKG's look so different from person to person? Are the differences caused by intrinsic heart irregularity, or by imperfect placement of electrode leads, or by respiration, or by the impossibility of perfectly placing the electrodes in the presence of wide anatomic differences from patient to patient? To investigate these questions, we use rubber mask principles in making quantitative comparisons of EKG waveforms from patient to patient. Typical results are shown in Fig. 17.

In Fig. 17a, single beat lead-1 QRS's are taken from two patients (John and Mike, same age group) and aligned with respect to each other as best possible in time and baseline (dc level). An 88 per cent mean-square error existed. Errors of this type generally run from 50 to 150 per cent.

The m.s.e. in QRS matching between the two patients was reduced substantially by comparing a synthetic projection from six leads of one patient (John) with lead 1 of the other patient (Mike). The result is shown in Fig. 17b. The mean-square error was 32.3 per cent.

Even closer matching was obtained by incorporating a small amount of time-stretching, limiting this always to within $\pm$ 10 per cent. The time durations of the salient EKG phenomena are frequent-

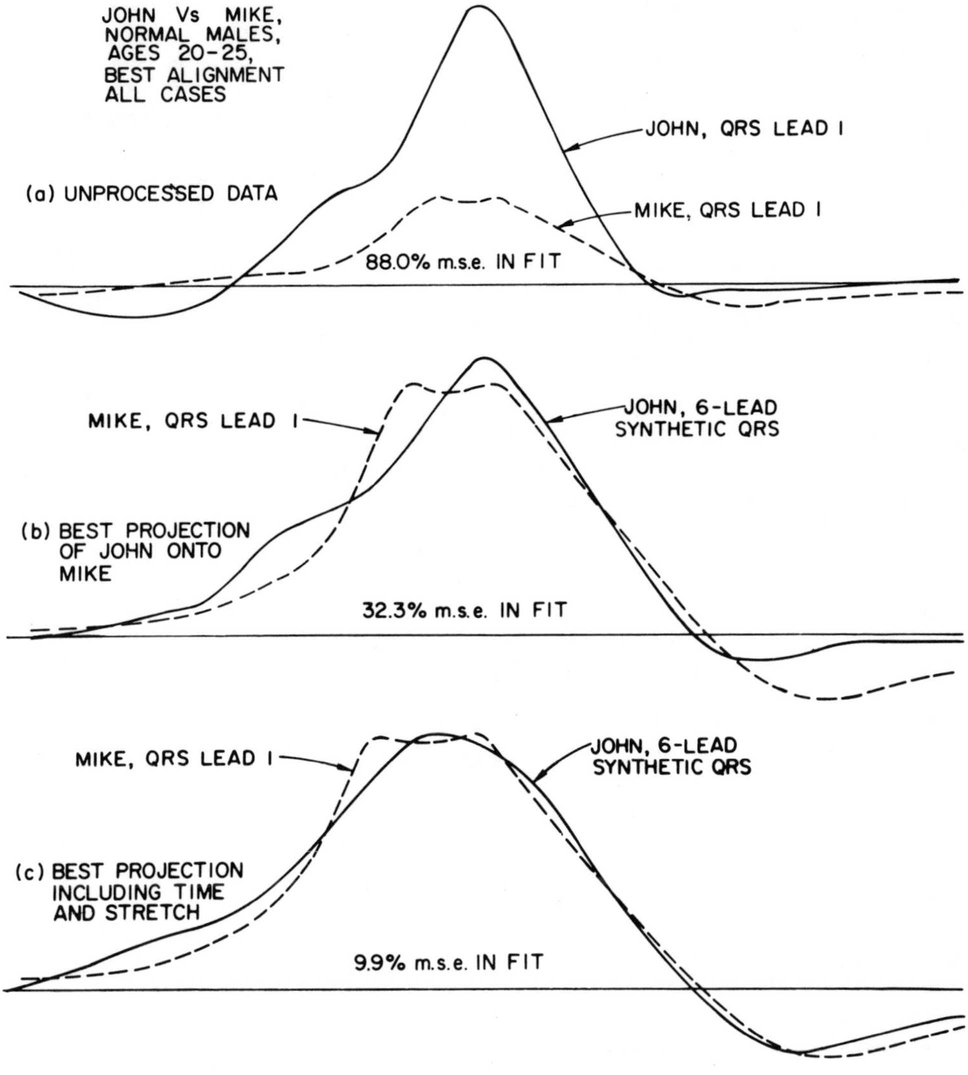

Fig. 17. Mike Lead 1 vs. Natural and Synthetic QRS John (Best Projection/Best Projection and Time Stretch).

ly used by cardiologists in the detection of disease. So the time base has not been stretched unduly. The effect of best time stretching along with best alignment and best projection is demonstrated in Fig. 17c. Here, the mean-square error was 9.9 per cent. It is significant to note that the mean-square error was reduced about ninefold in this example by projecting and stretching.

A tentative conclusion from this experiment and from many others like it is that there is far greater similarity in the intrinsic cardiac electrical activity than appears in comparing the EKG of one patient with that of another based on corresponding anatomical leads in accord with current clinical practice.

By the use of the rubber-mask principle as applied here to the QRS waveform, it is conceivable that a new set of $\pm 2\sigma$ bounds for normals could be derived that would be an order of magnitude tighter than those illustrated in Fig. 13.

Eliminating or reducing gross variations from QRS to QRS in the above described manner, the technique may place in evidence subtle local deviations in waveshape which may prove to be invaluable in diagnosis.

6. CONCLUSION

This paper has presented a set of examples in which the principle of rubber masks has been used in pattern measurement and analysis. Specific applications to the analysis of human chromosome images, chromatographic recordings, electrocardiogram waveforms, and electroencephalogram waveforms have been illustrated.

Previous methods of pattern analysis and recognition have in a very broad sense involved (a) template matching and (b) feature detection and classification. The rubber mask approach is based on template matching and it has many of the advantages of template matching. It does not require the measurement of special features, the design and choice of which tend to be very problem oriented and not very general. It makes direct use of incoming and stored pattern data. The rubber mask incorporates a promising method for overcoming the difficulties associated with template matching, since flexible templates can in many cases be tolerant to the imperfections and distortions that occur in natural pattern data.

One drawback of the rubber mask approach derives from the large computational requirements of the iterative stretching and fitting process. Further work is needed to simplify the algorithms in order for these techniques to enjoy wide applicability.

Acknowledgements-I would like to thank Mrs. MABEL ROCKWELL for editing this paper and for her many useful suggestions on the organization and technical presentation. Thanks also go to Mr. SIDHARTHA MAITRA who assisted in editing. So many people have significantly contributed to this research that their assistance is acknowledged by topic:

Chromosome analysis-The problem was posed by Dr. LEONARD HAYFLICK of the Department of Medical Microbiology, Stanford University Medical School, and Mr. MICHAEL LICCARDO. The experiments were done by Mr. ROBERT MELEN, assisted by students LAWRENCE MARPLE, FRANCISCO OLIVERA and NELSON CHANG. The author is grateful to Dr. KENNETH CASTLEMAN of the Jet Propulsion Laboratory for chromosome data, and for discussions of this subject with myself and my students. Helpful discussions with Dr. MORTIMER MENDELSOHN and Mr. LEE LITTLEPAGE of the Lawrence Radiation Laboratory are gratefully acknowledged. Dr. ANN MITCHELL, Department of Pediatrics, Stanford University, has given much insight into the banding-pattern approach and its problems.

Chromatogram analysis-The problem was posed by Drs. HOWARD SUSSMAN and KLARA EFRON of the Department of Pathology, Stanford University Medical School. In addition, discussions of this work were held with Dr. DAVID KORN, Chairman of Pathology, and his thinking has had significant effect on the technical approach. The experiments were done by students DAVID NEUHOFF, DOUGLAS BLAYNEY and STANLEY TENOLD. Deconvolution analysis was done by STEVE WERNECKE.

K-wave analysis-The problem of K-wave recognition was posed by Dr. VINCENT ZARCONE of the Department of Psychiatry of the Stanford University Medical School. His help and encouragement and that of Dr. WILLIAM DEMENT were invaluable to the progress of the experiments performed by Mr. WILLIAM GIBSON.

EKG analysis-The problem was posed by the author, and the experiments were performed by students SIDHARTHA MAITRA, STEVEN ZUCKER and DAVID WEST. The assistance of Drs. JOBST von der GROEBEN, RICHARD CROW and CHARLES WEAVER in making EKG data available for this study and for technical discussion of computer analysis of EKG waveforms over a period of years is gratefully acknowledged.

REFERENCES

1. Matched filter issue, IRE Trans. Information Theory IT-6, 309-417 (June 1960).
2. K. STEINBUCH and U.A.W. PISKE,"Learning matrices and their applications,"IEEE Trans. Electronic Computers EC-12, 846-862 (1963).

3. W.H. HIGHLEYMAN, "Linear decision functions, with application to pattern recognition," Proc. IRE 50, 1501-1514 (1962).
4. J.S. KOFORD and G.F. GRONER, "The use of an adaptive threshold element to design a linear optimal pattern classifier," IEEE Trans. Information Theory IT-12, 42-50 (1966).
5. T.M. COVER and P.E. HART, "Nearest neighbor pattern classification," IEEE Trans. Information Theory IT-13, 21-27 (1967).
6. G.L. FISCHER, JR., D.K. POLLOCK, B. RADACK and M.E. STEVENS (Eds.), Optical Character Recognition. Spartan Books, Washington, D.C. (1962).
7. L.N. KANAL (Ed.), Pattern Recognition. Thompson, Washington, D.C. (1968).
8. C.N. LIU and G.L. SHELTON, JR., "An experimental investigation of a mixed-font print recognition system," IEEE Trans. Computers EC-15 (6), 916-925 (1966).
9. Denver Conference. "A proposed standard system of nomenclature of human mitotic chromosomes," Ann. Human Genetics 24, 319-324 (1960)
10. T. CASPERSSON, L. ZECH, C. JOHANSSON and E.J. MODEST, "Identification of human chromosomes by DNA-binding fluorescent agents," Chromosoma 30, 215-227. (1970).
11. T. CASPERSSON and L. ZECH, "Chromosome identification by fluorescence." Hospital Practice, 51-62 (September 1972).
12. M.E. DRETS and M.W. SHAW, "Specific banding patterns of human chromosomes,"Proc. Nat. Acad. Sci. U.S.A. 68 (9), 2073-2077 (1971).
13. T. FLEISCHMANN, T. GUSTAFSSON, C.H. HAKANSSON and A. LEVAN, "Computer-display of the chromosomal fluorescence pattern," Heriditas 68, 325-328 (1971).
14. P.B. HAMILTON, "Ion exchange chromatography of amino acids," Anal. Chem. 35 (13), 2055-2064 (1963).
15. H.M. GLADNEY, B.F. DOWDEN and J.D. SWALEN, "Computer-assisted gas-liquid chromatography,"Anal. Chem. 41 (7), 823-828 (1969).
16. C.D. SCOTT, D.D. CHILCOTE and W. WILSON PITT, JR., "Method for resolving and measuring overlapping chromatographic peaks by use of an on-line computer with limited storage capacity," Clin. Chem. 16 (8), 637-642 (1970).
17. H. DAVIS, P.H. DAVIS, A.L. LOOMIS, E.N. HARVEY and G. HOBART, "Electrical reactions of the human brain to auditory stimulation during sleep," J. Neurophysiol. 2, 500-514 (1939).
18. L.C. JOHNSON and W.E. KARPAN, "Autonomic correlates of the spontaneous K-complex," Psychophysiology 4 (4), 444-452 (1968).
19. J.F. SASSIN and L.C. JOHNSON, "Body motility during sleep and its relation to the K-complex," Exp. Neurol. 22 (1), 133-144 (1968).
20. G. BREMER, "Detection of the K-complex in electroencephalograms," A thesis presented to the Graduate Council of the University of Florida, University of Florida (1970).
21. G.F. BREMER, J.R. SMITH and I. KARACAN, "Detection of the K-complex in electroencephalograms," IEEE Trans. Bio-med. Engng

17, 314-323 (1970)

22. J.R. SMITH and I. KARACAN, "EEG sleep stage scoring by an automatic hybrid system," Electroenceph. Clin. Neurophysiol, 31, 231-237 (1971).
23. Grant's Clinical Electrocardiography: The Spatial Vector Approach, Second edition revised by J.R. BECKWITH, McGraw-Hill, New York (1970).
24. L.A. GEDDES and L.E. BAKER, Principles of Applied Biomedical Instrumentation. Wiley, New York (1968).
25. H.C. BURGER, Heart and Vector: Physical Basis of Electrocardiography. Philips Technical Library, Eindhoven, Netherlands (1968).
26. E. FRANK, "An accurate, clinically practical system for spatial vectorcardiography," Circulation 13, 737-749 (1956).
27. C.S. WEAVER, J. von der GROEBEN and H.G. GLAZE, "Collecting and processing vector electrocardiograms," Stanford Electronics Laboratories, SU-SEL-66-122 (December 1966).
28. C.S. WEAVER, J. von der GROEBEN, P.E. MANTEY, C.A. COLE, JR., J.W. FITZGERALD and R.W. LAWRENCE, "Digital filtering with applications to electrocardiogram processing," IEEE Trans. Audio and Electroacoustics AU-16 (3), 350-391 (1968).

SELECTED READING

Chromosome Image and Banding Pattern Analysis

1. D. RUTOVITZ, "Centromere finding: Some shape descriptors for small chromosome outlines," Machine Intelligence 5, 435-562 (1970).
2. C.J. HILDITCH, "A system of automatic chromosome analysis," Automatic Interpretation and Classification of Images, A. GRASSELLI (ed.) Academic, New York (1969).
3. S. STONE, L. LITTLEPAGE and B. CLEGG, Second report on the chromosome scanning program at the Lawrence Radiation Laboratory, Pattern Recognition Studies S.P.I.E. Seminar Proc., vol. 18 (1969).
4. F. RUDDLE, S. SMITH, R. LEDLEY and M. BELSON, "Replication-precision study of manual and automatic chromosome analysis," Ann. N.Y. Acad. Sci. 157 (art 1), 400-423 (1969).
5. K. PATON, "Automatic chromosome identification by the maximum likelihood method," Ann. Human Genetics 33, 174-184 (1969).
6. P. NEURATH and K. ENSLEIN, "Human chromosome analysis as computed from arm lengths measurements," Cytogenetics 8, 337-354 (1969).
7. M. MENDELSOHN, D. HUNGERFORD, B. MAYALL, B. PERRY, T. CONWAY and J. PREWITT, "Computer-oriented analysis of human chromosomes-II." Ann. N.Y. Acad. Sci. 157 (art 1), 376-392 (1969).
8. C.J. HILDITCH and D. RUTOVITZ, "Chromosome recognition,"

Ann. N.Y. Acad. Sci. 157 (art 1), 339-364 (1969).

9. H. FREY, "An interactive computer program for chromosome analysis," Computers and Biomedical Research 2, 274-290 (1969).

10. J.W. BUTLER, M.K. BUTLER and B. MARCZYNSKA, "Automatic analysis of 835 Mormoset Spreads,"Ann. N.Y. Acad. Sci. 157 (art 1), 424-437 (1969).

11. G. GALLUS, N. MONTANARO and G. MOCCACARO, "A problem of pattern recognition in the automatic analysis of chromosomes; locating the centromere," Computers and Biomedical Research 2, 187-197 (1968).

12. Chicago Conference: Standardization in Human Cytogenetics. "Birth Defects: Original Article Series, II:2." The National Foundation, New York (196).

13. P. NEURATH, B. BALOUZIAN, T. WARMS, R. SERBAGI and A. FALEK, "Human chromosome analysis-an optical pattern recognition problem," Ann. N.Y. Acad. Sci. 128, 1013-1028 (1966).

14. L. PENROSE et al., "The London Conference on the normal human karyotype," Cytogenetics 2, 264-268 (1963).

15. M.A. BENDER and M.A. KASTENBAUM, "Statistical analysis of the normal human karyotype," Am. J. Human Genetics 21 (4), 322-351 (1969).

16. C.W. GILBERT and S. MULDAL, "Measurement and computer system for karyotyping human and other cells," Nature N.B. 230, 203-207 (April 1971).

17. A. KLINGER, A. KOCHMAN and N. ALEXANDRIDIS, "Computer analysis of chromosome patterns: Feature encoding for flexible decision making," IEEE Trans. Computers C-20 (9), 1014-1022 (1971).

18. M. MENDELSOHN et al., "Computer oriented analysis of human chromosomes I-photometric analysis of DNA content," Cytogenetics 5, 223-242 (1966).

19. S.R. PATIL, S. MERRICK and H.A. LUBS, "Identification of each human chromosome with a modified giemsa strain," Science 173, 821-822 (1971).

20. W. SCHNEDL, "Banding patterns of human chromosomes," Nature N.B. 233, 93-94 (15 September 1971).

21. T. CASPERSSON, L. ZECH, C. JOHANSSON and E.J. MODEST, "Identification of human chromosomes by DNA-binding fluorescing agents," Chromosoma 30, 215 (1970).

22. T. CASPERSSON, L. ZECH, and C. JOHANSSON, "Analysis of the human metaphase chromosome set by aid of DNA-binding fluorescent agents," Expl. Cell Res. 62, 490 (1970).

23. A. MÖLLER, H. NILSSON, T. CASPERSSON and G. LOMAKKA, "Identification of human chromosome regions by aid of computerized pattern analysis," Expl. Cell Res. 70, 475 (1972).

24. T. CASPERSSON, G. GAHRTON, J. LINDSTEN and L. ZECH, "identification of the Philadelphia chromosome as a number 22 by quinacrine mustard fluorescence analysis," Expl. Cell Res. 63, 238 (1970).

25. T.CASPERSSON, G. LOMAKKA and L. ZECH, "The 24 fluorescence patterns of the human metaphase chromosomes-distinguishing

characters and variability," Hereditas 67, 89 (1972).
26. T. CASPERSSON, G. LOMAKKA and A. MOLLER, "Computerized chromosome identification by aid of the quinacrine mustard fluorescence technique," Hereditas 67, 103 (1971)
27. H.J. EVANS, K.E. BUCKTON and A. T. SUMNER, "Cytological mapping of human chromosomes: Results obtained with quinacrine fluorescence and the acetic-saline-giemsa techniques," Chromosoma 35, 310-325 (1971).
28. W. SCHNEDL, "Analysis of the human karyotype using a reassociation technique," Chromosoma 34, 448-454 (1971).
29. G. MANOLOV, Y. MANOLOVA and A. LEVAN, "The fluorescence pattern of the human karyotype,"Hereditas 69, 273-286 (1971).
30. W. UNAKUL, R.T. JOHNSON, P.N. RAO and T.C. HSU, "Giemsa banding of interphase HeLa chromosomes," J. Cell Biol. 55, 264a (1972).
31. J.G. GALL and M. L. PARDUE, "Nucleic acid hybridization in cytological preparations," Methods in Enzymology, Nucleic Acids Part, D.L. GROSSMAN and K. MOLDAVE (Eds.) Vol. 21, pp. 470-480 (1971).
32. F.E. ARRIGHI and T.C. HSU, "Localization of heterochromatin in human chromosomes," Cytogenetics 10, 81-86 (1971).
33. B. DUTRILLAUX and J. LEJEUNE, C.R. Acad. Sci., Paris 272, 2638-2640 (1971).
34. M.E. DRETS and M.W. SHAW, "Specific banding patterns of human chromosomes," Proc. Nat. Acad. Sci. U.S.A. 68 (9), 2073-2077 (1971).
35. M. SEABRIGHT, "A rapid banding technique for human chromosomes." Lancet 2, 971-972 (1971).
36. Section of Cell Biology, M.D. Anderson Hospital and Tumor Institute, Mammalian Chromosomes Newsl. 13, 21-47 (1972).

Analysis of Spectrograms and Chromatograms

1. D.G. LUENBERGER, "Resolution of mass spectrometer data," Stanford Electronics Laboratories, Stanford, Calif., Tech. Report. SEL-64-129 (TR 6451-1) (November 1964)
2. L.R. SYNDER, "A rapid approach to selecting the best experimental conditions for high-speed liquid column chromatography," Part 1, J. Chromatog. Sci. 10, 200-212 (1972).

Computer Analysis of EKG Waveforms

1. R. HELM, "An accurate lead system for spatial vectorcardiography," Am. Heart Jl. 53, 415 (1957).
2. H.R. WARNER, A.F. TORONTO, L.G. VEASEY and R. STEPHENSON, "A mathematical approach to medical diagnosis, application to congenital heart disease,"J.A.M.A. 3, No. 177, 177-183 (1961).

3. C.A. STEINBERG, S. ABRAHAM and C.A. CACERES, "Pattern recognition in the clinical electrocardiogram," IRE Trans. Bio-medical Electronics BME-9, 23-30 (1962).
4. L. STARK, M. OKAJIMA and G.H. WHIPPLE, "Computer pattern recognition techniques; electrocardiographic diagnosis,"Commun. ACM 5, 527-532 (10 October 1962).
5. H.V. PIPBERGER, "Use of computers in interpretation of electrocardiograms,"Circulation Res. 11, 555 (1962).
6. T.Y. YOUNG and W.H. HUGGINS, "Computer analysis of electrocardiograms using a linear regression technique," IEEE Trans. Bio-medical Engineering BME-11, 60-67 (July 1964).
7. D.F. SPECHT, "Vectorcardiographic diagnosis using the polynominal discriminant method of pattern recognition," IEEE Trans. Bio-medical Engineering BME-14 (2), 90-95 (1967).
8. J.P. BROWN, D.B. FRANCIS, T.W. CALVERT and R.L. LONGINI, "Compensating the VCG for anatomic variations of individuals," Proc Annual Conf. on Engineering in Biology and Medicine, Boston, Mass. (November 1967).
9. J.P. BROWN, T.W. CALVERT, R.L. LONGINI and E.W. HECKERT,"Normalizing the VCG to facilitate diagnosis,"Proc. Annual Conf. on Engineering in Biology and Medicine, Houston Texas (1968).
10. A.A. LANGER, R.L. LONGINI and E.W. HECKERT, "Body compensator system for VCG's,"Proc. 8th Int. Conf. on Medical and Biological Engineering, Chicago, Illinois (1969).
11. R. GAMBOA, J.D. KLINGEMAN and H.V. PIPBERGER, "Computer diagnosis of biventricular hypertrophy from orthogonal electrocardiograms,"Circulation 39, 72-82 (January 1969).
12. D.B. GESELOWITZ and O. H. SCHMITT, "Electrocardiography," Bio-medical Engineering, H.P. SCHWANN (ed) McGraw-Hill, New York (1969).
13. J. von der GROEBEN, J.G. TOOLE and C.S. WEAVER, "Vectorcardiographic analysis with the aid of a small digital computer," Actuelle Probleme der Vektorkardiographie, R. WENGER (ed.) GEORG THIEME, Stuttgart (1968).
14. W.P. HOLSINGER, K.M. KEMPUER and M.H. MILLER, "A QRS preprocessor based on digital differentiation," IEEE Trans. Bio-medical Engineering BME-18 (3), 212-217 (May 1971).
15. D.B. GESELOWITZ, "Use of the multipole expansion to extract significant features of the surface electrocardiogram," IEEE Trans. Computers C-20 (9), 1086-1089 (September 1971).
16. R. PLONSEY, "Capability and limitations of electrocardiography and magnetocardiography," IEEE Trans. Bio-medical Engineering BME-19 (3), 239-244 (May 1972).

Other Work with Flexible Templates and Parametrized Descriptions of Patterns

1. R.G. CASEY and G. PURDY, "Moment normalization of handprinted characters," IBM Research, Yorktown Heights, New York: RC 2666

(14 October 1969).
2. H.J. BREMERMANN, "Pattern recognition by means of deformable prototypes," a talk presented at the 1971 Workshop on Pattern Recognition, Anaheim, Calif., October 27, 1971; abstracted on p. 547, IEEE Trans. Systems, Man and Cybernetics SMC-2 (4), (September 1972).
3. C.T. ZAHN and R.Z. ROSKIES, "Fourier descriptors for plane closed curves," IEEE Trans. Computers C-21, 269-281 (1972).
4. T. CASPERSSON and L. ZECH, "Chromosome identification by fluorescence," Hospital Practice 7 (9), 51-62 (1972).

General Bio-Medical Pattern Recognition

1. R.S. LEDLEY and L.S. ROTOLO, "Application of pattern recognition to biomedical problems," Automatic Interpretation and Classification of Images, A. GRASSELLI (ed.), pp. 323-362, Academic Press, New York (1969).
2. C.J. HILDITCH, "A system of automatic chromosome analysis," Ibid 363-390, (1969).
3. M. PFEILER, "Image transmission and image processing in radiology," Ibid 399-416 (1969).
4. R.S. LEDLEY, "Automatic pattern recognition for clinical medicine," Proc. IEEE 57 (11), 2017-2035 (1969).
5. E.E. GOSE, "Introduction to biological and mechanical pattern recognition," Methodologies of Pattern Recognition, S. WATANABE (ed.) pp. 203-253. Academic Press, New York (1969).
6. E.E. GOSE, J.W. BACUS and L.V. ACKERMAN, "A comparison of some computer-measured and human-measured pattern recognition properties," J. Cybernetics 1 (4), 68-74 (1971).
7. J.W. BACUS and E.E. GOSE, "Leukocyte pattern recognition," IEEE Trans. Systems, Man, and Cybernetics SMC-2 (4) 513-525 (1972).
8. R. LEDLEY, F. RUDDLE, J. WILSON, M. BELSON and J. ALBARRAN, "The case of touching and overlapping chromosomes" Pictorial Pattern Recognition, G. CHENG et al. (eds.) pp. 87-97. Thompson (1968).
9. R. LEDLEY, "Automatic pattern recognition for clinical medicine," Proc. IEEE 57, 2017-2035 (1969).
10. K. PATON, "An automatic method for finding metaphase spreads," Pictorial Pattern Recognition, G. CHENG et al. (eds.) pp. 135-146. Thompson (1968).
11. R. LEDLEY, M. LEGATOR and J. WILSON, "Automatic determination of mitotic index," Pictorial Pattern Recognition, G. CHENG et al. (eds.) pp. 99-103. Thompson (1968).
12. J.W. BUTLER, M.K. BUTLER and A. STROUD, "Automatic classification of chromosomes-11, and Automatic classification of chromosomes-111,"Data Acquisition and Processing in Biology and Medicine, K. ENSLEIN (ed.) pp. 45-57 and pp. 21-37. Pergamon, Oxford (1965 and 1968).

13. R. LEDLEY, "FIDAC: Film input to digital automatic computer and associated syntax directed pattern recognition programming system," Optical and Electro-Optical Information Processing, J. TIPPETT et al. (eds.) Chapter 33, pp. 591-612. MIT Press (1965).
14. R. LEDLEY, "High speed automatic analysis of biomedical pictures,"Science 146, 216-223 (9 October 1964).
15. K. ENSLEIN and P. W. NEURATH, "Augmented stepwise discriminent analysis applied to two classification problems in the biomedical field," Computers and Biomedical Research 2, 568-581 (1969).
16. D. RUTOVITZ, "Pattern recognition," Royal Statistical Society 12a, 504-530, series A (1966).
17. R. STEFANELLI and A. ROSENFELD, "Some parallel thinning algorithms for digital pictures," J. ACM 18 (2) 255-264 (1971).
18. J. HILDITCH and D. RUTOVITZ, "Chromosome recognition," Ann. N.Y. Acad. Sci. 157, 339-364 (1969).
19. M. INGRAM, P.E. NORGREN and K. PRESTON, JR., "Automatic differentiation of white blood cells," Image Processing in Biological Sciences, D.M. RAMSEY (ed.) Univ. Calif. Press, Berkeley, Calif. (1968).
20. M. INGRAM and K. PRESTON, JR., "Automatic analysis of blood cells," Scient. Am. 223, 72-82 (1970).
21. M.L. MENDELSOHN, B.H. MAYALL, J.M.S. PREWITT, R.C. BOSTROM and W.G. HOLCOMB, "Digital transformations and computer analysis of microscopic images," Advances in Optical and Electron Microscopy, V.E. COSLETT (ed.) pp. 77-150. Academic, New York (1968).
22. J.M.S. PREWITT and M.L. MENDELSOHN, "A general approach to image analysis by parameter extraction," Proc. Computers in Radiology, Chicago (1966).
23. "The analysis of cell images,"Ann. N.Y. Acad. Sci. 128, 1035-1053 (1966).
24. T. GOLAB, R.S. LEDLEY and L.S. ROTOLO, "FIDAC-Film input to digital computer," Pattern Recognition 3, 123 (1971).

THE "RUBBER-MASK" TECHNIQUE-II

PATTERN STORAGE AND RECOGNITION

Bernard Widrow

Stanford Electronics Laboratories

Stanford, California 94305, U.S.A.

ABSTRACT

This paper briefly summarizes much of the work in pattern recognition to date, and relates the rubber mask technique to previous work. A scheme for incorporating flexible-mask methods into a proposed pattern recognition and memory system is presented. A discussion based on some facts and on some conjecture of the human eye/brain system and how it recognizes patterns, possibly by flexible matching, is also presented.

INTRODUCTION

THE PREVIOUS paper (1) presented a set of examples in which the principle of rubber masks was used in pattern measurement and analysis. The purpose of the present paper is to show how this principle might be used in pattern-recognition and information-storage systems. A further purpose is to speculate on how the functioning of a rubber-mask pattern-recognition system could compare with the functioning of the natural eye/brain system.

Many of the papers and books on pattern recognition that have appeared in the literature during the past decade have begun with a description of the general pattern-recognition system shown in Fig. 1. This representation has proven to be sufficiently general to cover almost any recognition scheme that has been devised.(2)

The preprocessor in Fig. 1 derives its input signal (usually a vector) from an array of photocells, or equivalently from a

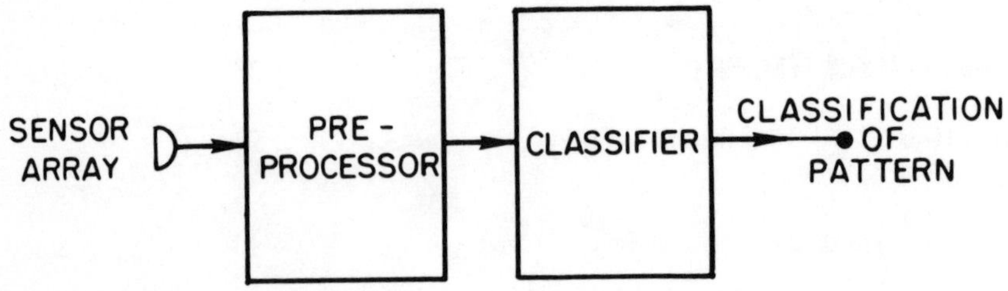

PRE-PROCESSOR:

- STRAIGHT THROUGH (e.g. ADALINE/MADALINE)
- RANDOM SCRAMBLE (e.g. PERCEPTRON)
- FEATURE DETECTOR (e.g. IN FINGERPRINT IDENTIFICATION: RIDGE COUNT, LOOP DETECT, ETC)
- PROPERTY MEASUREMENT (e.g. INTEGRAL GEOM.)
- PICTURE PARSER
- HEURISTIC FEATURES

⋮

CLASSIFIER:

- HYPERPLANES (e.g. MADALINE)
- POLYNOMIAL DISCRIMINANT (SPECHT)
- MATCHED FILTERS (e.g. LEARNING MATRIX)
- NEAREST NEIGHBOR RULES (COVER ET AL)
- CLUSTERING (ISODATA, ETC.)
- PROPERTY LIST
- PICTURE GRAMMAR
- FEATURE LIST

⋮

Fig. 1. Typical Forms . . Classical System.

scanning image dissector for image processing; or from a microphone for speech recognition; or from some other sensory source. The preprocessor outputs go to the classifier, which may be a trainable adaptive system or which may be some form of signal separator whose design is based on a priori knowledge. The classifier outputs are coded representations of the pattern classifications.

A representative (but non-exhaustive) list of schemes for preprocessing is presented in Fig. 1. Among the preprocessing schemes, the straight-through scheme involves no preprocessing at all. The random-scramble scheme performs a type of preprocessing which is generally designed for specific expected pattern features but does perform some measure of non-specific preprocessing. The feature-detection schemes are designed to look for and enumerate or quantify certain salient pattern "landmarks"; while property-measurement schemes measure pattern parameters (by statistical means or otherwise) which are generally insensitive to size, rotation, and translation. Heuristic feature detection and picture parsing are primarily concerned with pattern segmentation and context (spatial or temporal) or with the way in which the picture parts are interrelated or interspersed.

Also shown in Fig. 1 is a representative list of classifier schemes that have appeared in the literature. Hyperplane and matched-filter classifiers are in the same family, since they segment the pattern vector space with piecewise hyperplanar boundaries which can articulate as required to approximate higher-degree surfaces. All of these schemes lend themselves to adaptive or learning procedures for adjusting their parameters, with the learning process based on sets of identified reference or "training" patterns. Nearest-neighbor rules are easily implemented as learning schemes; as more training vectors are obtained, more identified points will be available in pattern-vector space with which to compare unknown vectors to be classified. Nearest-neighbor rules classify with boundaries that are also piecewise hyperplanar once the training data are assimilated. Clustering techniques involve a form of learning with unidentified input vectors. Pattern vectors that are in some sense close to one another and at the same time distant from all other vectors are clustered or associated. There are many ways to accomplish this automatically. The result is a set of vectors that are identified with respect to one another. The property-list classifier also forms separating boundaries in a vector space; the vector components are taken or encoded from measured pattern properties. Typical properties might be length, width, mass, color, or possibly the types of measures obtained by integral geometry. The feature-list classifier is a check list with which pattern features can be compared (as to their presence or absence) for classification purposes. Rules are established for the classifier tolerance to missing features and to "false-alarm" features. The picture-grammar classifier uses the feature-list principle in essence. In addition it uses contextual information derived from the picture parser.

The preprocessor and classifier systems described above, whether adaptive or non-adaptive, map pattern vectors into classification-space vectors, after which they implement boundaries for final pattern separation. The preprocessor/classifier scheme represents what we might call the classical approach to pattern recognition.

The next section considers the basic nature of the overall pattern-recognition problem, and attempts to present a new proposal for its eventual solution.

2. A RUBBER -MASK APPROACH TO PATTERN RECOGNITION

The fundamental problem in pattern recognition may be illustrated by the following example:

A person is now standing before you and you're trying to decide whether you've seen that person's face before. The face might have been seen previously with a slightly different shadow, with a different perspective, at a different distance, or with the head rotated at a somewhat different angle. With regard to all these aspects, the face has never been seen before quite as it is at present. Mentally, what one seems to do is to take previously encountered images and by distorting them in some reasonable way, determine whether a projection could be found that would give a good match between the remembered image and the actual image currently seen.

It is evident from the above example that the problem consists in recognizing the essential pattern that underlies and is common to the various views of a particular face that one is able to remember. This problem, the human eye/brain system seems to solve quite well. It is this problem that the rubber-mask approach addresses directly. R.L. GREGORY in his book Eye and Brain [3] points out that:

> ...perception is not determined simply by the stimulus patterns; rather it is a dynamic searching for the best interpretation of the available data...

In the field of automatic pattern recognition, classical methods have been tried with some success on specific problems. But no comprehensive approach has developed that has the possibility of solving general problems such as facial recognition, handwriting recognition, speech recognition, and others, in a manner analogous to that of the human eye/brain system in level

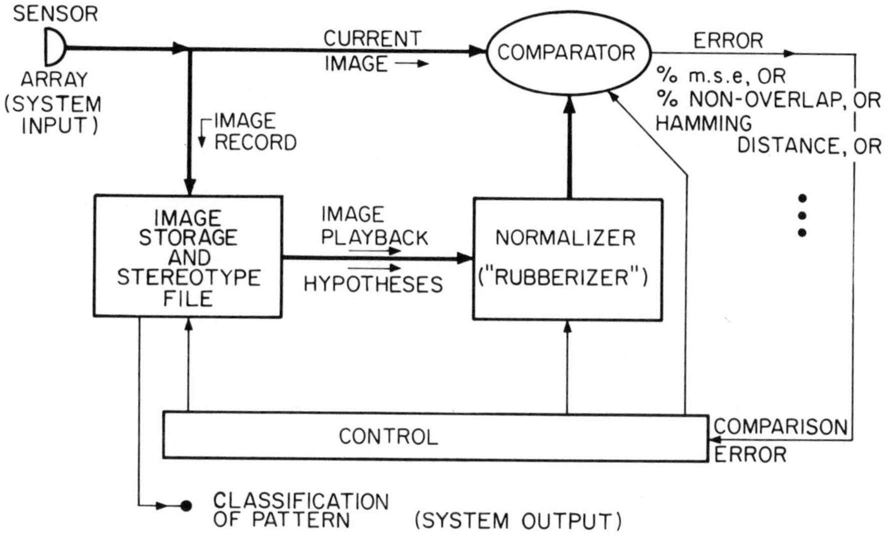

Fig. 2. "Rubber Mask" Pattern Recognition and Patter Memory System.

of performance and means of function.

It is proposed that automatic machinery be built (or programmed) to perform pattern recognition tasks in a natural way using the principle of rubber masks. The system diagrammed in Fig. 2 could represent a significant step in this direction. The sensor array feeds current image signal vectors into a comparator. Everything from the sensor array also goes into the image storage and stereotype file. From this store, patterns are fed through a normalizer ("rubberizer") and then compared against the current image. What is involved is a hypothesis-testing process. "Hypotheses" from the stereotype file are stretched and compared against unidentified input data. Each stereotype can be regarded as a hypothesis to be adjusted by the rubberizer for best fit to the current input image. The hypothesis that gives the best fit and/or requires the least stretching to achieve best fit is accepted. First and second runners-up would ususally be recalled to resolve conflicts.

The functions of the various boxes of Fig. 2 are represented in Fig. 3 as follows: The comparator compares, fixates, windows, and tests hypotheses. By "to fixate" we mean to focus the center of the window (the field of view) on a particular portion of the image. The rubberizer rotates, translates, scales, distorts, shades, projects in perspective, outlines, edge-detects, etc.

The stereotype file contains short-term and long-term memory, records image-vector inputs, and plays back recorded images. The short-term memory stores distorted images during the hypothesis testing phase; the long-term memory stores input images. Image features are used for associative addressing for record and playback. The control box stimulates the whole system to make it go.

Since new templates may be formed and recorded from current input data as the system is exposed to its environment, a form of learning associated with the pattern-recognition process is possible. That is to say, the <u>intrinsic</u> pattern of the function being studied may gradually become apparent as template formation progresses. An example is found in the EEG and EKG studies in Ref. (1).

The proposed rubber-mask pattern-recognition system is intended to represent a practical approach to pattern recognition and pattern memory based on template storage, template stretching, and template matching. The system is organized to perform sequences of hypothesis tests, and to recognize and measure pattern images. Each stored template is a hypothesis available for comparison with the current data. Alternate hypotheses could be ranked in terms of closeness of fit and extent of stretch required to achieve best fit.

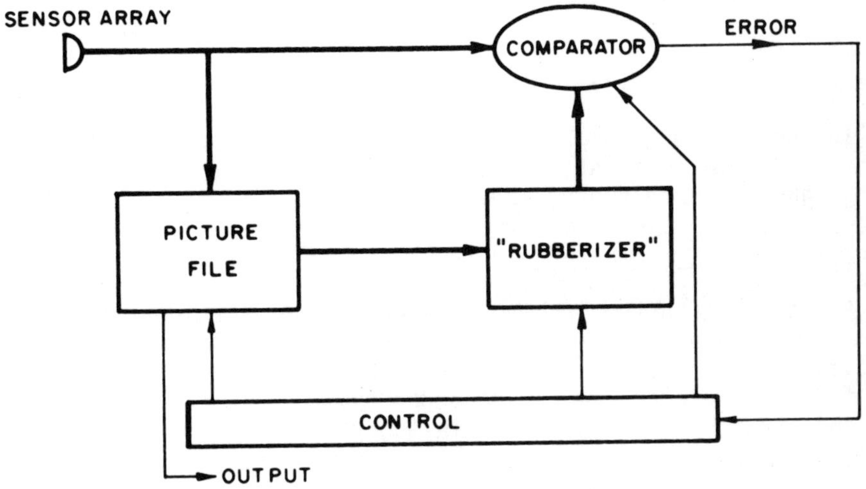

COMPARATOR : COMPARES, FIXATES, WINDOWS, TESTS HYPOTHESES.

NORMALIZER ("RUBBERIZER") : ROTATES, TRANSLATES, SCALES, DISTORTS, SHADES, PROJECTS IN PERSPECTIVE, OUTLINES, EDGE-DETECTS,···

PICTURE FILE : CONTAINS SHORT-TERM/LONG-TERM MEMORY, RECORDS PICTURE INPUTS, PLAYS BACK RECORDED IMAGES; COULD USE IMAGE FEATURES FOR ASSOCIATIVE ADDRESSING FOR RECORD AND PLAYBACK.

CONTROL : STIMULATES PICTURE FILE TO PRODUCE IMAGE HYPOTHESES; ACTUATES RUBBERIZER TO OPTIMIZE IMAGE FIT; ACTUATES COMPARATOR TO TRY VARIOUS FIXATIONS AND WINDOWS; IMPLEMENTS CLASSIFICATION RULE MINIMIZING ERROR.

Fig. 3. "Rubber Mask" System Functions.

3. SOLVING RECOGNITION PROBLEMS WITH THE RUBBER-MASK SYSTEM

In this section, we illustrate the functioning of the system of Fig. 3 by describing how it might address a practical problem, that of facial recognition. Two approaches are possible. One is based on the use of natural templates, i.e. templates derived from observation of nature. The other is based on the use of man-made (artificial) standard templates.

(a) Natural Templates

In the system of Fig. 2, many facial templates could be stored in the file. These might consist of actual photographs, digitized photographs, scanned photographs as in closed-circuit TV, or coded photographs. Whatever the form in which each image is stored, it would be derived from nature in some manner.

For a start on the problem of recognizing an unidentified image of an entire face, the eyes, sub-images, could be sought out and identified by template matching, using the rubberizer to obtain best fit to the remembered or recorded eyes by aligning, synthesizing, and stretching, as was done in the case of matching EKG waveforms in the previous paper.[1] For general usage in the system of Fig. 2, many eye templates from different people, taken under different lighting conditions, different perspectives, different amounts of rotation and translation, could be stored in the file. These would be direct images from nature, before any stretching or other distortion had been performed by the rubberizer to fit the remembered (or otherwise available) sample image. Separation between the eyes might be included in the matching process, i.e. an eye template could consist of a pair of eyes in various spatial relationships.

Such a procedure could be very powerful. For example, in the initial step a defocused image could be compared with a defocused template for a first crude elimination of totally unsuitable templates (pop eyes vs squinted eyes; almond-shaped eyes vs round eyes; and so on.) Allowing the comparison to sharpen after the first-stage elimination, the remaining candidate eye templates could be stretched and aligned to best-fit the data images-as a result of which, several likely individuals might well be selected from among hundreds of "possibles."

Next, the study might proceed to another important facial area-the mouth. Following the same procedure, several persons out of hundreds could possibly be selected as corresponding to an unidentified image on the basis of the mouth and its characteristic expression.

After tying down some good candidate examples of the eye and the mouth, the investigation might proceed to consideration of the nose, the ears, the hairline, the chin, and so on. True, the most likely sub-images as selected in each category are noted in terms of one or more candidate individuals who possess these sub-images; but the selected sub-images are ultimately viewed in combination. The entire unknown image is compared and matched to entire stored facial images made up of the selected sub-images arranged in various dimensional relationships to determine which candidate gives the best match. For example, when a candidate pair of eyes has been selected, this pair must be tried in an assembly of sub-images to simulate the remembered face.

A simple segmentation algorithm for the system of Fig. 2 having fairly general applicability in image analysis and recognition would look for regions within the image which contain significant fractions of the total area-regions which are rich in optical texture and which are isolated and self-contained. Such an algorithm would automatically choose the eyes, the mouth, the nose, and so on for isolation and study in facial analysis; and would eventually assemble these in various relationships for final recognition.

(b) Artificial Templates

Another approach to facial recognition by means of rubber masks would utilize stored artificial templates. For eye recognition, a small number of man-made stereotypes could be stored, corresponding to a reasonable range of directly-viewed eye shapes(as would be used in a police identification kit).(4) The eyes of an unknown facial image could be given a numerical classification in terms of the identification number of the eye template that best fits when best-stretched or projected. Separations between the eyes of a pair could be included in the analysis. Many people out of hundreds would have the same eye classification number; but the inclusion of eye separation might reduce the number of candidates.

If a small number of artificial stereotypes were stored for each sub-image, then the sub-images of any unknown facial image of interest could be assigned numerical identifications from flexibly fitting the stored artificial templates. From the point of view of data reduction, each face of interest could be approximately but efficiently stored, not with an actual picture, but with a set of numbers listed by sub-image. It should be noted that relative positions of the salient portions of the facial image-not only spacing between the eyes, but also distances from the eyes to nose and mouth, etc.-are also significant facial features that could be measured by rubber-mask techniques.

From the point of view of pattern recognition, an unknown

face would be recognized by comparing the list of sub-images and their assembly dimensions for the unknown face against the corresponding attributes of previously encountered known facial images.

4. IMPLEMENTATION OF RUBBER-MASK PATTERN-STORAGE AND RECOGNITION SYSTEM

Development of the complete system of Fig. 3, involving derivation of algorithms and their mathematical properties, is a long term goal of this research. Work is under way on the rubberizer and on the comparator. This work is being done in the context of the problems described in the previous paper, namely chromosome analysis, chromatogram measurement, and EEG and EKG analyses.

The image and stereotype file is a very complicated operation and little has been accomplished toward its realization so far. In the examples of the previous paper, the stereotype storage function has been very simple. In chromosome analysis, one merely stores the Denver Standards. In chromatogram analysis, one stores (or generates) a gaussian function. In EKG analysis, one would ultimately store a relatively small set of normal waveforms and also small sets of waveforms to typify each disease state of interest. None of these storage functions compares, however, to that which develops in facial recognition using perhaps thousands of natural faces (or their components) as templates.

All of the functions of the system of Fig. 3 are mechanistic and should eventually be realizable. It is clear, however, that some of these operations require large amounts of computer time in their implementation if present general-purpose equipment is used. In the future, it may in fact be necessary to develop completely new kinds of computing machinery to perform such algorithms.

5. HYPOTHESIS TESTING AND PATTERN MATCHING IN THE EYE/BRAIN SYSTEM

We have implied that emulating (even in the most remote degree) the pattern-recognition performance of the human eye/brain system is an ultimate goal of the rubber-mask system. The latter involves primarily hypothesis fitting (stretching, adjusting parameters), hypothesis testing (checking the fit), and trying alternative hypotheses. Such a process of sequential hypothesis testing seems to be a very natural one; it appears to be central to the functioning of the human mind. The ability to match one intricate pattern to another in spite of limited but significant rotation, translation, scale change, projection distortion, etc., is a remarkable attribute of the human eye/brain system.

That when the eye sees an image, the mind forms alternative hypotheses as to the nature of the perceived object is readily demonstrated when viewing one of the many classic optical-illusion images that have appeared in the literature. One famous example is the Necker cube.[3] One can stare at the picture of this seeming cube of wire and see it clearly in a certain spacial context-when all of a sudden the mind "flips" and the cube now appears to be inside out. Both hypotheses of what is seen are equally acceptable to the mind, which seems to like to test alternative hypotheses. Usually, one possibility fits the data better than all the rest. Sometimes, as is the case with illusions, more than one alternative best fits; or it may be that the hypothesis that seems to fit best is completely incorrect.

Interesting statements on seeing, perception, and hypothesis testing are given by GREGORY.[3]

> ...Perception involves going beyond the immediately given evidence of the senses: this evidence is assessed on many grounds and generally we make the best bet, and see things more or less correctly. But the senses do not give us a picture of the world directly; rather they provide evidence for checking hypotheses about what lies before us. Indeed, we may say that a perceived object is a hypothesis, suggested and tested by sensory data...When a perceptual hypothesis-a perception-is wrong we are misled as we are misled in science when we see the world distorted by a false theory.

The idea that intricate pattern-to-pattern comparisons can be made by the eye/brain system is suggested by the work of B. JULESZ[5] of the Bell Telephone Laboratories. In his experiments, two related patterns, constituting a "random dot stereogram," are shown simultaneously to the left and the right eye, respectively; and three-dimensional images are perceived by the eye/brain system in establishing the mutual connection between the images. The patterns consist of random dots. Individually, they appear in every respect to be feature-free.

As a simple example,* consider a planar area covered uniformly with white snow. Apply many black dots in a random array on top of the snow. Now look down on the scene with both eyes. Each eye sees the same random black dots on the white background.

---

*The reader should not blame Dr. Julesz for this illustration or for the implications drawn from it. The author assumes full responsibility!

The dot array appears to be identical to both eyes. It is true that the respective images are actually shifted with respect to each other because the eyes are physically displaced-but nevertheless, the brain perceives a single image of dots on a plane.

Next, imagine another planar area covered with fresh snow-undotted this time. In the midst of this area place a white polar bear. The bear is invisible, since he is white on white. Now put black dots on the snow and (oops!) on the polar bear. The observer's eye will see an image of black dots on a white background. The right eye will not simply see the same dot image as that of the left eye, since the dots on the polar bear will be displaced relative to the dots on the background because the polar bear "sticks up" from the snowy plane. The eye/brain then perceives the scene as a dotted polar bear sticking up from a dotted white background.

In this imaginary experiment, each eye sees only random dot patterns. Because of the relationship between the left-eye and the right-eye pattern, however, the eye/brain perceives a three-dimensional image a three-dimensional object made of dots-placed against a dotted background.

Julesz has generated random dot patterns by computer to represent three-dimensional objects such as cylinders, pyramids, and other geometrical shapes. Such patterns can be generated for sticking-up objects by starting with a random dot pattern for the left eye, and generating from it a pattern for the right eye by translating the dots that were within the original object silhouette by a small distance to the left relative to the background dots. (The silhouette of the object can never be seen in the individual dot patterns.) Vacated areas on the background are filled in by arbitrarily placing dots there. Overlapping background areas have their dot placements pre-empted by the object dot placements.

Remarkable three-dimensional effects are observed by viewing Julesz's random dot patterns.[5] The left-eye image and the right-eye image are addressed exclusively to the left eye and to the right eye, respectively, by using color-filter glasses-green for the left eye, red for right eye-to view two superposed images printed in the two colors. Julesz had obtained even more spectacular results with random dot patterns by using polaroid glasses with projected polarized-light images.

Since the individual random-dot patterns are apparently feature-free, one might possibly conclude from such experiments either that the eye/brain does not use features, or that if it does use them, they are not completely necessary for object recognition. Such a conclusion may in fact be correct; but the experiments do not necessarily lead to it. The differences between the left-eye and right-eye patterns contain the object shape and could thus contain its features. One point is clear: the eye/brain is capable of making very intricate pattern-to-pattern comparisons, apparently using direct pattern data in full detail. Imagine the detail contained in the random dot patterns!

In a seminar presented at Stanford University several years ago, Dr. Julesz described experiments that were performed with an eidetic subject (i.e. one having a "photographic memory"), If the right eye is covered, and the left-eye image is presented to the left eye, a random-dot array is perceived by the subject. A day later, the left eye is covered, and the right-eye image is presented to the right eye. When this is done with an ordinary subject, only a random-dot pattern is perceived. When the same experiment is performed with an eidetic subject, the image appears to him to be raised from the plane of the background.

After hearing of this stunning experiment, the following question was posed: Was the subject's head clamped in place on the two occasions, with the dot images precisely supported and lighted, to insure identical conditions for the two image sightings? The answer was no. On both occasions the dot pictures were hand held, and lighting was not critically maintained. Even if the pattern at the second sighting was rotated by as much a $\pm 10^{\circ}$ of the original orientation, and the distance of the pattern was changed as much as ± 10 per cent of the original distance, and if aspect angle was allowed a tolerance similar to that of the rotation angle, the three-dimensional image was easily perceived by the eidetic subject. An implication of this experiment is that the eye/brain is capable of matching very intricately-related patterns in spite of limited amounts of relative rotation, translation, scale change, aspect change, and so on.

Experiments along these lines with other kinds of distortions (twisting, differential stretching, etc.) would be very useful. The left-eye pattern could be a random dot array. The right-eye pattern could contain the dotted object (properly disposed relative to the left-eye pattern) including the desired distortion. Ordinary test subjects could be used, with both patterns presented to the eyes simultaneously. The interesting question is, how much distortion is permissible in the right-eye image relative to the left-eye image before the subject loses three-dimensional perception of the image?

The three-dimensional visual effect may well provide a medium for the experimental study of how the eye/brain matches one pattern against another and how it tests alternative hypotheses. The results of such experiments would give ideas on the types of functions that might be built into an automatic pattern-recognition system.

6. CONCLUSION

A complete memory and pattern-recognition system is suggested based on the rubber-mask concept. Three important functions are needed: Hypothesis testing (pattern matching); pattern stretching; and pattern memory. There is some evidence that these functions can be implemented in automatic equipment, and it is speculated that the same functions are performed in a natural way in the eye/brain system.

One of the earliest thinkers to look at perception from the point of view of hypothesis formation and testing was Plato. He believed that man is endowed with a sense of beauty and perfection, and that man perceives natural objects as imperfect versions of perfect geometric forms such as straight lines, circles and so on. Plato's models, which he called ιδεας (archetypes or perfect forms) were all in the mind and were by definition perfect-whether they represented perfect circularity, perfect equality, perfect justice, or any other idealized standard never achieved in the natural universe.

The basic character of geometric proofs was evidently known to Plato. For instance, he was undoubtedly familiar with Pythagoras' proof that the square on the hypothenuse of a right triangle is equal to the sum of the squares on the other two sides-a proof that had to be (and has to be today) based on purely mental concepts of the perfect circle, the perfect right angle, and so on. No such perfect forms could in the olden days be traced in the dust with a stick-nor can they today be drawn on vellum with compass and ruler nor reproduced by a computerized system of graphics!

It is interesting to read from PLATO'S Republic[6] his ideas on hypotheses and idealizations. As reported by Plato, Socrates (into whose mouth Plato put many of his own theories) is engaged in a dialogue with his pupil Glaucon.

> ...You are aware that students of geometry, arithmetic, and the kindred sciences assume the odd and the even and the figures and three kinds of angles and the like in their hypotheses, which they and everybody are supposed to know, and therefore they do not deign to give any account of them either to themselves or others; but they begin with them, and to on until they arrive at last, and in a consistent manner, at their conclusion?
>
> Yes, he said, I know.
>
> And do you not know that although they make use of the visible forms and reason about them, they are thinking

> not of these, but of the ideals which they resemble; not of the figures which they draw, but of the absolute square and the absolute diameter, and so on-the forms which they draw or make, and which have shadows and reflections in water of their own, are converted by them into images, but they are really seeking to behold the things themselves, which can only be seen with the eye of the mind?
>
> That is true.
>
> Republic, Book VI (Jowett trans.)

Acknowledgments-I would like once again to thank Mrs. MABEL ROCKWELL for editing this paper, many useful suggestions on the organization and technical presentation, and for bringing the thoughts of Plato into the discussion of rubber masks. SIDHARTHA MAITRA assisted in the editing, and his help is gratefully acknowledged.

REFERENCES

1. B. WIDROW,"The "rubber mask" technique I: Pattern measurement and analysis," Pattern Recognition 5, 175-197 (1973).
2. C.J.W. MASON, "Pattern Recognition Bibliography", IEEE Systems, Man, and Cybernetics, Group Newsletter, October 1970, December 1970, February 1971, June 1971.
3. R.L. GREGORY, "Eye and Brain: The Psychology of Seeing", World University Library. McGraw-Hill, New York (1966).
4. Police identification kits are manufactured by the Identi-Kit Company which claims for its product in an advertisement on page 39 of Law and Order, September 1972:...It allows you to go to the witness or victim of a crime and develop a composite facial likeness of a suspect within minutes. Manual overlay of six to nine facial components becomes a complete line drawing without the use of complicated machines, photographs or the services of an artist. Also, Identi-Kit composites are easily reproduced and automatically code themselves for instant transmission to other Identi-Kit users all over the world....
5. B. JULESZ, Foundations of Cyclopean Perception. University of Chicago Press, Chicago (1971).
6. Plato's Republic, Book VI, Jowett Translation, Random House, Vintage Paperbacks.

SELECTED READING

Adaline/Madaline

1. B. WIDROW and M.E. HOFF, "Adaptive switching circuits", WESCON Convention Rec., Institute of Radio Engineers, N.Y. Pt. 4, pp. 96-104 (1960).
2. B. WIDROW, "Generalization and information storage in networks of Adaline 'neurons'," Self Organizing Systems, M.C. YOVITZ, G.T. JACOBI and G.D. GOLDSTEIN (eds) pp. 435-461. Spartan, Washington, D.C. 1962.
3. K. STEINBUCH and B. WIDROW, "A critical comparison of two kinds of adaptive classification networks", IEEE Trans. Electronic Computers EC-14 (5), 737-740 (1965).
4. J.S. KOFORD and G.F. GRONER, "The use of an adaptive threshold element to design a linear optimal pattern classifier,"IEEE Trans. Information Theory IT-12, 42-50 (1966)
5. C.H. MAYS, "Effects of adaptation parameters on convergence time and tolerance for adaptive threshold elements", IEEE Trans. Electronic Computers (short notes) EC-13, 465-468 (1964).
6. N.J. NILSSON, Learning Machines. McGraw-Hill, New York (1965).
7. B. WIDROW et al., "Adaptive antenna systems," Proc. IEEE 55, 2143-2159 (1967).

Perceptron

1. F. ROSENBLATT, "The perceptron, a theory of statistical separability in cognitive systems", Report VG-1196-G-1, Cornell Aeronautical Lab., Buffalo, New York (1958).
2. F. ROSENBLATT, Principles of Neurodynamics, and the Theory of Brain Mechanisms. Spartan Books, Washington, D.C. (1962).
3. H.D. BLOCK, "The perceptron: A model for brain functioning". I, Rev. Mod. Phys. 34 (1), 123-135 (1962).
4. H.D. BLOCK, B.W. KNIGHT and F. ROSENBLATT, "Analysis of a four-layer series-coupled perceptron." II, Rev. Mod. Phys. 34 (1), 135-142 (1962).
5. N.J. NILSSON, Learning Machines. McGraw-Hill, New York (1965).
6. G. NAGY, "The state of art in pattern recognition", Proc. IEEE 56 (5), 836-862 (1968).
7. M.L. MINSKY and S. PAPERT, Perceptrons: an Introduction to Computational Geometry. M.I.T. Press, Cambridge, Mass. (1969).

Matched Filters and Learning Matrices

1. "Matched filter issue", IRE Trans. Information Theory IT-6, 309-417 (1960).

2. T. KAILATH, "Adaptive matched filters", Mathematical Optimization Techniques, R. BELLMAN (ed), pp. 109-140. University of California Press (1963).
3. K. STEINBUCH and U.A.W. PISKE, "Learning matrices and their applications", IEEE Trans. Electronic Computers EC-12 (5), 846-862 (1963).
4. H. KAZMIERCZAK and K. STEINBUCH, "Adaptive systems in pattern recognition," IEEE Trans. Electronic Computers EC-12 (5), 822-835 (1963).

Linear and Polynomial Discriminants

1. W.H.HIGHLEYMAN, "Linear decision functions, with application to pattern recognition", Proc. IRE 50, 1501-1514 (1962).
2. T.W. ANDERSON, Introduction to Multivariate Statistical Analysis. Wiley, New York (1965).
3. J.S. KOFORD and G.F. GRONER,"The use of an adaptive threshold element to design a linear optimal pattern classifier", IEEE Trans. Information Theory IT-12, 42-50 (1966).
4. D. F. SPECHT, "Generation of polynomial discriminant-functions for pattern recognition", IEEE Trans. Electronic Computers EC-16 (1), 308-319 (1967).
5. D.F. SPECHT, "Vectorcardiographic diagnosis using the polynomial discriminant method of pattern recognition", IEEE Trans. Biomedical Electronics BME-14 (2), 90-95 (1967).

Nearest Neighbour Rules

1. T.M. COVER and P. E. HART, "Nearest neighbor pattern classification", IEEE Trans. Information Theory IT-13 (1), 21-27 (1967).
2. T.M. COVER, "Estimation by the nearest neighbor rule", IEEE Trans. Information Theory IT-14 (1), 50-55 (1968).

Integral Geometry

1. L.A. SANTALO, Introduction to Integral Geometry. Hermann, Paris (1953).
2. A.B.J. NOVIKOFF, "Integral geometry as a tool in pattern perception", Principles of Self-Organization, VON FORESTER and ZOPF (eds), pp. 347-368, Pergamon, Oxford (1962).
3. G. TENERY, "Pattern recognition function of integral geometry", IEEE Trans. Military Electronics MIL-7, 196-199 (1963).
4. E. WONG and J.A. STEPPE, "Invariant recognition of geometrical shapes", Methodologies of Pattern Recognition, S. WATANABE (ed), pp. 535-543. Academic, New York (1969).

Pattern Classification by Clustering/ Learning Without a Teacher

1. F. MOSTELLER, "On pooling data", J. Am. Stat. Assoc. 43 (242), 231-242 (1948).
2. T.T. TANIMOTO and D.J. ROGERS, "A computer program for classifying plants", Science 132 (3434), 1115-1118 (1960).
3. D.B. COOPER and P.W. COOPER, "Nonsupervised adaptive signal detection and pattern recognition", Inform. Control 7 (3), 416-444 (1964).
4. R.E. BONNER, "On some clustering techniques", IBM J. Res. Dev. 8 (1), 22-32 (1964).
5. R.L. MATTSON and J.E. DAMMANN, "A technique for determining and coding subclasses in pattern recognition problems", IBM J. Res. Dev. 9 (4), 294-302 (1965).
6. G.H. BALL and D.J. HALL, "Isodata, a novel method of data analysis and pattern classification", Stanford Research Institute, Menlo Park, Calif. (April 1965).
7. J. SPRAGINS, "Learning without a teacher", IEEE Trans. Information Theory IT-12 (2), 223-230 (1966).
8. S.C. FRALICK, "Learning to recognize patterns without a teacher", IEEE Trans. Information Theory IT-13 (1), 57-64 (1967).
9. W.C. MILLER, "A modified mean square error criterion for use in unsupervised learning", Stanford Electronics Laboratories, Stanford, Calif., Tech. Rept. SEL-67-066 (TR 6778-2) (August 1967).
10. YA.Z. TSYPKIN, "Self learning - what is it?" IEEE Trans. Automatic Control AC13 (6), 608-612 (1968).
11. C.G. HILBORN, JR. and D.G. LAINIOTIS, "Unsupervised learning minimum risk pattern classification for dependent hypotheses and dependent measurements", IEEE Trans. Systems Science and Cybernetics SSC-5 (2), 109-115 (1969).
12. A.N. MUCCIARDI and E.E. GOSE, "An automatic algorithm and its properties in high-dimensional spaces," IEEE Trans. Systems, Man and Cybernetics SMC-2 (2), 247-254 (1972).

Linguistic Approach to Pattern Recognition

1. R.A. KIRSCH, "Computer interpretation of English text and picture patterns", IEEE Trans. Electronic Computers EC-13, 363-376 (1964).
2. R.S. LEDLEY, "High-speed automatic analysis of biomedical pictures", Science 146, 216-223 (1964).
3. N. CHOMSKY, Aspects of the Theory of Syntax. M.I.T. Press (1965).
4. R. NARASIMHAN, "Syntax-directed interpretation of classes of pictures", Comm. ACM 9, 166-173 (1966).
5. M.B. CLOWES, "Transformational grammars and the organization of pictures", Automatic Interpretation and Classification of

Images, A. GRASSELLI (ed), pp. 43-76. Academic, New York (1969).
6. R. NARASIMHAN, "On the description, generation, and recognition of classes of pictures", Automatic Interpretation and Classification of Images, A. GRASSELLI (ed), pp. 1-41. Academic, New York (1969).
7. M. EDEN, "Handwriting generation and recognition", Recognizing Patterns, P.A. KOLERS and M. EDEN (eds) pp. 138-154. M.I.T. Press, Cambridge, Mass. (1968).
8. W.F. MILLER and A.C. SHAW,"Linguistic methods in picture processing - a survey", Fall Joint-Computer Conference, pp. 279-290 (December 1968).
9. A. ROSENFELD, "Chapter on picture description and picture languages," Picture Processing by Computer, pp. 167-184. Academic, New York (1969).
10. J. HOPCROFT and J. ULLMAN, Formal Languages and their Relation to Automata. Addison-Wesley, Reading, Mass. (1969).
11. A.C. SHAW, "Parsing of graph-representable pictures", J. ACM 17 (3), 453-481 (1970).
12. H.C. LEE and K.S. FU, "A stochastic syntax analysis procedure and its application to pattern recognition," IEEE Trans. Comp., (July 1972).
13. "Special Issue on Syntactic Pattern Recognition", Pattern Recognition 4, No. 1 (January 1972).

Fingerprint Recognition

1. WILLIAM DIENSTEIN, Technics for the Crime Investigator. Charles C. Thomas (1952).
2. A. GRASSELLI, "On the automatic classification of fingerprints", Methodologies of Pattern Recognition, S. WATANABE (ed.), pp. 253-315. Academic, New York (1969).
3. L.S. PENROSE, "Dermatoglyphics", Scient. Am. pp. 72-84 (December 1969).
4. "Proc. 1st and 2nd Natl. Symp. Law Enforcement Science and Technology," Thompson/Academic Press and IITRI. Articles
   (a) JOHN E. GAFFNEY, JR. et al., "Fingerprint encoding and matching procedures for automated recognition systems", pp. 401-407.
   (b) C.B. SHELMAN, "Machine extraction of ridge endings from a noisy fingerprint", pp. 409-416.
   (c) VINCENT V. HORVATH and CHARLES D. DOYLE, "Recent developments in fingerprint recognition using coherent optical processing", pp. 417-426.
   (d) BERNARD M. VAN EMDEN, "Advanced computer based fingerprint automatic classification technique" (FACT), pp. 493-505.
   (e) M.D. FREEDMAN and E.D. HIETANEN, "Application of parallel neighborhood logic to fingerprint processing", pp. 507-509.
   (f) RICHARD W. SCHWARTZ, "System considerations in automated fingerprint classification", pp. 511-515.

Human Vision

1. R.L. GREGORY, Eye and Brain, the Psychology of Seeing. World University Library, McGraw-Hill, New York (1966).
2. E.R. HILGARD and G.H. BOWER, Theories of Learning, 3rd ed. Appleton-Century-Crofts, New York (1966).
3. P.C. DODWELL, Visual Pattern Recognition. Holt, Rinehart and Winston, New York (1970).
4. T.N. CORNSWEET, "Measuring movements of the retinal image with respect to the retina", Biomedical Sciences Instrumentation, W.E. MURRAY and P.F. SALISBURY (ed). vol. 2. Plenum, New York (1964).
5. N.H. MACKWORTH, "A stand camera for line-of-night recording", Perception and Psychophysics 2 (1967).
6. N.H. MACKWORTH and J.S. BRUNER, "How adults and children search and recognize pictures", Human Development 13, 149-177 (1970).

Books on Pattern Recognition

1. G.L. FISCHER, JR., D.K. POLLOCK, B. RADACK and M.E. STEVENS (eds), Optical Character Recognition. Spartan Books, Washington (1962).
2. G. SEBESTYEN, Decision-Making Process in Pattern Recognition. Macmillan, New York (1962).
3. D.K. POLLOCK, C.J. KOESTER and J.T. TIPPETT, Optical Processing of Information. Spartan Books, Washington, D.C. (1963).
4. N.J. NILSSON, Learning Machines. McGraw-Hill, New York (1965).
5. L.UHR, Pattern Recognition. Wiley, New York (1966).
6. K.S. FU, Sequential Methods in Pattern Recognition and Machine Learning. Academic, New York (1968).
7. L.N. KANAL (ed), Pattern Recognition. Thompson, Washington (1968).
8. A. ROSENFELD, Picture Processing by Computer. Academic, New York (1969).
9. M.S. WATANABE, Methodologies of Pattern Recognition. Academic, New York (1969).
10. A. GRASSELLI (ed), Automatic Interpretation and Classification of Images. Academic, New York (1969).
11. J. MENDEL and K. S. FU, Adaptive Learnings and Pattern Recognition: Theory and Applications. Academic, New York (1970).
12. N. BONGARD, Pattern Recognition. Spartan Books, New York (1970).
13. YA. Z. TSYPKIN, Adaptation and Learning in Automatic Systems. Academic, New York (1971).
14. K.S. FU (ed) Pattern Recognition and Machine Learning. Plenum, New York (1971).
15. O. GRUSSER and R. KLINKE (eds), Pattern Recognition in Biological and Technical Systems. Springer, Berlin, New York

(1971).
16. K. FUKUNAGA, Introduction to Statistical Pattern Recognition. Academic, New York (1972).

Optical Character Recognition (OCR)

1. G.L. FISCHER, JR., D.K. POLLOCK, B. RADACK and M.E. STEVENS (eds), Optical Character Recognition. Spartan Books, Washington, D.C. (1962).
2. C.N. LIU and G.L. SHELTON, JR., "An experimental investigation of a mixed-font print recognition system", IEEE Trans. Computers EC-15 (6), 916-925 (1966).
3. G. NAGY, "Preliminary investigation of techniques for automated reading of unformatted text," IBM Res. RC 1867, Yorktown Heights, New York (30 June 1967).
4. T. TRICKETT, "The design of a standard type font for optical character recognition," Honeywell Comput. Jl. 3 (1), 2-11 (1969).
5. "Special issue on Optical Character Recognition", Pattern Recognition 2, No. 3 (September 1970).
6. J.R. PARKS, "A multi-level system of analysis for mixed font and hand-blocked printed character recognition", Automatic Interpretation and Classification of Images, A. GRASSELLI (ed), pp. 295-322. Academic, New York (1969).

LEARNING TEXTURE INFORMATION FROM SINGULAR PHOTOGRAPHS AND ITS APPLICATION IN DIGITAL IMAGE CLASSIFICATION

S. J. Dwyer, J. K. Chang, R. W. McLaren, G. S. Lodwick
University of Missouri
Columbia, Missouri

INTRODUCTION

An area which has rapidly gained interest in the last few years is that of remote sensing by imagery.[1,2] Certainly the successful mission of the ERTS has contributed to an expanding interest and increased potential.[3] The potential uses of such remotely sensed data cover a variety of applications, including land use studies, crop quality, accurate map making, and evaluating natural resources. In utilizing this data in a particular application, a significant problem is the amount of data that must be processed to obtain specific information or features pertinent to the application. This is inherent in image or visual information; when this is multiplied by a large number of the available images, the problem of obtaining details or small features from such images is significant. This is particularly true when some set of features is to be used in a parametric pattern recognition scheme. However, there are many instances when one is interested in identifying an area or region which may encompass many details, but is small compared with an overall image and represents a defined entity. Then, if one is interested in identifying or classifying such a region, its gross characteristics must be represented while ignoring small details that may vary significantly from one sample observation to another. One approach to this problem is to use texture information.[4,5] This visual property can be considered as a "fine structure" distributed randomly over a small region. This paper concerns the use of texture information in classifying small regions in an overall digitized photograph. The classification scheme uses parameters learned from training observations. Training observations are represented by small regions or sub-images having known or defined classifications.

This paper will describe a method for utilizing limited representative regions of a defined texture class to generate a training sample. Finally, texture information from the training samples will be used in a Bayes decision rule to distinguish among classes of timberland.

TEXTURE INFORMATION

Texture is a visual property of an image; it corresponds to a "graininess" or fine structure that varies widely for different materials, e.g., glass, sand, grass, or soil. Texture can be viewed as an elemental pattern structure randomly or uniformly distributed over a small region representing a defined object or surface. This view leads one to describe texture by statistical properties or features. The choice of such features is based on providing a unique representation of texture over a region declared to be of a particular class. Here, a digitized image having n x n elements and r gray levels is considered. A sub-region of the image is identified and a square array of m x m elements just contained in this sub-region is then used to generate learning observations, each observation being a sub-image of size q x q, $q \ll m$. The following "statistical" features were used to describe the texture of the observation.[6] Let the number of gray levels be r = 16. Also, let P(i), i = 0,...,15 be the first-order probabilities of the 16 gray levels, estimated from the sample observations. The average value for the observations is defined by,

$$\mu = \sum_{i=0}^{15} (i)^2 P(i) \tag{1}$$

This represents the first feature. The second feature is similar to variance,

$$v = \sum_{i=0}^{15} (i)\, P(i) - \mu^2 \tag{2}$$

The next features are, skewness, S,

$$S = \frac{\sqrt{B_1}\,(B_2 + 3)}{2(5B_2 - 6B_1 - 9)} \quad , \tag{3}$$

where, $B_1 = C_3{}^2/C_2{}^3$, $B_2 = C_4/C_2{}^2$, and kurtosis, K,

$$K = C_4/C_2{}^2 - 3 \ , \tag{4}$$

wherein, C_t is the t th central moment,

$$C_t = \sum_{i=0}^{15} (i - \mu)^t P(i) \tag{5}$$

The ramaining 20 features are related to second-order properties of the elements in the texture pattern as follows.

Let P_z denote the second-order probability matrix which consists of 16 x 16 elements as follows:

$$P_z = \begin{bmatrix} P_z(0,0) & P_z(0,1) & \cdots & P_z(0,15) \\ P_z(1,0) & P_z(1,1) & \cdots & P_z(1,15) \\ \cdot & & & \\ \cdot & & & \\ \cdot & & & \\ P_z(15,0) & P_z(15,1) & \cdots & P_z(15,15) \end{bmatrix}, \tag{6}$$

$$z = 1, 2, \ldots, 8.$$

The element P_1 (k,h) of the associated probability matrix specifies the probability that neighboring image points f (i,j) and f(i+1,j) of the texture class have gray values k and h respectively. That is,

$$P_1(k,h) = Pr\{ f(i,j) = k,\ f(i + 1,j) = h\} \tag{7}$$

For other values of z, the following neighboring image points are used in the previous equation.

z	image points
1	f(i,j), f(i+1,j)
2	f(i,j), f(i+1,j+1)
3	f(i,j), f(i,j+1)
4	f(i,j), f(i-1,j+1)
5	f(i,j), f(i-1,j)
6	f(i,j), f(i-1,j-1)
7	f(i,j), f(1,j-1)
8	f(i.j), f(i+1,j-1)

(8)

For example, one has,

$$P_5(k,h) = \Pr\{ f(i,j) = k,\ f(i-1,j) = h\} \qquad k = 0,1,\ldots,15; \quad h = 0,1,\ldots,15. \tag{9}$$

The four second-order joint probability matrices are defined as follows:

<u>Horizontal Joint Probability Matrix P_H</u>

$$P_H = (P_1 + P_5)/2 \tag{10}$$

<u>Vertical Joint Probability Matrix P_V</u>

$$P_V = (P_3 + P_7)/2 \tag{11}$$

<u>Left Diagonal Joint Probability Matrix P_{LD}</u>

$$P_{LD} = (P_4 + P_8)/2 \tag{12}$$

<u>Right Diagonal Joint Probability Matrix P_{RD}</u>

$$P_{RD} = (P_2 + P_6)/2 \tag{13}$$

The five measurements called energy, entropy, correlation, local homogeneity, and inertia which are derived for each of the previous joint probability matrices, P, are then defined as follows:

$$E(P) = \sum_{k=0}^{15} \sum_{h=0}^{15} p(k,h)^2 \text{ , the \underline{energy} ;} \tag{14}$$

$$H(P) = \sum_{k=0}^{15} \sum_{h=0}^{15} \{-p(k,h) \log p(k,h)\} \text{ , the \underline{entropy};} \tag{15}$$

$$C(P) = \frac{\sum_{k=0}^{15} \sum_{h=0}^{15} (k-M)(h-M)\, p(k,h)}{\sum_{k=0}^{15} (k-M)^2 \sum_{h=0}^{15} p(k,h)} \text{ , the \underline{correlation};} \tag{16}$$

where

$$M = \sum_{k=0}^{15} k \sum_{h=0}^{15} p(k,h) \; ; \qquad (17)$$

$$L(P) = \sum_{k=0}^{15} \sum_{h=0}^{15} \frac{1}{1+(k-h)^2} p(k,h), \text{ the } \underline{\text{local homogeneity;}} \qquad (18)$$

and finally,

$$I(P) = \sum_{k=0}^{15} \sum_{h=0}^{15} (k-h)^2 \, p(k,h), \text{ the } \underline{\text{moment of inertia}}. \qquad (19)$$

Thus, these five features are defined for each of the joint probability matrices defined in Eqs. (10) - (13), yielding 4 x 5 = 20 features.

GENERATION OF A TRAINING SAMPLE

Each of the descriptors or features will vary in value over individual sub-image observations sampled from the same texture class. It is desired to use these features in a parametric decision scheme, e.g., a Bayes decision rule. This decision rule utilizes the mean vector and covariance matrix for each defined pattern class when the feature vector X is assumed to be multivariate normal. These parameters will be estimated from a training sample for each class. Such a training sample consists of a finite number of sub-images, each derived from a larger image generated from a region of a texture class. In setting-up such a training sample, however, practical limitations were encountered. First, the basic procedure to generate a training sample involved partitioning a single image into training observation sub-images, say 16 x 16 in size; this provides flexibility in processing the training sample to estimate the texture descriptors over a given class and in selecting the size of the sub-image and thus the training sample size. However, a single total image, particularly of the type dealt with here-aerial photographs, does not represent a single texture but rather a distribution of contiguous regions of differing texture classes. Furthermore, often the texture class whose descriptors are to be learned is represented by only a few regions of sufficient size having the same texture class. For these reasons, the following procedure was employed to generate a training sample. Using a single image having a region whose texture is well-defined for a given class is identified and a rectangular image is then just fitted within the region (and at least 10 times the area of the expected size of the sub-image serving as a training observation). Then, this image is repeatedly concatenated with itself spatially to just fill the raster size of the original entire image. Now, there are a number of factors or design parameters to consider. One is the size of the sub-image that forms one observation in the training sample. As the size of this image decreases, the variation in the descriptors calculated from one sub-image to another increases, which could degrade recognition performance. At the same time, the training sample size increases, which should improve performance. A compromise would be evaluated in terms of classification results. A compromise employed here was to use a sub-image of size 16 x 16 when the overall image was 256 x 256, yielding 256 observations for the training sample. The square region used to generate an overall image of the same texture class was chosen as being at least 10 times the area of the 16 x 16 sub-image to maintain "randomness" in the sample observations. Other design parameters include the overall image size (often limited by equipment) and perhaps the size of an optimal feature subset.

CLASSIFICATION

Pattern classification of the image classes was implemented by a digital computer using Bayes decision rule. Bayesian decision rules are discussed in more detail in (7), for example. Let the conditional probability density function of the feature vector X given class ω_i be assumed to be multivariate normal,

$$f(X/\omega_i) = (2\pi)^{-\frac{n}{2}} \quad |\Sigma_i|^{-1/2} \exp\{-\tfrac{1}{2}\,(X-\underline{\mu}_i)'\ \Sigma_i^{-1}\ (X-\underline{\mu}_i)\}\ , \qquad (20)$$

$i = 1,2,\ldots,m.$

where $\underline{\mu}_i$ is the mean vector and Σ_i is the covariance matrix for class ω_i, $i = 1,\ldots,m$, and X is the feature vector $X = [x_1 \ldots x_n]'$. For each class, ω_i, $\underline{\mu}_i$ and Σ_i are estimated from the corresponding training sample. Then, for a given digitized image to be processed, for each non-overlapping 16 x 16 sub-image, the same size as the sub-images forming the training samples, the features are extracted (up to 24 of them). This forms the feature vector X. Then, Bayes decision rule is applied. Assign X to class k if,

$$(X-\underline{\mu}_i)'\ \Sigma_i^{-1}(X-\underline{\mu}_i) - \log\,(P_i^{\,2}/|\Sigma_i|) \qquad (21)$$

is minimum for i = k.

In this expression, P_i is the prior probability of class ω_i. In this way, each small region in the overall (aerial) photograph can be identified, thus, for example locating certain resources in a given area covered by the photograph.

APPLICATION AND RESULTS

The proposed automatic procedure for using texture information extracted from digitized images in classification has been applied to the identification of areas in an aerial photograph as one of several classes of timberland. The location and identification of different types of timber areas is of practical importance in Forestry and Conservation. At present the location and identification of different types of timber areas have been done manually from the aerial photograph. The computer-aided approach outlined previously can offer speed, consistency of results, flexibility, and for a situation involving a large number of photographs, decreased cost.

Four classes or timber areas are identified here, reproductive, poles, saw timber and mixed. Reproductive timber refers to an area where trees are mostly less than five inches in diameter at branch height (dbh), poles refer to those from six to twelve inches dbh, saw timber, twelve inches and over dbh, while a mixed timber area consists of trees of mixed sizes. In addition to such timber areas, one is also interested in locating and identifying fields, bodies of water, and the like.

The computer-aided procedure for locating and identifying timber areas consists of the following steps. First, an aerial photograph is digitized by scanning the transparency of the photograph using an image dissector camera. The resolution of the scanned image is 256 x 256 pixels, each representing 256 shades of gray (8 bits). However, in the extraction of texture information, the use of only 16 shades of gray proved to be adequate in terms of classification results and at the same time, reduced computation requirements.

The second step in the computer-aided procedure is the extraction of texture information from a square sub-image of predetermined size, 4 x 4, 8 x 8, or 16 x 16, of the digitized aerial photograph. In the present study, 16 x 16 sub-images were used, so that the entire digitized aerial photograph is partitioned into 256 non-overlapping contiguous square areas. For each area, the 24 features (described previously) are calculated.

The third step is to classify each of the 256 sub-images into one of the defined texture classes based on the 24 dimensional feature vector. Classification is implemented here by a Bayes

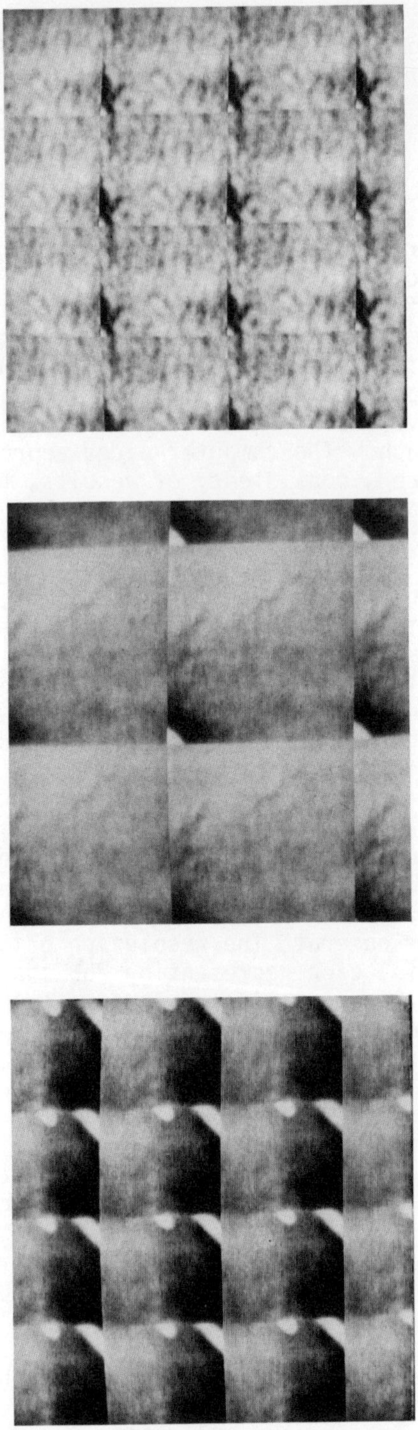

a. Reproductive b. Poles c. Saw Timber

Fig. 1. Images Used to Generate Training Samples.

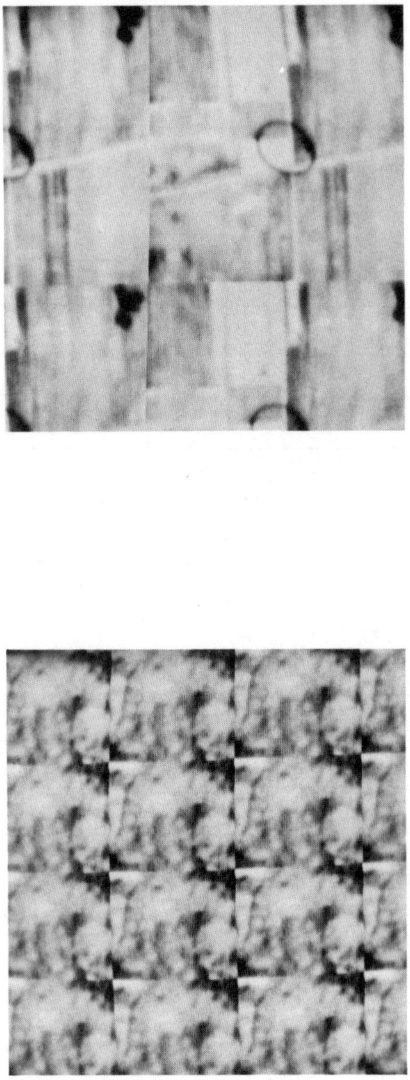

d. Mixed

e. Field

Fig. 1. (cont'd.).

Table I. Training Results

		Class Assigned by Decision Rule					
		F	S	M	P	R	% Correct
True Class	F	256	0	0	0	0	100
	S	0	256	0	0	0	100
	M	0	0	256	0	0	100
	P	0	0	0	256	0	100
	R	0	0	0	1	255	99.6

F = Field S = Saw Timber M = Mixed P = Poles R = Reproduction

Table II. Jackknife Results

		Class Assigned by Decision Rule					
		F	S	M	P	R	% Correct
True Class	F	251	0	2	2	1	98.1
	S	0	232	24	0	0	90.6
	M	3	0	253	0	0	98.8
	P	5	0	0	249	2	97.3
	R	3	0	2	4	247	96.5

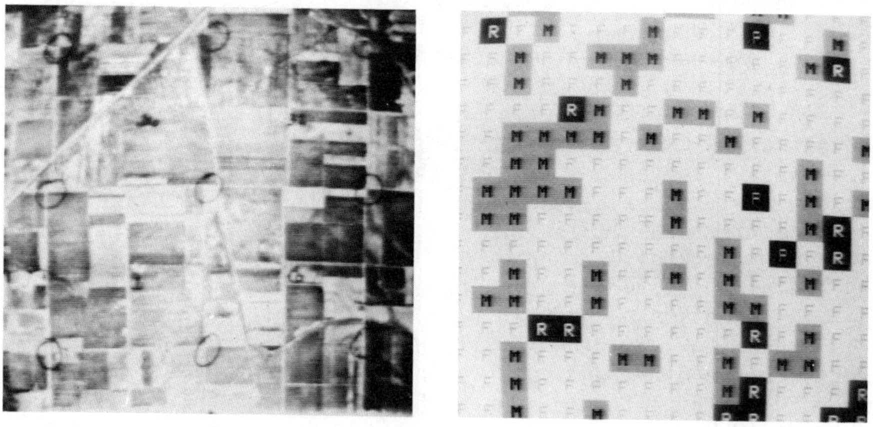

a. Digitized Image

b. Classification of Regions

Fig. 2. Classification of Regions Over an Entire Image

decision rule assuming that the texture features are multivariate normally distributed for each texture class. Decision parameters for each class are estimated from training samples as previously described. Finally, classification results for the 256 sub-images are displayed on a television display device in the form of assigning a letter for each sub-image representing the decision.

Figures (1a) to (1e) show the digitized training image for each of the four classes of timberland plus field before sampling. As described earlier, each of these images was generated by first extracting a rectangular image of size at least 64 x 64 just contained in a region clearly identified as representing a defined (texture) class. Then, this image was repeatedly written, non-overlapping, to fill the entire 256 x 256 raster. Then, for each of these training images, 256 (16 x 16) training observations were formed, from which the basic statistics of the feature vector could be estimated for the decision rule.

Results for dicriminating among the five distinct texture classes are summarized in the confusion matrices of Tables I and II. Table I summarizes results when the training sample observations are used as test observations. This accounts for the nearly perfect results. Table II, however, summarizes results based on a jackknife procedure in which approximately 10% of the training sample set was separated and used as a test sample, the remaining 90% being used to estimate decision parameters. This table shows average results over repeated jackknife experiments, when a different 10% of training sample observations are separated as a test sample on each of the 10 experiments. These results are very encouraging.

As a further test and as a preliminary study in directly applying the method, the texture classification technique was applied to an entire digitized image consisting of different textures. The digitized image and a display of classified regions of it by the proposed method are shown in Figures (2a) and (2b).

SUMMARY AND CONCLUSIONS

A method has been proposed for the computer-aided classification of regions in an aerial photograph. This procedure utilizes texture information for discrimination among classes. Texture has been viewed as an elemental pattern distributed randomly over a small region. Some 24 features are defined and used to describe texture. Classification is implemented by a Bayesian decision rule, assuming normality. Decision parameters are estimated from training samples, each generated from a single region of an aerial photograph, the region representing a single texture class. The basic procedure is applied to the identification of 5 distinct textures corresponding to 5 classes of timberland (including fields). The results obtained thus far are very encouraging and indicate the possibility of applying such a computer-oriented procedure to other photographs, particularly to aerial photographs or to those returned by satellite in order to locate and identify other land features.

REFERENCES

1. R.K. Holz, Editor, The Surveillant Science: Remote Sensing of the Environment, Houghton Mifflin, 1973.
2. Proceedings of International Symposia on Remote Sensing of Environment, Ann Arbor, University of Michigan, Institute of Science and Technology, Willow Run Laboratories, 1969-1973.
3. R.N. Colwell, "Remote Sensing as an Aid to the Management of Earth Resources", American Scientist, Vol. 61, March/April, 1973, pp. 175-183.
4. B.S. Lipkin and A. Rosenfeld, Editors, Picture Processing and Psycho pictorics, Academic Press, 1970, p. 289-308, p. 287, pp. 347-370.
5. P.H. Stoloff, "Detection and Scaling of Statistical Differences Between Visual Textures", Perception and Psychophysics, Vol. 6(6A), 1969.
6. D.A. Ausherman, "Texture Discrimination Within Digital Imagery", Ph.D. Thesis, Electrical Engineering, University of Missouri, Dec., 1972.
7. K.S. Fu, Sequential Methods in Pattern Recognition and Machine Learning, Academic Press, 1968.

A THEORY OF CHARACTER RECOGNITION BY PATTERN MATCHING METHOD

Taizo Iijima
Tokyo Inst. of Tech.
Tokyo, Japan

Hiroshi Genchi and Ken-ichi Mori
Toshiba Research and Devel. Ctr.
Kawasaki, Japan

1. INTRODUCTION

Character recognition is a categorizing process of unknown input pattern into one of known finite number of character categories. Various practical methods of realizing this process has been devised. Pattern matching method is one of the most commonly used techniques in which the similarity of input pattern is tested with the reference pattern of each category.

The problem of pattern matching is how to quantitatively measure the similarity between two patterns. Most conventional methods have been experimentally devised such as direct optical pattern matching where correlated optical transparency output is measured by photo-multipliers,(1) sampled point reference pattern matching where the sample points are selected by statistical calculation of pattern data, similarity measurement defined by the distance or the direction cosine between two vectors in figure space representing patterns.(2)

In this paper the similarity measure is theoretically studied based on the basic theory of figures developed by one of the authors (Iijima.)(3)- (5)

Next, preprocessing method is studied so as to alleviate the effect of noises accompanying with the input pattern for practical character readers. The input pattern is defined in a generalized form including a "blurring parameter." The blurring operation has the merit of suppressing noises, but the demerit of decreasing the similarity measure. To overcome this difficulty, a new preprocessing method of canonicalization is defined.

Thirdly, the mathematical structure of the similar pattern set of the same category class is studied. An augmented similarity measure based on the structure of the pattern set, "the multiple similarity measure", is proposed. It is shown that this multiple similarity is a proper measure for recognizing character patterns accompanied with noises.

2. MATHEMATICAL STRUCTURE OF SIMILARITY MEASURE

Let a character pattern be expressed by the light energy distribution function of positional vector r on a hyperplane S_∞. Let a reference pattern be denoted by f_0 (r) and an input pattern by f(r). Then the conventional similarity measure of two patterns is defined by

$$s(f_0,f) = \frac{(f_0, f)}{\| f_0 \| \, \|f\|} \tag{1}$$

where (f_0, f) and $\|f\|$ mean the inner product of two patterns and the norm of pattern f, respectively. The range of the value of $s(f_0, f)$ is obtained from Schwartz's inequality as:

$$-1 \leq s(f_0, f) \leq 1. \tag{2}$$

Therefore by testing whether

$$1 - s(f_0, f) < \theta \tag{3}$$

holds for a sufficiently small constant θ, the decision can be made whether f(r) belongs to the category of $f_0(r)$.

To study the characteristics of the similarity measure $s(f_0, f)$, we introduce the generalized pattern $f(r, T(\tau))$ for a given pattern f(r) defined by the use of "blurring parameter" as follows.

$$f(r, T(\tau) = \int_{S_\infty} G(r - r', T(\tau) - T(\tau_0))f(r')\, dr', \qquad (\tau_0 < \tau < \infty) \tag{4}$$ [†]

where
$$G(r, T) = \frac{1}{4\pi \sqrt{|T|}} \exp(-\frac{1}{4} r \cdot T^{-1} \cdot r)., \tag{5}$$

and T_0 is a symmetric dyadic such that

---

†The light energy of the pattern f is assumed to diminish where r tends to infinity.

$$\lim_{|r| \to \infty} \exp(\epsilon r \cdot r) f(r, T(\tau)) = 0, \quad (\epsilon > 0).$$

$$\left.\begin{aligned} T_0 &= \frac{1}{2K} \int_{S_\infty} (r - a)(r - a)^T f(r)\, dr, \\ a &= \frac{1}{K} \int_{S_\infty} r\, f(r)\, dr, \\ K &= \int_{S_\infty} f(r)\, dr. \end{aligned}\right\} \qquad (6)$$

τ_0 is a constant such that

$$\tau_0 = \sqrt{|T_0|} \quad . \qquad (7)$$

When the blurring operator τ tends to τ_0, the generalized pattern approaches to f(r).

$$\lim_{\tau \to \tau_0} f(r, T(\tau)) = f(r), \ (T(\tau_0) = T_0). \qquad (8)$$

The larger τ is, the more blurred is the generalized pattern.

The inner produce of two patterns (f,f') with the blurring parameter set to τ_1 is defined as

$$(f,f') = \int \frac{f(r,T_1)\ f'\ (r,T_1)}{G(r - a,\ T_1)}\ dr, \qquad (9)$$

where $T_1 = T(\tau_1)$.

The norm of a figure is defined

$$\|f\| = \sqrt{(f,\ f)} \quad . \qquad (10)$$

If $f(r,\ T(\tau)) = f'(r,\ T(\tau))$, the similarity measure obviously becomes unity, $s(f,f) = 1$.

When the similarity measure between two patterns is evaluated, the center of gravity, a, the secondary moment about the center of gravity, T_1, must be normalized for both patterns. This operation is called preprocessing of pattern for pattern matching.

3. BLURRING OPERATION AND CANONICALIZATION OF PATTERN

It has been widely believed that the blurring of patterns would be undesirable if two similar patterns, such as capital B and numeral 8, should be discriminated. It has therefore been accepted that patterns should be sharpened as much as possible. However, it has been known that the blurring operation of input patterns contributes to suppressing noises accompanying the pattern.[6]

(A) Suppression of random noises in input patterns.
(B) Reduction of the effect of positional deviation to the similarity measure by increasing the value of the blurring parameter.
(C) Reduction of the number of sampling points of input patterns.

No matter what merits have the blurring operation, a fatal demerit prevents it from being adopted in practical character reader. The demerit is expressed in the following relation.

$$\lim_{\tau_1 \to \infty} s(f,f') = 1. \qquad (11)$$

This states that any two patterns that belong to different character categories tend to be identical when they are extensively blurred.

To overcome this difficulty, we propose another preprocessing method, named canonicalization. Consider a transformation of a pattern $f(r, T(\tau_1))$ to $g(r, T(\tau_1))$ defined by

$$f(r, T(\tau_1)) = g(r, T(\tau_1)) + K \cdot G(r - a, T(\tau_1)), \qquad (12)$$

where K and a are the same as defined previously.

The generalized figure $f(r, T(\tau))$ is expanded by the fundamental figure function system derived by Iijima[4]†

$$f(r, T(\tau)) = KG(r - a, T(\tau)) \sum_{\ell=0}^{\infty} \left(\frac{\sigma}{\sqrt{\tau}}\right)^{\ell} \kappa_\ell \left\{ \sum_{m=0}^{\ell} \upsilon_m^{(\ell)} \, q^*_{m\ell-m}\left(\frac{U^{-1}(r-a)}{\sqrt{\tau}}, I\right)\right\} \qquad (13)$$

†The fundamental figure function system $\{q^*_{mn}\}$ is the orthonormal function system satisfying the basic equation of figure, where H_m is the mth order Hermitian function.

$$q^*_{mn}\left(\frac{r}{\sqrt{\tau}}, I\right) = \frac{1}{\sqrt{m!\,n!}} H_m\left(\frac{x}{\sqrt{2\tau}}\right) H_n\left(\frac{y}{\sqrt{2\tau}}\right).$$

Some of the absolute feature coefficients κ_ℓ are:

$$\kappa_0 = 1, \ \kappa_1 = \kappa_2 = 0, \text{ and } \kappa_4 = 1$$

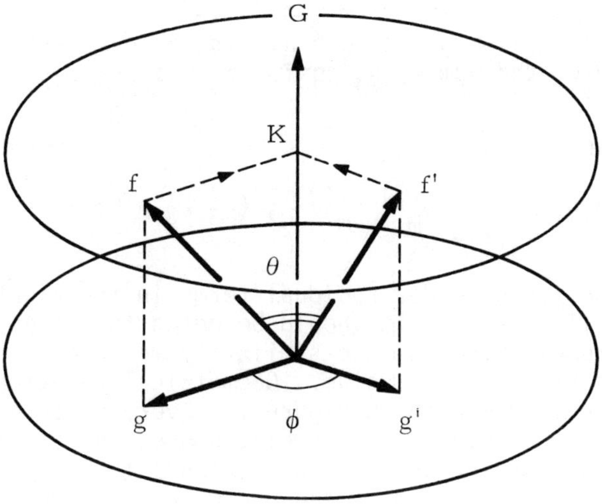

$$\begin{cases} s(f,f') = \cos\theta \\ S(g,g') = \cos\phi \end{cases}$$

Fig. 1. Illustration of canonicalization in three dimensional space

σ, κ_ℓ, and $\upsilon_m^{(\ell)}$ are constants independent of τ, called figure scale, absolute feature coefficients, and feature distribution factors, respectively.

Using this expansion and the above transformation, the similarity measure between two patterns g' and g is obtained as

$$s(g,g') = \frac{\sum_{\ell=3}^{\infty} \left(\frac{\sigma}{\tau_1}\right)^{\ell} \kappa_\ell \kappa'_\ell \sum_{m=0}^{\ell} \upsilon_m^{(\ell)} \upsilon_m'^{(\ell)}}{\sqrt{\sum_{\ell=3}^{\infty} \left(\frac{\sigma^2}{\tau_1}\right)^{\ell} \kappa_\ell^2} \sqrt{\sum_{\ell=3}^{\infty} \left(\frac{\sigma^2}{\tau_1}\right)^{\ell} \kappa_\ell'^2}} . \qquad (14)$$

τ_1 growing to infinity,

$$\lim_{\tau_1 \to \infty} s(g,g') = \sum_{m=0}^{3} \upsilon_m^{(3)} \upsilon_m'^{(3)} , \quad (\kappa_3 \kappa'_3 \neq 0) \qquad (15)$$

If $\kappa_3\kappa'_3 = 0$, the higher order residual terms in the right side of equation (15) will remain. It should be noted that even if input pattern is extremely blurred, the similarity measure of canonicalized patterns has a finite value. Geometrical relation of the canonicalization is depicted in Figure 1. Every pattern f comes close to G when it is blurred extremely; hence, the similarity measure of two patterns f and f', cos θ, tends to unity. Whereas the canonical pattern g which is the projected pattern of f to a plane perpendicular to G surely varies its norm by the blurring; but the similarity measure of g and g', corresponding to f and f' respectively, cos θ^*, keeps invariant by the blurring. The combination of the blurring operation and the canonicalization suppresses the noises inherently accompanying input patterns and augments the reading performance of a practical character reader. The blurring operation is easily implemented by defocusing the optical system in the scanner of an optical character reader.

4. MATHEMATICAL STRUCTURE OF SIMILAR PATTERN SET

In the pattern matching method discussed above, a single reference pattern is prepared for each category class where the most similar reference pattern to input pattern is sought. However, a real distribution of similar pattern sets of a category class is so prolate[6] that the conventional pattern matching method cannot attain satisfactory recognition performance required from practical applications because it assumes the distribution is ideally equi-directional.

In this section a mathematical structure of the similar pattern set of a category class is introduced so that a more augmented pattern matching method may be derived.

Let us consider a set of known canonicalized pattern set $\{g_0(r, T(\tau); \beta)\}$ which are similar to each other and belong to the same character category, where β means an indexing parameter of the set in an indexing space D. Let $w(\beta)$ denote the occurrence probability density of these patterns:

$$\int_D w(\beta)\, d\beta = 1, \qquad w(\beta) \geq 0. \tag{16}$$

Then every canonicalized pattern $g_0\,(r, T(\tau_1); \beta)$, is transformed to $k(r, T(\tau); \beta)$ as follows;

$$k(r, T(\tau_1); \beta) = \frac{\|\bar{g}_0\|^2}{(\bar{g}_0, g_{0\beta})}\, g_0(r, T(\tau_1); \beta), \tag{17}$$

where

$$\bar{g}_0(r, T(\tau_1)) = \int_D w(\beta) g_0(r, T(\tau_1); \beta) d\beta\ , \tag{18}$$

$$(\bar{g}_0, g_{0\beta}) = (\bar{g}_0(r, T_1), g_0(r, T_1; \beta))\ . \tag{19}$$

Because each pair of the members of the set $\{g_{0\beta}\}$ is similar, $(\bar{g}_0, g_{0\beta})$ has a positive value. If the occurrence probability density $w(\beta)$ is also transformed to $v(\beta)$ as

$$v(\beta) = \frac{(\bar{g}_0\ g_{0\beta})}{\|\ \bar{g}_0\|^2}\ w(\beta)\ , \tag{20}$$

then from equations (16) and (18),

$$\int_D v(\beta)\, d\beta = 1,\ v(\beta) \geq 0. \tag{21}$$

So, hereafter, $\{k(r, T(\tau_1); \beta\}$ abbreviated as $\{k_\beta\}$ is considered without confusion as the similar pattern set itself, and $v(\beta)$ is supposed to be the occurrence probability density of $k(r, T(\tau_1); \beta)$. Then following relations are obtained.

$$\overline{k}(r, T(\tau_1)) = \int_D v(\beta)\, k(r, T(\tau_1);\, \beta)\, d\beta$$

$$= \overline{g}_0(r, T(\tau_1)), \tag{22}$$

$$(\overline{k}, k_\beta) = \frac{\|\overline{g}_0\|^2}{(\overline{g}_0, g_{0\beta})} (\overline{k}, g_{0\beta}) = \|\overline{k}\|^2 . \tag{23}$$

Consider a functional defined by

$$J[\phi] = \int_D v(\beta) \frac{(k_\beta, \phi)^2}{\|\phi\|^2}\, d\beta \tag{24}$$

This functional represents the mean square value of the coordinate components of a pattern k_β in the direction of the functional coordinate axis $\phi/\|\phi\|$. From the Schwartz's inequality, for an arbitrary pattern $\phi(r, T(\tau))$, the following relation always holds,

$$(k_\beta, \phi)^2 \leqq \|k_\beta\|^2 \cdot \|\phi\|^2 . \tag{25}$$

Then

$$0 \leqq J[\phi] \leqq \int_D v(\beta) \| k_\beta \|^2\, d\beta . \tag{26}$$

Inequality (26) shows the value of $J[\phi]$ is positive and bounded. Therefore, the problem of obtaining the functional coordinate axis that has the largest component in the sense of average mean to the set of similar patterns can now be formulated as a problem of obtaining the solution that maximizes the value of $J[\phi]$.

Let a variation of equation (24) be taken as,

$$\delta J[\phi] = 0, \tag{27}$$

then the following integral equations hold,

$$\int_D \frac{K(r, r', T(\tau_1))}{G(r', T(\tau_1))} \phi_n(r', T(\tau_1))\, dr'$$

$$= \lambda_n \phi_n (r, T(\tau_1)), \quad (n = 0,1,2,\ldots) , \tag{28}$$

where

$$\left.\begin{aligned} K(r, r', T(\tau_1)) &= \int_D v(\beta)\, k(r, T(\tau_1);\beta)\, k(r', T(\tau_1);\beta)\, d\beta \\ \lambda_n &= J[\phi_n]. \end{aligned}\right\} \quad (29)$$

It is well known that λ_n and $\phi_n(r, T(\tau_1))$ are called the eigenvalues and eigenfunctions (mode functions) respectively.

To analytically solve the integral equations in (28) is difficult but the mean pattern $\overline{k}(r, T(\tau_1))$ of the similar pattern set can be proved to be the first eigenfunction corresponding to the maximum eigenvalue. So, hereafter the maximum eigenvalue λ_0 and the corresponding eigenfunction ϕ_0 are set such that

$$\lambda_0 = \|\overline{k}\|^2 \ ,$$

$$\phi_0(r, T(\tau_1)) = \frac{\overline{k}(r, T(\tau_1)}{\|\overline{k}\|} \ . \quad (30)$$

This is valid even in the practical application for recognizing noisy patterns because the norm of the noise factor does not exceed the norm of the mean pattern under the preprocessing with blurring and canonicalization operation described in section 3. The upper bound of $J[\phi]$ can be narrowed as

$$0 \leq J[\phi] \leq \|\overline{k}\|^2 \leq \int_D v(\beta)\, \|k_\beta\|^2\, d\beta \ . \quad (31)$$

The mathematical structure of the distribution of the similar pattern set of a category class is proved to be expressed with $\{\phi_n\}$ and $\{\lambda_n\}$. The similar patterns distribute on the functional coördinate axes of mode functions and the expanse of the distribution in the direction of each axis is proportional to λ_n. The prolateness of the distribution depends directly on the magnitude of λ_n.

5. MULTIPLE SIMILARITY METHOD

A logical way to improve the conventional pattern matching method is to use the whole information of the known similar pattern set $\{k_\beta\}$ instead of only the mean pattern $\overline{k}$ of the set. In this section, a new definition of the similarity measure between an input pattern and the similarity pattern set of a category class will be proposed.

First consider the decomposition of k_β by the mode functions $\{\phi_n\}$. Any pattern $k(r, T(\tau_1);\beta)$ in a similar pattern set can be expanded using the mode functions $\{\phi_n\}$ as,

$$k(r, T(\tau_1);\beta) = \sum_{n=0}^{\infty} b_n(\beta)\, \phi_n(r, T(\tau_1)), \tag{32}$$

where the expansion coefficients $b_n(\beta)$ are determined as

$$b_n(\beta) = (k_\beta, \phi_n), \quad (n = 0,1,2,\ldots). \tag{33}$$

From equations (23) and (30), $b_0(\beta)$ equals to $\|k\|$, then equation (32) can be rewritten as

$$k(r, T(\tau_1);\beta) = \bar{k}(r, T(\tau_1)) + \sum_{n=1}^{\infty} b_n(\beta)\, \phi_n(r, T(\tau_1)). \tag{34}$$

The new definition of the similarity measure between a canonicalized input pattern $g(r, T(\tau_1))$ and the similar pattern set $\{k_\beta\}$ of a character category are proposed as

$$S^2(k_\beta, g) \equiv \int_D v(\beta) \frac{(k_\beta, g)^2}{\|k_\beta\|^2\, \|g\|^2}\, d\beta . \tag{35}$$

Using the form of similarity measure described in section 2, the above definition could be rewritten as,

$$= \int_D v(\beta) \frac{\|k_\beta\|^2}{\|\bar{k}\|^2} \{s(k_\beta, g)^2\, d\beta\} \tag{36}$$

It may be understood the new definition is a kind of mean square of the similarity measure $s(k_\beta, g)$ between input pattern g and every similar pattern k_β that is known to belong to a specified category.

If all member patterns $\{k_\beta\}$ of a similar pattern set are equal to its mean pattern k, then it is clear that

$$S^2(\bar{k}, g) = s^2(\bar{k}, g). \tag{37}$$

So $S^2(k_\beta, g)$ may be considered as a generalized form of the square of conventional similarity measures. $S(k_\beta, g)$ is called the multiple similarity measure, and the pattern matching method that uses this measure as the multiple similarity method.

From equation (31), the value of the multiple similarity measure has the range of

$$0 \leqq S^2(k_\beta, g) \leqq 1. \tag{38}$$

Specifically, when an input pattern g equals the mean pattern $\overline{k}$, $S^2(k_\beta, g)$ has the maximum value

$$S^2(k_\beta, \overline{k}) = 1. \tag{39}$$

To make calculation easy, the multiple similarity measure defined by equation (35) should be transformed using equation (17) to

$$S^2(k_\beta, g) = \sum_{n=0}^{\infty} \frac{\lambda_n}{\lambda_0} \frac{(\phi_n, g)^2}{\|g\|^2} . \tag{40}$$

Considering that the value of λ_n decreases rapidly to zero as n increases, the sum in the above equation can be truncated at not-so-large n terms.

Now, because the mode functions $\{\phi_n\}$ determined from the similar pattern set of a category do not necessarily form a complete system to any input pattern $g(r, T(\tau_1))$, the expansion of g with the mode functions may include the residual term $h(r, T(\tau_1))$.

$$g(r, T(\tau_1)) = \sum_{n=0}^{\infty} b_n\phi_n(r, T(\tau_1)) + h(r, T(\tau_1)), \tag{41}$$

$$b_n = (g, \phi_n) \quad , \quad (n = 0,1,2,\ldots). \tag{42}$$

where $h(r, T(\tau_1))$ is perpendicular to $\{\phi_n\}$,

$$(h, \phi_n) = 0, \qquad (n = 0,1,2,\ldots) . \tag{43}$$

Using equation (42), the norm of $g(r, T(\tau_1))$ becomes

$$\|g\|^2 = \sum_{n=0}^{\infty} b_n^2 + \|h\|^2 .$$

Substitution of equations (41) and (43) into equation (40), leads to a relation on the multiple similarity measure,

$$S^2(k_\beta, g) = \frac{\sum_{n=0}^{\infty} \frac{\lambda_n}{\lambda_0} b_n^2}{\sum_{n=0}^{\infty} b_n^2 + \|h\|^2} \qquad (44)$$

$$= \left[1 - \frac{\|h\|^2}{\sum_{n=0}^{\infty} b_n^2 + \|h\|^2} \right] \left[1 - \frac{\sum_{n=1}^{\infty} \left(\frac{\lambda_0 - \lambda_n}{\lambda_0}\right) b_n^2}{\sum_{n=0}^{\infty} b_n^2} \right] .$$

Meanwhile the conventional similarity measure of the pattern matching method can be expressed in the same form as

$$s^2(\overline{k}, g) = \frac{\|\overline{k}\|^2}{\sum_{n=0}^{\infty} b_n^2 + \|h\|^2}$$

$$= \left[1 - \frac{\|h\|^2}{\sum_{n=0}^{\infty} b_n^2 + \|h\|^2} \right] \left[1 - \frac{\sum_{n=1}^{\infty} b_n^2}{\sum_{n=0}^{\infty} b_n^2} \right] \qquad (45)$$

The first term of the right side of both equations (44), (45) mean the effect of the residual term in the expansion (41) to the similarity values. This is identical for both definitions of the similarity measure. However, if an input pattern differs from the exact mean pattern $\overline{k}$, the similarity measure is directly affected to reduce the recognition power as shown in the second term of the right side of equation (45). Whereas in the multiple similarity measure, the effect is moderated so far as an input pattern varies within the range of distribution of the similar pattern set as shown in equation (44). For this reason, the multiple similarity method is more powerful and reliable as a recognition principle than the conventional pattern matching method.

6. CONCLUSION

In this paper, the conventional pattern matching method is analyzed in detail based on the structure of patterns to break through the difficulty inherent to the method. Proposed are two ways that considerably augment recognition capability.

(1) The blurring operation combined with the canonicalization of input patterns supresses noises accompanying the patterns. The noises, such as random noise, stroke width variation or positional deviation of input patterns prevented the conventional pattern matching method from being applied to a wider range of practical character reader.

(2) A generalized definition of the similarity measure, "the multiple similarity measure," is derived from the analysis of mathematical structure of the similar pattern set of the same category class. It is shown that this multiple similarity measure is a more reliable measure for recognizing character patterns accompanied with noises than the conventional one.

A practical optical page character reader ASPET/71 is developed by the authors using the theory of the blurring operation with canonicalization and the multiple similarity method.(7) ASPET/71 has a capabiltiy of reading alphanumeric characters and symbols of OCR-B font with a speed of 2,000 characters per second. The most outstanding capability of ASPET/71 is its accuracy and reliability of reading performance even to scratchy or smudgy character images printed by high speed printers as predicted by the theory described in this paper.

REFERENCES

(1) W.J. Hannan, "The RCA Multi-font Reading Machine." Optical Character Recognition, Spartan Books, 1962, pp 3-14.

(2) J.R. Parks, "Automatic Recognition of Low-quality Printed Characters using Analogue Techniques," Proc. of Inst. Radio, Engrs. Vol. 55, pp 67-80, Aug. 1967.

(3) T. Iijima, "Basic Equation of Figure and Observational Transformation" Systems, Computers & Controls, Vol 2, No. 4, pp 70-77, 1971.

(4) T. Iijima, "Basic Theory on Feature Extraction for Figures," Systems, Computers & Controls, Vol. 3, No. 1, pp 32-39, 1971.
(5) T. Iijima, "Basic Theory on the Structural Recognition of Figure," Trans. IECE, Vol. 55-D, No 8, pp 499-506, 1972 (in Japanese).
(6) T. Iijima, H. Genchi and K. Mori, "A Theoretical Study of the Pattern Identification by Matching Method," Proc. of First USA-JAPAN Computer Conference, pp 42-48, Oct. 1972.
(7) T. Iijima and K. Mori, "A New Generation of OCR-ASPET/71," J. Of Nikkei Electronics, No. 30, pp 66-80, May, 1972 (in Japanese).

INDEX